FOCUS ON QUANTUM GRAVITY RESEARCH

FOCUS ON QUANTUM GRAVITY RESEARCH

DAVID C. MOORE
EDITOR

Nova Science Publishers, Inc.
New York

For permission to use material from this book please contact us:
Telephone 631-231-7269; Fax 631-231-8175
Web Site: http://www.novapublishers.com

Library of Congress Cataloging-in-Publication Data

Focus on quantum gravity research / David C. Moore (editor).
p. cm.
Includes bibliographical references and index.
ISBN 1-59454-660-6
1. Quantum gravity--Research. I. Moore, David C.
QC178.F63 2006
530.14'3--dc22 2005028206

Published by Nova Science Publishers, Inc. ✢ New York

CONTENTS

PREFACE

Quantum gravity is the field of theoretical physics attempting to unify the theory of quantum mechanics, which describes three of the fundamental forces of nature, with general relativity, the theory of the fourth fundamental force: gravity. The ultimate goal is a unified framework for all fundamental forces: a theory of everything. This new book examines state-of-art research in this field.

Tracing the history of the universe, spacetime in its early epoch would be totally fluctuating quantum mechanically, and there would be no classical spacetime to describe dynamics. The models of quantum gravity chapter 1 is going to discuss are aimed to describe such states with background-metric independence and transitions to the conventional spacetime dynamics. The first model chosen is a renormalizable theory of gravity developed on the basis of the conformal gravity in four dimensions. In this theory the metric tensor field is written as a product of the conformal factor and the traceless tensor, and the conformal mode is treated non-perturbatively while the traceless tensor field is handled as a perturbation. The renormalized coupling of the traceless mode indicates the asymptotic freedom. It implies that at very high energies beyond the Planck mass scale, quantum fluctuations of the conformal field are dominated as a consequence of the background-metric independence. The second model discussed is a numerical method for quantizing gravity carried on the simplicial spacetime manifold. The theory treats dynamics non-perturbatively within the limited space allowed in a cluster of computers. It is significant that these two theories seem to be in the same universality class so far as the critical exponents are compared. It means that a conformal invariant spacetime can be mapped to a simplicial manifold with varying connectivity of simplices. The asymptotic freedom also implies that there exists a dynamical scale where the effective coupling of traceless mode diverges, and the conformal symmetry loses its validity. About this point the universe is expected to make a transition from the quantum spacetime to the classical Einstein spacetime, and fluctuations of the conformal field would be frozen to the classical background. The primordial power spectrum obtained from the two point correlation function of scalar curvature shows a significant character of the theory which is expected to be observed cosmologically. The sharp fall off of the angular power spectra at low multipole components of anisotropies in the cosmic microwave background is a topic discussed.

Much research has been done in the latter years on the subject of Lorentz violation induced by Quantum Gravity effects. On the theoretical side it has been shown that both Loop Quantum Gravity and String Theory predict that Lorentz violation can be induced at an energy near to the Planck scale. On the other hand, most of the experimental results in the latter years, have confirmed that the laws of physics are Lorentz invariant at low energy with very high accuracy. The inclusion of one- and two-loop contributions from a Lorentz violating Lagrangian dramatically change the above picture: the loop momenta run into the Planck scale and above and from the "divergent" terms finite Lorentz violating contributions of order one arise. These can be suppressed through suitable counterterms in the Lagrangian, originating a strong fine tuning problem. A brief discussion of these issues and their possible influence in future research follows in chapter 2.

It is shown by the author in chapter 3 that if gravitons are super-strong interacting particles and the low-temperature graviton background exists, the basic cosmological conjecture about the Dopplerian nature of redshifts may be false. In this case, a full magnitude of cosmological redshift would be caused by interactions of photons with gravitons. A new dimensional constant which characterizes one act of interaction is introduced and estimated. Non-forehead collisions with gravitons will lead to a very specific additional relaxation of any photonic flux. It gives a possibility of another interpretation of supernovae 1a data - without any kinematics. Of course, all of these facts may implicate a necessity to change the standard cosmological paradigm. Some features of a new paradigm are discussed here, too. A quantum mechanism of classical gravity based on an existence of this sea of gravitons is described for the Newtonian limit. This mechanism needs graviton pairing and "an atomic structure" of matter for working it, and leads to the time asymmetry. If the considered quantum mechanism of classical gravity is realized in the nature, then an existence of black holes contradicts to Einstein's equivalence principle. It is shown that in this approach the two fundamental constants - Hubble's and Newton's ones - should be connected between themselves. The theoretical value of the Hubble constant is computed. In this approach, every massive body would be decelerated due to collisions with gravitons that may be connected with the Pioneer 10 anomaly. It is shown that the predicted and observed values of deceleration are in good agreement. Some unsolved problems are discussed, so as possibilities to verify some conjectures in laser-based experiments.

The quantum gravity is formulated based on the principle of local gauge invariance. The model discussed in chapter 4 has local gravitational gauge symmetry and gravitational field is represented by gauge field. Then gravitational gauge interactions of pure gravitational gauge field is studied. It is found that the field equation of gravitational gauge field is just the Einstein's field equation. Path integral quantization of the theory is discussed in the paper. The quantum gauge gravity discussed in this paper is a perturbatively renormalizable quantum gravity, which is one of its most important advantage. In leading order approximation, the gravitational gauge theory gives out classical Newton's theory of gravity. In classical level, it gives out Einstein's general relativity. Another important advantage of the quantum gauge theory of gravity is that it can explain both classical tests of gravity and quantum effects of gravitational interactions, such as gravitational phase effects found in COW experiments and gravitational shielding effects found in Podkletnov experiments.

Quantum gravity effects of zeroth order in the Planck constant are investigated in the framework of the low-energy effective theory. In chapter 5, a special emphasis is placed on establishing the correspondence between classical and quantum theories, for which purpose transformation properties of the $\hbar^0$-order radiative contributions to the effective gravitational field under deformations of a reference frame are determined. Using the formalism of canonical transformations, it is shown that the one-loop contributions violate the principle of general covariance, in the sense that the quantities which are classically invariant under such deformations take generally different values in different reference frames. In particular, variation of the scalar curvature under transitions between different reference frames is calculated explicitly. Furthermore, the long-range properties of the two-point correlation function of the gravitational field are examined. Using the Schwinger-Keldysh formalism it is proved that this function is finite in the coincidence limit outside the region of particle localization. In this limit, the leading term in the long-range expansion of the correlation function is calculated explicitly. It shown also that in the case of a macroscopic gravitating body, the terms violating general covariance, and the field fluctuation are both suppressed by a factor *1/N*, where *N* is the number of particles in the body. This leads naturally to a macroscopic formulation of the correspondence between classical and quantum theories of gravitation. As an application of the obtained results, the secular precession of a test particle orbit in the fluctuating gravitational field is determined.

Chapter 6 extends the Gross-Perry-Yaffe approach of hot flat space instability to Minkowski space. This is done by examining the massless gravitational perturbation in a saddle point approximation of the partition function. In particular, the authors show that a nonconformal instability appears in the transverse-traceless tensor sector (massless gravitons) if a Schwarzschild wormhole background is considered. The appearance of an instability in the whole manifold is here interpreted as a black hole pair creation. The instability of flat space via black hole pair creation is further explored evaluating the Casimir energy to one loop in a Hamiltonian approach. A variational approach associated to a Sturm-Liouville problem is used to evaluate the Casimir energy by means of trial wave functionals of Gaussian form. The Casimir energy is here interpreted as an induced cosmological constant associated to the Sturm-Liouville eigenvalue. To handle with divergences which appear to one loop approximation, they use a zeta function regularization. A renormalization and a renormalization group equation are involved to handle with ultra-violet divergences.

Canonical quantization of gravity requires knowledge about the representation theory of its constraint algebra, which is physically equivalent to the algebra of arbitrary 4-diffeomorphisms. All interesting lowest-energy representations are projective, making the relevant algebra into a four-dimensional generalization of the Virasoro algebra. As described in chapter 7, such diffeomorphism anomalies are invisible in field theory, because the relevant cocycles are functionals of the observer's trajectory in spacetime. The multi-dimensional Virasoro algebra acts naturally in the phase space of arbitrary histories, with dynamics playing the role of first-class constraints. General relativity is regularized by expanding all fields in Taylor series around the observer's trajectory, and truncating at some fixed order. This regularized but manifestly general-covariant theory is quantized in the history phase space, and dynamics is imposed afterwards, in analogy with BRST quantization. Infinities arise when the regularization is removed; it is presently unclear how these should be dealt with.

Some reasonings are presented in chapter 8 that the problem of a singularity in general relativity with the problem of freezing of the 5th dimension can be connected. It is shown that some solutions in the 5D Kaluza-Klein gravity with the cross section in the Planck region have a region ($\approx l_{pl}$) where the metric signature changes from { +,-,-,-,- \} to { -,-,-,-,+ \}. The idea is discussed that such switching can not occur following some general point of view that the Planck length is the minimal length in the nature and consequently the physical quantities can not change very quickly in the course of this length. In this case the dynamic of the G_{55} metric component should be frozen.

The new first order, rheonomic, κ-supersymmetric formalism recently introduced by us for the world-volume action of the $D3$ brane is extended to the case of $D5$ branes in chapter 9. This extension requires the dual formulation of the Free Differential Algebra of type IIB supergravity in terms of 6-form gauge potentials which was so far missing and is given here. Furthermore relying on our new approach we are able to write the $D5$ world volume action in a manifestly SL(2,ℝ) covariant form. This is important in order to solve the outstanding problem of finding the appropriate boundary actions of $D3$-branes on smooth ALE manifolds with twisted fields. The application of our results to this problem is however postponed to a subsequent publication.

In: Focus on Quantum Gravity Research
Editor: David C. Moore, pp. 1-47
ISBN 1-59454-660-6

Chapter 1

Background Free Quantum Gravity and Cosmology

Ken-ji Hamada[a,b], Shinichi Horata[c], and Tetsuyuki Yukawa[c]
[a]Institute of Particle and Nuclear Studies,
KEK, Tsukuba 305-0801, Japan
[b]Department of Particle and Nuclear Physics,
The Graduate University for Advanced Studies (Sokendai),
Tsukuba 305-0801, Japan
[c]Hayama Center for Advanced Research,
The Graduate University for Advanced Studies (Sokendai),
Hayama 240-0193, Japan

Tracing the history of the universe, spacetime in its early epoch would be totally fluctuating quantum mechanically, and there would be no classical spacetime to describe dynamics. The models of quantum gravity we are going to discuss are aimed to describe such states with background-metric independence and transitions to the conventional spacetime dynamics.

The first model we choose is a renormalizable theory of gravity developed on the basis of the conformal gravity in four dimensions. In this theory the metric tensor field is written as a product of the conformal factor and the traceless tensor, and the conformal mode is treated non-perturbatively while the traceless tensor field is handled as a perturbation. The renormalized coupling of the traceless mode indicates the asymptotic freedom. It implies that at very high energies beyond the Planck mass scale, quantum fluctuations of the conformal field are dominated as a consequence of the background-metric independence. The second model we plan to discuss is a numerical method for quantizing gravity carried on the simplicial spacetime manifold. The theory treats dynamics non-perturbatively within the limited space allowed in a cluster of computers. It is significant that these two theories seem to be in the same universality class so far as the critical exponents are compared. It means that a conformal invariant spacetime can be mapped to a simplicial manifold with varying connectivity of simplices.

The asymptotic freedom also implies that there exists a dynamical scale where the effective coupling of traceless mode diverges, and the conformal symmetry loses its validity. About this point the universe is expected to make a transition from the quantum spacetime to the classical Einstein spacetime, and fluctuations of the conformal field would be frozen to the classical background. The primordial power spectrum obtained from the two point correlation function of scalar curvature shows a significant character of the theory which is expected to be observed cosmologically. The sharp fall off of the angular power spectra at low multipole components of anisotropies in the cosmic microwave background is a topic we wish to discuss.

1.1 Introduction

The discovery of the cosmic microwave background radiation (CMB) by Penzias and Wilson [1] has proven that the universe was once a melting pot of high energy reactions at its early stage, and since then it has been cooling as it expands. Valuable information on physical processes taking place in the expanding universe has been recorded in the CMB as tiny anisotropies. Angular power spectra of the anisotropies, recently observed by the Cosmic Background Explorer (COBE) [2] and the Wilkinson Microwave Anisotropy Probe (WMAP) [3], are, roughly speaking, projection of the history for the period between its birth to the present. It gives us a amazing hope that if we believe the idea of inflational universe [4, 5] describing a extremely rapid expansion without global thermalization, the long-distance correlation in observed anisotropies can provide the information about dynamics of the period before the universe grew to the Planck scale. We are now on the stage of revealing and verifying the quantum aspect of universe. This article aims to provide models of quantum gravity which have the possibility to understand the observed spectrum naturally without adding any artificial field by hands [6].

After passing many theoretical as well as experimental tests, the general theory of relativity [7] has been established as the fundamental theory of gravity capable of describing the evolution of universe. Many questions in the mysteries of universe have arose expecting that the theory can give answers, such as, how the universe is created, why it is expanding, how it will evolve in the future, and so on. In order to answer these questions, we need to know the initial condition of the universe that would be set at the Planck time. However, it is the period when quantum mechanics rules the law of universe, and it is beyond the power of the classical general relativity. If we wish to apply the Einstein theory for the Planck scale phenomena, it has fatal difficulties, such as the black-hole singularity and divergences in the canonical quantization procedure. Existence of these problems makes us wonder whether the Planck scale is the upper limit of the energy beyond which nature cannot be described in terms of physical laws. It is our ultimate task to construct the quantum theory which breaks the wall of Planck scale and creates a new frontier.

The black-hole singularity originates from the fact that the gravitational force is attractive between all matters, and once the gravitational force takes over other repulsive forces, matters collapse to a point to form a black hole, whenever they contain over a certain critical mass. In order to remove the singularity, we need a repulsive force in gravity. We naively expect such a force may exist since the universe has been expanding from its beginning against a lot of mass it contains. The cosmological constant gives such effect, but the present value is known to be too small to inflate it at the beginning. Since the energy required in the inflation is about the Planck mass scale, it is rather natural to expect a repulsive force of quantum mechanical origin.

The black-hole singularity of a microscopic size will be much influenced by quantization. For a massive particle with the mass heavier than the Planck scale the ordinary point-like description becomes inadequate, because any elementary particle with such a mass is considered to become a black-hole. If there were such a particle, the Compton wave length will be smaller than its Schwarzshild radius. Then, most of information of a particle would be lost, since it cannot be extracted outside the horizon, and we are forced to abandon the standard description of a particle with the mass beyond the Planck mass.

The key to resolve the singularity problem is to make use of the conformal invariance which the theory possesses as a consequence of the background-metric independence, since the invariance simply means that there is no fixed scale and no special point in space. In the very early universe spacetime fluctuations are so great that geometry loses its classical meaning, and the background-metric independent picture will emerge.

At low energies the theory should reduce to the Einstein theory of relativity in which the classical metric acquires a physical significance for measuring time and distance. We expect that there would be a kind of phase transition between the quantum universe with background-metric independence and the semi-classical universe of gravitons on a classical background effectively describing the Einstein theory. Accordingly the theory should possess a dynamical scale which divides those two phases.

At present we do not experience the conformal invariant spacetime, where a standard description of particle motion becomes inadequate. However, there is a possibility to observe the instance of the transition occurred at a certain time in history of the universe, since we can trace the past guided by the known physical laws as far as the classical spacetime exists.

How can one construct a model capable of describing such a dynamics? As a guiding principle we require the following three basic requirements on the theory:

- background-metric independence,
- finiteness,

- four dimensional spacetime.

At first we assume the background-metric independence as the fundamental symmetry. Since quantization can be performed by the functional integral over the metric tensor field, general coordinate transformation invariance means that the partition functional does not depend on the choice of a special background.

The finiteness condition is required because any physically meaningful theory should give quantitative predictions, or should have methods to compute with finite numbers for proper observables. In quantum gravity the meaning of finiteness is twofold. One is the appearance of spacetime singularity like a black-hole, and the other is a theoretical requirement known as the renormalizability. In the functional integral method if there is a singular configuration, the integral becomes ill-defined and the renormalizability will be lost. Study of the renormalization in quantum theory of gravity has a long history, some of which will be discussed in the next section.

There exist models of quantum gravity which are defined in space with dimensions higher than four. The introduction of extra dimensions is approximately equivalent to adding extra fields coupled to gravity from the view point of a four dimensional theory. The extra space dimensions have an advantage to formulate a theory in a unified way for interactions including gravity. On the other hand we need to explain why the present spacetime is four dimensional. Of course it could be tested experimentally in future whether there exists any extra dimension, but there is no evidence at present to doubt that the spacetime dimensions are four.

In order to be background-metric independent, the theory should be quantized non-perturbatively. We are going to take two approaches which have started explosively in 1990's mainly in the context of string theory known as the Liouville theory and the matrix model. One is a renormalizable quantum theory based on the conformal gravity in four dimensions. The theory is formulated in such a way that the conformal mode in the metric field is treated exactly, while the traceless tensor mode is handled as a perturbation. This treatment is justified by the asymptotic-free behavior of traceless mode. It implies that the conformal invariance becomes significant at very high energies where the coupling strength becomes small.

Another is a four-dimensional simplicial quantum gravity based on the dynamical triangulation. This is a formulation adopting the background-metric independence as the first principle. The path integral over metric function is replaced by the summation over all spacetime configurations numerically in a simplicial manifold. In the recent analysis it has been recognized that these two methods belong to the same universality class.

In this article, we review two models and add some new developments and numerical results, especially those which are related to cosmology. In next section, we give a short summary on the renormalizable model. The following section 3 is devoted to the review of the simplicial quantum gravity. The last section is the application of theories to the cosmological problems, especially on the anisotropy of

CMB.

1.2 Renormalizable Quantum Gravity

Since it has been recognized that any attempt to quantize Einstein gravity perturbatively [8, 9] cannot be succeeded [10], most of researchers in this field feel the necessity of quantizing gravity in either a non-perturbative way or others. Historically, the divergence problem are tackled introducing four-derivative terms in the action of gravity by many authors [11, 27, 13], because the gravitational coupling constant becomes dimensionless, and at the same time we can avoid the unbounded problem of the action. Afterwards the four-derivative gravity has been one of preferred quantum theory of gravity for those who wish to realize the background-metric independence [14, 15, 16, 17, 18, 19, 20, 21, 22].

However, a few problems have remained to be solved in existing four-derivative gravity models. One of them is the renormalizability. Although these models seem to be renormalizable in view of the power-counting, it does not imply genuine renormalizability at all. The second is the asymptotical non-free problem of the R^2 action even at the one-loop level, when we choose the correct sign to make the action bounded below. The third problem is unitarity, because of the higher derivative terms which create undesired negative-metric modes.

Through studies of 2D quantum gravity [23, 24, 25], it has become clear that all of these problems arise simply because the formulation does not correctly take into account the background metric independence. We here propose that the conformal gravity without R^2 term is a possible candidate for the theory free from these problems [20]. The conformal invariance forces us change the aspect of spacetime at high energies above the Planck scale, where the background metric does not have the meaning in measuring the distance any longer, and the ordinary picture of particles propagating on the background spacetime is inadequate. Consequently, this requires a new prescription to deal with negative-metric modes which can be carried out by making use of the conformal symmetry [17, 21, 22].

1.2.1 Brief Summary of The Model

The model we consider here is defned by the action,

$$\begin{aligned} I \; = \; \int d^4x\sqrt{-g}\Big\{ & -\frac{1}{t^2}C^2_{\mu\nu\lambda\sigma} - bG_4 + \frac{M_P^2}{2}R - \Lambda \\ & -\frac{1}{4}Tr\left(F^2_{\mu\nu}\right) - \frac{1}{2}\left(\partial^\mu X\partial_\mu X + \frac{1}{6}RX^2\right) + \cdots \Big\}, \end{aligned} \qquad (1.2.1)$$

where we write the reduced Planck mass as $M_P = 1/\sqrt{8\pi G}$, and the cosmological constant as Λ. There are two conformal invariant actions in four dimensions: the square of the Weyl tensor and the Euler term. The Weyl tensor is defined by

$$C_{\mu\nu\lambda\sigma} = R_{\mu\nu\lambda\sigma} - \frac{1}{2}g_{\mu[\lambda}R_{\sigma]\nu} + \frac{1}{2}g_{\nu[\lambda}R_{\sigma]\mu} + \frac{1}{6}g_{\mu[\lambda}g_{\sigma]\nu}R, \tag{1.2.2}$$

where the anti-symmetrized product is denoted as $a_{[\mu}b_{\nu]} = a_\mu b_\nu - a_\nu b_\mu$. This term represents the field strength of traceless modes, and t is the dimensionless coupling constant. The Euler density G_4 in four dimensions is given by

$$G_4 = R^2_{\mu\nu\lambda\sigma} - 4R^2_{\mu\nu} + R^2. \tag{1.2.3}$$

The constant b is introduced to renormalize divergences proportional to G_4.

The third term is the Einstein action, which will dominate at energies below the Planck scale. $F_{\mu\nu}$ is a gauge field strength, and X is a scalar field. Both have conformal coupling required by the integrability condition discussed below. In addition to these fields, we consider fermions, which are conformally invariant in any dimensions.

This model is expected to be renormalizable at higher orders, and the beta function of the renormalized coupling constant t_r indicates asymptotic freedom. This implies that at high energies beyond the Planck scale, the coupling constant gets small, and accordingly configurations with the vanishing Weyl tensor will dominate. Then, the singular configulation like a black hole at which the Riemann-Christoffel curvature tensor is divergent is excluded quantum mechanically, while it cannot be excluded in the Einstein theory because such a configuration has the vanishing scalar curvature so that its quantum weight in the path integral is unity.

The partition function is defined by the functional integral over the metric tensor field $g_{\mu\nu}$. We here decompose the metric field into the conformal mode ϕ, the traceless mode $h^\lambda{}_\nu$ and the background metric $\hat{g}_{\mu\lambda}$ as

$$g_{\mu\nu} = e^{2\phi}\bar{g}_{\mu\nu}, \tag{1.2.4}$$

and

$$\bar{g}_{\mu\nu} = \left(\hat{g}e^{th}\right)_{\mu\nu} = \hat{g}_{\mu\lambda}\left(\delta^\lambda{}_\nu + th^\lambda{}_\nu + \frac{t^2}{2}(h^2)^\lambda{}_\nu + \cdots\right), \tag{1.2.5}$$

where $tr(h) = h^\lambda{}_\lambda = 0$. The signature of the metric is taken as $(-1, 1, \cdots, 1)$. The contraction of the indices of $h^\mu{}_\nu$ is done by using the background metric. In the following, quantities with the hat and the bar on them are defined in terms of the metric $\hat{g}_{\mu\nu}$ and $\bar{g}_{\mu\nu}$, respectively.

The technique to treat diffeomorphism invariance is developed in the two dimensional theory at the end of 1980's [23, 24, 25], and extended later in four dimensions [14, 15, 16, 17, 18, 19, 20, 21, 22]. We change the path-integral measures from the diffeomorphism invariant measures to the practical measures defined on the background metric $\hat{g}_{\mu\nu}$. Consequently, in order to preserve the diffeomorphism invariance the Wess-Zumino action [27] S is necessary as the Jacobian, and the partition function is expressed as

$$Z = \int \frac{[d\phi dh dA dX]_{\hat{g}}}{\mathrm{Vol(diff.)}} \exp\{iS(\phi, \bar{g}) + iI(A, X, g)\}. \tag{1.2.6}$$

We note that the induced action S contains the kinetic term for the conformal field.

Dynamics of the conformal field is induced by the Wess-Zumino action which also takes care of the conformal anomalies [28, 29, 30]. Although the conformal anomaly breaks the conformal symmetry by a fixed background, it will be recovered when the conformal field is quantized. Therefore, the spacetime dynamics at very high energies is described completely by the conformal field theory (CFT_4).

We exclude the R^2 term, which is commonly introduced as the kinetic term of the conformal field, in the action because of the following reasons. Firstly, the coupling constant attached to this action does not show asymptotic freedom even at the one-loop level of the renormalization group equation. Secondly, in order that the functional integral over the conformal field is well defined, the effective action Γ should satisfy the integrability condition [31]. This condition requires that there should be no divergences proportional to R^2. The integrability condition also gives a constraint on the kinetic term of scalar field, while the Einstein-Hilbert action, the cosmological constant, and mass terms of matter fields do not have any effect from the condition. In this way, the integrability condition reduces a part of ambiguities in four-dimensional actions.

The beta function of the renormalized coupling t_r for the traceless mode has been calculated within the lowest order of perturbation to be $\beta_t = -\beta_0 t_r^3$ with

$$\beta_0 = \frac{1}{(4\pi)^2}\left\{\frac{1}{240}(N_\mathrm{X} + 3N_\mathrm{W} + 12N_\mathrm{A}) + \frac{197}{60}\right\}, \tag{1.2.7}$$

where N_X, N_W and N_A are the numbers of scalar fields, Weyl fermions and gauge fields added to the action, respectively. Solving the renormalization group equation, we obtain the running coupling constant α_G as

$$\alpha_\mathrm{G} = \frac{t_r^2(p)}{4\pi} = \frac{1}{2\pi\beta_0}\frac{1}{\log(p^4/\Lambda_\mathrm{QG}^4)}, \tag{1.2.8}$$

where the scale parameter $\Lambda_\mathrm{QG} = \mu \exp\{-1/2\beta_0 t_r^2(\mu)\}$ with μ being a renormalization mass scale, and p is a momentum defined in the physical space. The running coupling constant measures the degree of deviation from CFT_4. The new scale parameter Λ_QG represents the energy scale where description of the metric field in terms of the conformal mode and the traceless mode changes its appearance to conventional Einstein gravity [6].

A systematic method of renormalization based on the dimensional regularization is presented in Subsection 2.3. There, the effective action will be obtained for energies larger than $\Lambda_{\rm QG}$ as

$$\begin{aligned}\Gamma &= \int d^4x\sqrt{-g}\Biggl\{-\frac{\beta_0}{2}C_\mu{}^{\nu\lambda\sigma}\log\left(\frac{\Delta_4}{\Lambda_{\rm QG}^4}\right)C^\mu{}_{\nu\lambda\sigma} \\ &\quad -\frac{b_1}{(4\pi)^2}\frac{1}{8}E_4\frac{1}{\sqrt{-g}\Delta_4\left(1+\frac{2a_1}{\beta_0\log(\Delta_4/\Lambda_{\rm QG}^4)}\right)}\sqrt{-g}E_4 \\ &\quad +\frac{M_{\rm P}^2}{2}R-\Lambda+\mathcal{L}_{\rm Matter}\Biggr\}, \end{aligned} \tag{1.2.9}$$

where $\sqrt{-g}\Delta_4$ is a conformal invariant fourth order self-adjoint operator. In the case acting on a scalar field it is written as

$$\Delta_4 = \nabla^4 + 2R^{\mu\nu}\nabla_\mu\nabla_\nu - \frac{2}{3}R\nabla^2 + \frac{1}{3}(\nabla^\mu R)\nabla_\mu. \tag{1.2.10}$$

The effective action in the second line is a higher-order generalization of the non-local Riegert action [14] for the conformal anomaly. The quantity E_4 represents the modified Euler density,

$$E_4 = G_4 - \frac{2}{3}\nabla^2 R. \tag{1.2.11}$$

The Riegert action is the source of the kinetic term for conformal field. The coefficient b_1 in front of the logarithmic divergence proportional to the Euler term has been computed within the lowest order as

$$b_1 = \frac{1}{360}\left(N_{\rm X} + \frac{11}{2}N_{\rm W} + 62N_{\rm A}\right) + \frac{769}{180}. \tag{1.2.12}$$

The coefficient a_1 is expected to be a positive constant which will be computed later in Subsection 2.3.

As energies gets lower to the dynamical scale $\Lambda_{\rm QG}$, the running coupling $\alpha_{\rm G}$ will diverge and the higher derivative terms in the effective action disappear. There, the conformal invariance breaks down and the background metric appears as a physical measure of space and time. Beyond the energy smaller than this point the Einstein gravity govern the dynamics. The cosmological consequence of this transition will be discussed in Section 4.

Reduction of the traceless field to the ordinary graviton field and the classical background also occurs at energy about the $\Lambda_{\rm QG}$. While the conformal field dominates in determining the classical background fluctuations of the traceless field will propagate as either $1/p^2M(p^2)$ or $1/M(p^2)$, where $M(p^2) = M_{\rm P}^2 + 2\beta_0 p^2\log(p^4/\Lambda_{\rm QG}^4)$. The negative-metric mode propagates as $1/M(p^2)$, which does not have a real pole. Also there is no tachyon pole in case of the ordering of scales $M_{\rm P} > \sqrt{4\beta_0/e}\Lambda_{\rm QG}$ satisfied, which is a cosmologically natural guess. Thus, the physical pole of the propagator $1/p^2$ is regarded to be the graviton, and the undesired ghost modes contained in the four-derivative gravitational fields will be shifted to a pair of complex poles [27]. In the quantum field theory a state decaying from a real-pole to a complex-pole is known to give measure-zero contributions [32, 33]. Since the function $M(p^2)$ becomes $M_{\rm P}^2$ at the $\Lambda_{\rm QG}$ scale, only the graviton pole becomes dynamical below this scale.

1.2.2 Conformal Invariance and Physical States

In a scale-invariant spacetime, we can not have the ordinary particle picture any longer such as propagating on a classical spacetime. Physical states are generated by conformal symmetry, and they are classified by the representation of conformal algebra. In this subsection, we discuss the construction and classification of physical states [21, 22].

2D Quantum Gravity Before discussing physical states in the four-dimensional theory, it is instructive to take a look at the case of two-dimensional quantum gravity [23, 24, 25]. In two dimensions the traceless field has no dynamical degrees of freedom, so that we can take the conformal gauge $h_{\mu\nu} = 0$. If we take this gauge, the corresponding Wess-Zumino action is given by the so-called Liouville/Polyakov action [23],[1]

$$S_{\rm L}(\phi, \hat{g}) = -\frac{b_{\rm L}}{4\pi} \int d^2x \sqrt{-\hat{g}} \left(\phi \hat{\Delta}_2 \phi + \hat{R}\phi \right), \qquad (1.2.13)$$

where $\sqrt{-g}\Delta_2 = \sqrt{-g}(-\nabla^2)$ is a conformally invariant operator in two dimensions. The coefficient has been computed as

$$b_{\rm L} = (25 - c)/6, \qquad (1.2.14)$$

for the case of gravity coupled with matter fields with the total central charge c.

In the conformal gauge [25, 26], the residual gauge degrees of freedom ξ^μ satisfy the conformal Killing equations: $\delta_\xi h_{\mu\nu} = \partial_\mu \xi_\nu + \partial_\nu \xi_\mu - \eta_{\mu\nu} \partial^\lambda \xi_\lambda = 0$. As the standard procedure to remove these degrees of freedom, so-called bc ghosts are introduced. Then, the system is described by a matter field, the conformal field governed by the Liouville action (1.2.13) with the coefficient (1.2.14), and the bc-ghosts.

Each field is expanded by modes represented on the cylindrical background $R \times S^1$, parameterized by $x^\mu = (\eta, \sigma)$ with $0 < \sigma < 2\pi$. The generator of the conformal algebra is given in terms of the stress tensor by $L_\xi = \int_{S^1} d\sigma \xi^\mu T_{\mu 0}$. Corresponding to the infinite number of conformal Killing vectors, $(e^{in(\eta+\sigma)}/2, e^{in(\eta+\sigma)}/2)$ and $(e^{in(\eta-\sigma)}/2, -e^{in(\eta-\sigma)}/2)$ with $n \in \mathbf{Z}$, there are infinite number of generators $L_n^\pm = e^{in\eta} \int_{S^1} d\sigma\, e^{\pm in\sigma}(T_{00} \pm T_{01})/2$ satisfying the reality condition $L_n^{\pm\dagger} = L_{-n}^\pm$.

[1] The Liouville action is given by integrating the conformal anomaly over the conformal field as $S_{\rm L} = -(b_{\rm L}/4\pi) \int d^2x \int_0^\phi d\phi \sqrt{-g} R$, with using the relation $\sqrt{-g}R = \sqrt{-\bar{g}}(2\bar{\Delta}_2\phi + \bar{R})$.

These generators form the conformal algebra, called the Virasoro algebra, without central charge,

$$\left[L_n^{\pm}, L_m^{\pm}\right] = (n-m)L_{n+m}^{\pm}, \tag{1.2.15}$$

where L_n^+ and L_m^- commute. In fact, the generator $L_n^{\pm}$ is composed of three generators: $L_n^{\mathrm{L}\pm}$ for the Liouville sectors with central charge $1+6b_{\mathrm{L}}$, $L_n^{\mathrm{M}\pm}$ for the matter sector with central charge c and $L_n^{bc\pm}$ for the ghost sector with central charge -26. The total central charge indeed vanishes. The Hamiltonian $H = L_0^+ + L_0^-$ is expanded as

$$H = \hat{p}^2 + b_{\mathrm{L}}/2 + \sum_{n=1}^{\infty}\left\{ n a_n^{+\dagger} a_n^+ + n a_n^{-\dagger} a_n^- \right\} + \cdots, \tag{1.2.16}$$

where $\hat{p}$ is the conjugate operator of $\hat{q}$ in the Liouville zero-mode $\phi_0 = \hat{q}/\sqrt{2b_{\mathrm{L}}}$, and $a_n^{\pm\dagger}(a_n^{\pm})$ creates (annihilates) eigenmode of the Liouville field. The dots denote contributions from the other fields. On the cylinder background, it is identified to be the dilatation generator counting the conformal weight of the state.

Conformally invariant vacuum defined by the condition, $L_n^{\pm}|\Omega\rangle = 0$, for $n \geq -1$ is uniquely determined as $|\Omega\rangle = e^{-b_{\mathrm{L}}\phi_0}|0\rangle$, where $|0\rangle$ is the standard Fock vacuum with vanishing the zero-mode momentum $\hat{p}$. The exponential factor is due to the background charge coming from the $\hat{R}\phi$ term in the Liouville action. Here, we use the Virasoro generators without the ghost sector, which is considered to be integrated out. The physical states are then spanned by states satisfying the following conditions in the Fock space created by $a_n^{\pm\dagger}$ and $\hat{q}$ on the conformally invariant vacuum:

$$H|\mathrm{phys}\rangle = 2|\mathrm{phys}\rangle \quad \text{and} \quad L_n^{\pm}|\mathrm{phys}\rangle = 0 \tag{1.2.17}$$

with $n \geq 1$. The eigenvalue 2 of the Hamiltonian indicates that the physical states has the conformal weight 2, so that its volume integral has the vanishing weight in two dimensions. If we consider full generators including the ghost sector, the physical state conditions are represented as the Wheeler-DeWitt equation such that all the Virasoro generators with $n \geq 0$ vanish. The ghost excitation is responsible for the shift in the Hamiltonian.

Let us take an example of the physical states satisfying conditions (1.2.17). When we write the primary state of the matter sector with the conformal weight 2Δ as $|\Delta\rangle = \Phi_\Delta^\dagger|0\rangle$, which satisfies the conditions $L_n^{\mathrm{M}\pm}|\Delta\rangle = 0$ for $n \geq 1$ [34], the physical state coupled to the primary field is given by

$$e^{\gamma_\Delta \phi_0}\Phi_\Delta^\dagger|\Omega\rangle. \tag{1.2.18}$$

Here, γ_Δ is determined by the Hamiltonian condition, which yields, $-(\gamma_\Delta - b_{\mathrm{L}})^2/2b_{\mathrm{L}} + b_{\mathrm{L}}/2 + 2\Delta = 2$, and thus we obtain

$$\gamma_\Delta = b_{\mathrm{L}}\left(1 - \sqrt{1 - \frac{4-4\Delta}{b_{\mathrm{L}}}}\right), \tag{1.2.19}$$

where we select the solution which approaches the canonical value $2 - 2\Delta$ in the classical limit, $b_{\rm L} \to \infty$. This state is called the dressed state. For each dressed state, there is a field operator $e^{\gamma_\Delta \phi}\Phi_\Delta(\eta, \sigma)$ such that the state is given by the limit: $\lim_{\eta \to i\infty} e^{-2i\eta} e^{\gamma_\Delta \phi}\Phi_\Delta(\eta, \sigma)|\Omega\rangle$. The field with $\Delta = 0$ is the term corresponding to the cosmological constant. The scaling dimension of the physical field is determined by γ_Δ, which will be used in Section 3.

Besides, there are more complicated states, the so-called discrete states [35, 36] and the states possessing a ring structure, called the grand ring [37]. These states are related to the W_∞ symmetry of two dimensional target space [38], which are not relevant in the later discussion, and we shall skip to consider them, here.

4D Quantum Gravity We follow the canonical quantization procedure for the 4-dimensional gravity. Since the coupling constant of traceless field vanishes at very high energies, we take traceless modes within the lowest order of perturbation. After carrying out a gauge-fixing for the traceless field, we obtain the Riegert action [14] as the lowest term of the Wess-Zumino action:[2]

$$S_1(\phi, \hat{g}) = -\frac{b_1}{(4\pi)^2} \int d^4x \sqrt{-\hat{g}} \left(2\phi \hat{\Delta}_4 \phi + \hat{E}_4 \phi \right), \qquad (1.2.20)$$

where the coefficient b_1 has been given in (1.2.12). We quantize the combined system, $S_1 + I$, as a classical action. The action I contains the kinetic term of the traceless field coming from the Weyl tensor term. The Einstein-Hilbert action is considered as a potential term, and they are disregarded considering we are in the high energy regime.

To quantize the model in practice, we need to specify the background metric. Owing to the conformal invariance, all models transformed by a conformal transformation into each other are equivalent. We here choose the $R \times S^3$ background because it has several advantages. Mode expansions of higher derivative fields become simple and the canonical commutation relations have diagonal forms, contrary to the case of flat background [39] in which there is an unusual time-dependence in the undesired mode and the commutaters in general becomes off-diagonal. Also, we can use tools developed in the $SU(2)$ representation theory [40], because the isometry group of S^3 is $SU(2) \times SU(2)$.

Equations of the motion for the traceless field are invariant under the gauge transformation $\delta_\xi h_{\mu\nu} = \hat{\nabla}_\mu \xi_\nu + \hat{\nabla}_\nu \xi_\mu - \frac{1}{2}\hat{g}_{\mu\nu}\hat{\nabla}^\lambda \xi_\lambda$. To fix the gauge, we impose the transverse conditions, $\hat{\nabla}_i h^i{}_0 = \hat{\nabla}_i h^i{}_j = 0$. Furthermore, we take the radiation gauge $h_{00} = 0$ using the residual gauge symmetry. Then, we obtain the following

[2]The Wess-Zumino action in four dimensions is given by integrating the conformal anomaly E_4 with respect to the conformal field as $S_1 = -(b_1/(4\pi)^2) \int d^4x \int_0^\phi d\phi \sqrt{-g} E_4$, where the relation, $\sqrt{-g}E_4 = \sqrt{-\bar{g}}(4\bar{\Delta}_4\phi + \bar{E}_4)$, is used. The above action is the ϕ-dependent part of the non-local Riegert action, $-(b_1/8(4\pi)^2) \int d^4x \sqrt{-g} E_4 (\sqrt{-g}\Delta_4)^{-1} \sqrt{-g} E_4$.

equations for the gravitational fields,

$$\begin{aligned}\left(\partial_\eta^4 - 2\Box_3\partial_\eta^2 + \Box_3^2 + 4\partial_\eta^2\right)\phi &= 0, \\ \left(\partial_\eta^4 - 2\Box_3\partial_\eta^2 + \Box_3^2 + 8\partial_\eta^2 - 4\Box_3 + 4\right)h^{ij} &= 0, \\ \left(\Box_3 + 2\right)\left(-\partial_\eta^2 + \Box_3 - 2\right)h^{0i} &= 0, \qquad (1.2.21)\end{aligned}$$

where h^{ij} and h^{0i} are called the transverse-symmetric-traceless mode and the transverse mode of the traceless field, respectively. η is the conformal time, and $\Box_3 = \hat{\nabla}^i\hat{\nabla}_i$ is the Laplacian on S^3. We take the radius of S^3 to be unity.

Dynamical fields are expanded in symmetric-traceless-transverse (ST^2) spherical tensor harmonics [41]. Expansion coefficients have the time dependence such as $e^{iE\eta}$, determined by the equations of motion, where E becomes the conformal weight for each mode. The ST^2 tensor harmonics of rank n are constructed and classified using the $(J+\varepsilon_n, J-\varepsilon_n)$ representation of the group $SU(2)\times SU(2)$ for each sign of the polarization index $\varepsilon_n = \pm n/2$ [21]. They, denoted by $Y^{i_1\cdots i_n}_{J(M\varepsilon_n)}$, are the eigenfunction of the Laplacian on S^3: $\Box_3 Y^{i_1\cdots i_n}_{J(M\varepsilon_n)} = \{-2J(2J+2)+n\}Y^{i_1\cdots i_n}_{J(M\varepsilon_n)}$, where J ($\geq n/2$) takes integer or half-integer values, and $M=(m,m')$ represents the multiplicity. Below, we use the parametrizations, $\varepsilon_0 = 0$, $\varepsilon_1 = y = \pm 1/2$, $\varepsilon_2 = x = \pm 1$, $\varepsilon_3 = z = \pm 3/2$ and $\varepsilon_4 = w = \pm 2$, for the tensor indices up to rank 4.

The space of the residual symmetry in the radiation gauge is slightly bigger than the space generated by the 15 conformal Killing vectors. In order to construct generators of the conformal algebra, we make the space of residual gauge symmetry equal to the space spanned by the conformal Killing vectors on $R\times S^3$. For this reason, we remove the lowest mode with $J=1/2$ in the transverse field satisfying the equation $(\Box_3+2)\,h^{0i}=0$. We call this choice as the radiation$^+$ gauge [21].

The Hamiltonian in the radiation$^+$ gauge on $R\times S^3$ is expanded as

$$\begin{aligned} H &= \frac{1}{2}\hat{p}^2 + b_1 + \sum_{J\geq 0}\sum_{M}\{2Ja^\dagger_{JM}a_{JM} - (2J+2)b^\dagger_{JM}b_{JM}\} \\ &\quad + \sum_{J\geq 1}\sum_{M,x}\{2Jc^\dagger_{J(Mx)}c_{J(Mx)} - (2J+2)d^\dagger_{J(Mx)}d_{J(Mx)}\} \\ &\quad - \sum_{J\geq 1}\sum_{M,y}(2J+1)e^\dagger_{J(My)}e_{J(My)}. \qquad (1.2.22)\end{aligned}$$

Here $a^\dagger_{JM}(a_{JM})$ and $b^\dagger_{JM}(b_{JM})$ are the positive-metric and the negative-metric mode creation (annihilation) operators of conformal modes, and $c^\dagger_{J(Mx)}(c_{J(Mx)})$ and $d^\dagger_{J(Mx)}(d_{J(Mx)})$ are those in the transverse-symmetric-traceless field, respectively. $e^\dagger_{J(My)}(e_{J(My)})$ is the negative-metric mode creation (annihilation) in the transverse field. The operator $\hat{p}$ is the conjugate momentum of $\hat{q}$ which appears in the zero-mode of the conformal field as $\phi_0 = \hat{q}/\sqrt{2b_1}$. As the two-dimensional case, the

Hamiltonian is the dilatation operator counting the number of the conformal weight for a state.

Corresponding to the conformal Killing vectors ξ^μ, there are 15 generators of the conformal algebra given in terms of the stress tensor for the combined system, $S_1 + I$, as

$$Q_\xi = \int_{S^3} d\Omega_3 \xi^\mu T_{\mu 0}, \tag{1.2.23}$$

where $d\Omega_3$ is the volume element of the unit S^3 with the volume, $\mathrm{Vol}(S^3) = 2\pi^2$.

The Hamiltonian is the generator of dilatation corresponding to the time translation along $\xi^\mu = (1,0,0,0)$. The Killing vectors on S^3 represented by $\xi^\mu = (0, \xi^i_R)$, satisfying equations $\hat{\nabla}^i \xi^j_R + \hat{\nabla}^j \xi^i_R = 0$ and $\partial_\eta \xi^i_R = 0$, give 6 generators of its isometry group $SU(2) \times SU(2)$, denoted by $R_{\underline{MN}}$ with the properties $R_{\underline{MN}} = -\epsilon_{\underline{M}} \epsilon_{\underline{N}} R_{-\underline{N}-\underline{M}}$ and $R^\dagger_{\underline{MN}} = R_{\underline{NM}}$, where the indices, $\underline{M}$ and $\underline{N}$, are the **4**-vectors on $SU(2) \times SU(2)$ and $\epsilon_{\underline{M}}$ is the phase factor $\epsilon_M = (-1)^{m-m'}$ for the **4**-vector. Corresponding to the conformal Killing vectors, $\xi^\mu/\sqrt{\mathrm{Vol}(S^3)} = (e^{i\eta} Y^*_{1/2\underline{M}}/2, -ie^{i\eta}\hat{\nabla}^i Y^*_{1/2\underline{M}}/2)$ and their complex conjugates, there are the $4+4$ generators of the special conformal transformations, which are denoted by $Q_{\underline{M}}$ and its hermite conjugate, respectively. These 15 generators form the closed algebra:

$$\begin{aligned}
\left[Q_{\underline{M}}, Q^\dagger_{\underline{N}}\right] &= 2\delta_{\underline{MN}} H + 2R_{\underline{MN}}, \\
\left[H, Q_{\underline{M}}\right] &= -Q_{\underline{M}}, \quad \left[H, R_{\underline{MN}}\right] = 0, \\
\left[Q_{\underline{M}}, R_{\underline{M}_1\underline{M}_2}\right] &= \delta_{\underline{M}\underline{M}_2} Q_{\underline{M}_1} - \epsilon_{\underline{M}_1}\epsilon_{\underline{M}_2}\delta_{\underline{M}-\underline{M}_1} Q_{-\underline{M}_2}, \\
\left[R_{\underline{M}_1\underline{M}_2}, R_{\underline{M}_3\underline{M}_4}\right] &= \delta_{\underline{M}_1\underline{M}_4} R_{\underline{M}_3\underline{M}_2} - \epsilon_{\underline{M}_1}\epsilon_{\underline{M}_2}\delta_{-\underline{M}_2\underline{M}_4} R_{\underline{M}_3-\underline{M}_1} \\
&\quad -\delta_{\underline{M}_2\underline{M}_3} R_{\underline{M}_1\underline{M}_4} + \epsilon_{\underline{M}_1}\epsilon_{\underline{M}_2}\delta_{-\underline{M}_1\underline{M}_3} R_{-\underline{M}_2\underline{M}_4}.
\end{aligned} \tag{1.2.24}$$

The rotation generator has the vanishing conformal weight and has a diagonal form for each mode labeled by J. This generator raises and lowers the $SU(2) \times SU(2)$ indicies $M = (m, m')$ with a fixed J. On the other hand, the generator for the special conformal transformation $Q_{\underline{M}}$ has the weight -1 (its conjugate has the weight 1), and it is the **4**-vector on $SU(2) \times SU(2)$. Then, this generator is given by a **4**-vector constructed as a proper combination of creation and annihilation modes with different weights by 1. For the conformal field sector [17], it has the form

$$\begin{aligned}
Q_{\underline{M}} &= \left(\sqrt{2b_1} - i\hat{p}\right) a_{\frac{1}{2}\underline{M}} + \sum_{J \geq 0} \sum_{M_1, M_2} \mathbf{C}^{\frac{1}{2}\underline{M}}_{JM_1, J+\frac{1}{2}M_2} \Big\{ \alpha(J) \epsilon_{M_1} a^\dagger_{J-M_1} a_{J+\frac{1}{2}M_2} \\
&\quad + \beta(J) \epsilon_{M_1} b^\dagger_{J-M_1} b_{J+\frac{1}{2}M_2} + \epsilon_{M_2} a^\dagger_{J+\frac{1}{2}-M_2} b_{JM_1} \Big\},
\end{aligned} \tag{1.2.25}$$

where $\alpha(J) = \sqrt{2J(2J+2)}$, and $\beta(J) = -\sqrt{(2J+1)(2J+3)}$. The first term is the zero-mode contribution which appears only in the conformal field sector. The

$\mathbf{C}$ function is the $SU(2) \times SU(2)$ Clebsch-Gordan coefficient defined by the integral of three products of scalar harmonics over S^3,

$$\mathbf{C}^{J_3 M_3}_{J_1 M_1, J_2 M_2} = \sqrt{\frac{(2J_1+1)(2J_2+1)}{2J_3+1}} C^{J_3 m_3}_{J_1 m_1, J_2 m_2} C^{J_3 m_3'}_{J_1 m_1', J_2 m_2'}, \tag{1.2.26}$$

where $C^{J_3 m_3}_{J_1 m_1, J_2 m_2}$ is the standard $SU(2)$ Clebsch-Gordan coefficient. The significant property of the generator $Q_{\underline{M}}$ is that this generator mixes the positive-metric mode and the negative-metric mode.

Similarly, generators for the traceless field sector are given by proper combinations of creation and annihilation operators of the transverse-symmetric-traceless and the transverse modes. They have more complicated forms, which have been derived in [21]. There exist cross terms of the positive-metric mode $c_{J(Mx)}$ and the negative-metric modes, $d_{J(Mx)}$ and $e_{J(My)}$, which shows that a closed algebra can be constructed only when both the positive-metric and the negative-metric modes are combined properly.

A conformally invariant vacuum annihilated by all the generators is uniquely given by $|\Omega\rangle = e^{-\sqrt{2b_1}\hat{q}}|0\rangle = e^{-2b_1\phi_0}|0\rangle$, where $|0\rangle$ is the standard Fock vacuum with the zero eigenvalue of $\hat{p}$. The exponential factor indicates the background charge coming from the $\hat{E}_4\phi$ term in the Riegert action. The physical states are spanned by the Fock space generated on the conformally invariant vacuum. They must satisfy the conformal invariance conditions [21, 22],

$$Q_{\underline{M}}|\text{phys}\rangle = 0 \tag{1.2.27}$$

and

$$(H-4)|\text{phys}\rangle = R_{\underline{MN}}|\text{phys}\rangle = 0, \tag{1.2.28}$$

where the eigenvalue 4 indicates that the subtraction of the conformal weight of the volume integral in four dimensions. The ghost fields in the radiation$^+$ gauge, which have 15 degrees of freedom, are decoupled and considered to be integrated out. If we consider the full generators including the ghost sector, the Hamiltonian condition has the Wheeler-DeWitt form of $H = 0$ [17, 22].

The physical state is now decomposed into two sectors: the conformal field and the traceless field. The matter field sector is not considered here. Each sector consists of the Hamiltonian eigenstates satisfying the $Q_{\underline{M}}$ condition (1.2.27). We first construct such states, and then impose the conditions (1.2.28) after combining all sectors.

In order to find states satisfying condition (1.2.27), we seek a creation operator that commutes with the generator $Q_{\underline{M}}$. Such an operator will provide a building block of physical states. We first consider the sector of the conformal field. Since there is no creation mode that commutes with $Q_{\underline{M}}$, we look for operators constructed in a bilinear form. Using the crossing properties of the $SU(2) \times SU(2)$ Clebsch-Gordan coefficients, we find two $Q_{\underline{M}}$-invariant combinations with the weight $H = 2L$

[22]:

$$
\begin{aligned}
S^{\dagger}_{LN} &= \chi(\hat{p})a^{\dagger}_{LN} + \sum_{K=\frac{1}{2}}^{L-\frac{1}{2}} \sum_{M_1,M_2} x(L,K)\mathbf{C}^{LN}_{L-KM_1,KM_2} a^{\dagger}_{L-KM_1} a^{\dagger}_{KM_2}, \\
\mathcal{S}^{\dagger}_{L-1N} &= \psi(\hat{p})b^{\dagger}_{L-1N} + \sum_{K=\frac{1}{2}}^{L-\frac{1}{2}} \sum_{M_1,M_2} x(L,K)\mathbf{C}^{L-1N}_{L-KM_1,KM_2} a^{\dagger}_{L-KM_1} a^{\dagger}_{KM_2} \\
&\quad + \sum_{K=\frac{1}{2}}^{L-1} \sum_{M_1,M_2} y(L,K)\mathbf{C}^{L-1N}_{L-K-1M_1,KM_2} b^{\dagger}_{L-K-1M_1} a^{\dagger}_{KM_2}, \qquad (1.2.29)
\end{aligned}
$$

for integers $L \geq 1$. The coefficients are given by

$$
x(L,K) = \frac{(-1)^{2K}}{\sqrt{(2L-2K+1)(2K+1)}} \sqrt{\begin{pmatrix} 2L \\ 2K \end{pmatrix} \begin{pmatrix} 2L-2 \\ 2K-1 \end{pmatrix}} \qquad (1.2.30)
$$

and $y(L,K) = -2\sqrt{(2L-2K+1)(2L-2K-1)}x(L,K)$. For any half-integer L, these functions vanish. The $\hat{p}$-dependent operators are given by $\chi(\hat{p}) = (\sqrt{2b_1} - i\hat{p})/\sqrt{2(2L-1)(2L+1)}$ and $\psi(\hat{p}) = -\sqrt{2}(\sqrt{2b_1} - i\hat{p})$.

By joining these operators using the $SU(2) \times SU(2)$ Clebsch-Gordan coefficients, we can construct the basis of $Q_{\underline{M}}$-invariant creation operators in the conformal field sector. Due to the crossing properties of the Clebsch-Gordan coefficients [40], any $Q_{\underline{M}}$-invariant creation operators can be expressed in such a fundamental form. Thus, these two types of $Q_{\underline{M}}$-invariant operators are expected to be the building blocks of physical states in the conformal field sector, which are summarized in Table 2.1.

Table 1.2.1: Building blocks in the conformal field sector.

rank of tensor index	0
creation op.	$S^{\dagger}_{LN}$ $\mathcal{S}^{\dagger}_{L-1N}$
weight ($L \in \mathbf{Z}_{\geq 1}$)	$2L$

The creation mode commuting with $Q_{\underline{M}}$ in the traceless field sector is only the lowest positive-metric mode, $c^{\dagger}_{1(Mx)}$, in the transverse-symmetric-traceless field. The $Q_{\underline{M}}$-invariant creation operators constructed as a bilinear form of the creation modes are rather complicated. We must consider such operators including the tensor indices up to rank 4. We do not here present the explicit forms of such operators, which have been constructed in [22]. They will give the building blocks for the

Table 1.2.2: Building blocks in the traceless field sector.

rank of tensor index	0	1	2	3	4
creation op.	$A^\dagger_{LN}$	$B^\dagger_{L-\frac{1}{2}(Ny)}$	$c^\dagger_{1(Nx)}$	$D^\dagger_{L-\frac{1}{2}(Nz)}$	$E^\dagger_{L(Nw)}$
	$\mathcal{A}^\dagger_{L-1N}$				$\mathcal{E}^\dagger_{L-1(Nw)}$
weight ($L \in \mathbf{Z}_{\geq 3}$)	$2L$	$2L$	2	$2L$	$2L$

traceless field sector listed in Table 2.2. Any $Q_{\underline{M}}$-invariant states are expected to be constructed from these building blocks.

We now construct the physical state satisfying all the conditions (1.2.27) and (1.2.28). Firstly, consider the states that depends only on the zero mode of the conformal field satisfying the $Q_{\underline{M}}$ invariance condition, which is given by $|p,\Omega\rangle = e^{ip\hat{q}}|\Omega\rangle = e^{ip\sqrt{2b_1}\phi_0}|\Omega\rangle$. This is the eigenstate of $\hat{p}$ with eigenvalue $p + i\sqrt{2b_1}$. The general state satisfying the conditions for $Q_{\underline{M}}$ and $R_{\underline{MN}}$ is constructed by acting those building blocks on the state $|p,\Omega\rangle$, with all the tensor indices contracted by using the $SU(2) \times SU(2)$ Clebsch-Gordan coefficients. This state is denoted by $\mathcal{R}_n(S^\dagger, \cdots)|p,\Omega\rangle$, where the operator $\mathcal{R}_n$ carries the conformal weight n. The Hamiltonian condition gives the equation $\left(p + i\sqrt{2b_1}\right)^2/2 + b_1 + n = 4$, so that p should have the purely imaginary value $-i\gamma_n/\sqrt{2b_1}$ with [15, 17, 21]

$$\gamma_n = 2b_1\left(1 - \sqrt{1 - \frac{4-n}{b_1}}\right). \tag{1.2.31}$$

Here, the solution that γ_n approaches the canonical value, $4-n$, in the large b_1 limit is selected. Scaling dimensions of these physical fields are determined by γ_n, which are defined when we compare them with numerical results for the simplicial quantum gravity in Section 3, and with the observed spectra for the CMB anisotropies in Section 4.

As examples, we show the lower n states up to 4 in the followings. The lowest weight state is

$$e^{\gamma_0\phi_0}|\Omega\rangle, \tag{1.2.32}$$

which corresponds to the cosmological constant $\sqrt{-g}$. For $n=2$, there is a gravitational state

$$\mathcal{S}^\dagger_{00}e^{\gamma_2\phi_0}|\Omega\rangle. \tag{1.2.33}$$

This corresponds to the scalar curvature $\sqrt{-g}R$. For $n=4$, there are three gravitational states,

$$\sum_{N,x}\epsilon_N c^\dagger_{1(-Nx)}c^\dagger_{1(Nx)}|\Omega\rangle, \quad \left(\mathcal{S}^\dagger_{00}\right)^2|\Omega\rangle, \quad \sum_N \epsilon_N S^\dagger_{1-N}S^\dagger_{1N}|\Omega\rangle. \tag{1.2.34}$$

The first state corresponds to the square of the Weyl tensor, $\sqrt{-g}C^2_{\mu\nu\lambda\sigma}$, and the second is the square of the scalar curvature, $\sqrt{-g}R^2$. The third is a diffeomorphism invariant state independent of the first two states.

The conformal field sector becomes unnormalizable because of the purely imaginary value of the zero-mode with the momentum p. This is a peculiar property of quantum gravity with the background-metric independence [26]. In such circumstances, there will be no spacetime we can recognize, and physical quantities like the S-matrix become inadequate, since they are always defined on a fixed classical spacetime of observers. As it will be discussed in Section 3, this model with conformal invariance seems to be in the same universality as the four-dimensional simplicial quantum gravity. There, the partition function is defined by a grand canonical ensemble with exponential weight proportional to the volume of the conformal field. It is shown that each Boltzmann weight is positive-definite, and there is no signal to violate the positivity of its ensemble in numerical simulations. Although our understanding of this subject is far from complete, we expect the positivity of the model will be clarified in future.

1.2.3 Renormalization

There are many regularization schemes to extract divergences, among which we show that the dimensional regularization [42] is a systematic and manifestly diffeomorphism invariant method to compute higher order corrections [20]. Renormalization is then carried out with great care concerning the conformal mode dependence. In this regularization scheme, the measure contributions to the action S is completely included between D and 4 dimensions, while the results are independent of how the measure is chosen.

Dimensional Regularization Let us first define the D-dimensional action. We shall consider quantum gravity coupled to QED with n_F fermions, because, QED is the simplest prototype of a wide range of field theories we meet in practice. The renormalizable D-dimensional action near four dimensions is written as

$$I = \int d^D x \sqrt{g} \left\{ \frac{1}{t^2} F_D + bE_D + \frac{1}{4} F_{\mu\nu}F^{\mu\nu} + \sum_{j=1}^{n_F} i\bar{\psi}_j \not{D} \psi_j - \frac{M_{\mathrm{P}}^2}{2} R + \Lambda \right\}. \quad (1.2.35)$$

Here, the Euclidean signature is taken. We can always go back to the Lorentz signature by the Wick rotation which we take in the following discussions.

In order to determine the actions F_D and E_D, we apply the argument given in Subsection 2.1, where we have considered the integrability condition for conformal variations of the effective action, by generalizing the condition to D dimensions. The conformal anomaly defined by a conformal variation of the effective action is a candidate for the counterterm, or the bare action. There are three conformal anomalies satisfying such a condition, in which F_D and E_D are provided as those reducing to the square of the Weyl tensor and the Euler density at four dimensions, respectively.

The action F_D is given by the square of the D-dimensional Weyl tensor,

$$F_D = R_{\mu\nu\lambda\sigma}R^{\mu\nu\lambda\sigma} - \frac{4}{D-2}R_{\mu\nu}R^{\mu\nu} + \frac{2}{(D-1)(D-2)}R^2. \tag{1.2.36}$$

This action gives the kinetic term of the traceless field. The quantity E_D is the D dimensional generalizations of the modified Euler density (1.2.11) defined by

$$E_D = G_D - \frac{4(D-3)^2}{(D-1)(D-2)}\nabla^2 R \tag{1.2.37}$$

with

$$G_D = G_4 + \frac{(D-3)^2(D-4)}{(D-1)^2(D-2)}R^2, \tag{1.2.38}$$

where G_4 is defined by the combination (1.2.3) used in four dimensions.

An important property of E_D is that its volume integral, which is the same as the volume integral of G_D, is expanded around $D = 4$:[3]

$$\int d^D x\sqrt{g}E_D = \sum_{n=0}^{\infty}\frac{(D-4)^n}{n!}S_n, \tag{1.2.39}$$

with

$$S_n(\phi,\bar{g}) = \int d^D x\sqrt{\bar{g}}\left\{2\phi^n\bar{\Delta}_4\phi + O(\phi^n)\right\}, \tag{1.2.40}$$

where $O(\phi^n) = \bar{E}_4\phi^n + \cdots$, and Δ_4 and E_4 are defined by expressions (1.2.10) and (1.2.11), respectively. The action for $n = 1$ corresponds to the Riegert action given previously.

Renormalization can be carried out by replacing the bare quantities in the action I with the renormalized quantities multiplied by the renormalization factors. To make the model finite, we need the following renormalization factors:

$$A_\mu = Z_3^{1/2}A_\mu^r, \quad \psi_j = Z_2^{1/2}\psi_j^r, \quad h_{\mu\nu} = Z_h^{1/2}h_{\mu\nu}^r. \tag{1.2.41}$$

The coupling constants of QED and the traceless field are renormalized as

$$e = Z_e e_r, \quad t = Z_t t_r. \tag{1.2.42}$$

[3]The integrable quantity satisfying this property is unique. This series has a resemblance to that of the scalar curvature expanded around near two dimensions: $\int d^D x\sqrt{g}R = \sum_{n=0}^{\infty}(D-2)^n S_n^{(2)}/n!$ with $S_n^{(2)}(\phi,\bar{g}) = \int d^D x\sqrt{\bar{g}}\left\{\phi^n\bar{\Delta}_2\phi + O(\phi^n)\right\}$, where $O(\phi^n) = \bar{R}\phi^n + \cdots$. This is unique possibility for 2D quantum gravity, and $S_1^{(2)}$ is the Liouville action discussed in previous subsections. The actions for $n > 1$ are generalized forms of the Liouville action. These actions will be necessary to compute anomalous dimensions of composite fields like cosmological constant in dimensional regularization.

The Ward-Takahashi identity holds even when QED couples with quantum gravity, so that $Z_e = Z_3^{-1/2}$. The renormalization procedure concerning b will be discussed below.

The most significant property of this model is that the renomalization factor for the conformal field is unity:

$$Z_\phi = 1. \tag{1.2.43}$$

This comes from the fact that we treat the conformal field non-perturbatively without introducing the coupling constant for this field. Hence, the verification of this relation is considered to be a consistency check of renormalizablity at higher orders.

Renormalization is carried out by expanding the renormalized quantities in the Laurent series, and regarding terms with non-negative powers of $D-4$ as kinetic terms or vertices, and those with negative powers of $D-4$ as counterterms. To begin with, let us consider the QED sector. Z_3 is expanded as

$$Z_3 = 1 + \frac{x_1}{D-4} + \frac{x_2}{(D-4)^2} + \cdots, \tag{1.2.44}$$

where the residues are given by functions of the renormalized couplings, e_r and t_r. Using this expression, the gauge field action is expanded as

$$\begin{aligned}
&\frac{1}{4}\int d^D x \sqrt{g} F_{\mu\nu} F^{\mu\nu} \\
&= \frac{1}{4} Z_3 \int d^D x \, e^{(D-4)\phi} \bar{F}^r_{\mu\nu} \bar{F}^{r\mu\nu} \\
&= \frac{1}{4}\int d^D x \biggl\{ \left(1 + \frac{x_1}{D-4} + \frac{x_2}{(D-4)^2} + \cdots \right) \bar{F}^r_{\mu\nu}\bar{F}^{r\mu\nu} \\
&\qquad + \left(D-4 + x_1 + \frac{x_2}{D-4} + \cdots \right) \phi \bar{F}^r_{\mu\nu}\bar{F}^{r\mu\nu} \\
&\qquad + \frac{1}{2}\left((D-4)^2 + (D-4)x_1 + x_2 + \cdots \right) \phi^2 \bar{F}^r_{\mu\nu}\bar{F}^{r\mu\nu} + \cdots \biggr\},
\end{aligned} \tag{1.2.45}$$

where $\bar{F}^r_{\mu\nu}\bar{F}^{r\mu\nu} = \bar{g}^{\mu\lambda}\bar{g}^{\nu\sigma}F^r_{\mu\nu}F^r_{\lambda\sigma}$. Thus, the residues, x_1 and x_2, are determined from the logarithmic and double logarithmic UV divergences proportional to the kinetic term of the gauge field. These residues induce the new vertices and counterterms of the form $\phi^n \bar{F}^{r2}_{\mu\nu}$.

Similarly, the action of the traceless field can be expanded in terms of the renormalized quantities using expression $\sqrt{g}F_D = e^{(D-4)\phi}\bar{F}_D$ and the renormalization factors Z_t and Z_h. In the exactly same manner as above, new vertices and counterterms of the type $\phi^n \bar{F}_D$ will be induced. Although the expression becomes more complicated, the expansion is straightforward.

The UV divergences related to the E_D term are renormalized by writing b as

$$b = \frac{1}{(4\pi)^2}\sum_{n=1}^{\infty} \frac{b_n}{(D-4)^n}. \tag{1.2.46}$$

Here, we can take $b_0 = 0$, because the lowest (tree) part, $b_0\sqrt{g}G_D$, does not contain the kinetic term for gravitational fields, so that it is not relevant dynamically. We then obtain the following Laurent expansion:

$$\begin{aligned} b\int d^Dx\sqrt{g}E_D & \\ = \frac{1}{(4\pi)^2}\int d^Dx\biggl\{ & \left(\frac{b_1}{D-4}+\frac{b_2}{(D-4)^2}+\cdots\right)\bar{G}_4 \\ & +\left(b_1+\frac{b_2}{D-4}+\cdots\right)\left(2\phi\bar{\Delta}_4\phi+\bar{E}_4\phi+\frac{1}{18}\bar{R}^2\right) \\ & +\frac{1}{2}\Big((D-4)b_1+b_2+\cdots\Big)\left(2\phi^2\bar{\Delta}_4\phi+\bar{E}_4\phi^2+\cdots\right)+\cdots\biggr\}. \end{aligned} \tag{1.2.47}$$

The residues b_n are determined by UV divergences proportional to $\bar{G}_4$. The coefficient, b_1, which corresponds to the logarithmic divergence, produces the kinetic term of the conformal field.

Let us carry out renormalization practically. We employ the Feynaman gauge for the traceless field, so that the propagator becomes

$$\frac{D-2}{2(D-3)}\frac{1}{k^4}I^H_{\mu\nu,\lambda\sigma}, \tag{1.2.48}$$

where $I^H_{\mu\nu,\lambda\sigma} = (\delta_{\mu\lambda}\delta_{\nu\sigma}+\delta_{\mu\sigma}\delta_{\nu\lambda})/2 - \delta_{\mu\nu}\delta_{\lambda\sigma}/D$ is the projection operator to the traceless field. The momentum k is defined on the flat background, not the physical space. This propagator is described by a spiral line in the Feynman diagrams.

The propagator for the conformal field, which is described by a solid line, is given by

$$\frac{(4\pi)^2}{4b_1}\frac{1}{k^4}. \tag{1.2.49}$$

The value, b_1, should be determined self-consistently from the residue of the logarithmic divergence proportional to $\bar{G}_4$, using this propagator itself and new induced vertices with b_1. As given in (1.2.52), the lowest term of b_1 becomes a constant independent of b_1 itself as well as the couplings, because the b_1-dependences in propagators cancel out those in vertices for corresponding diagrams. Therefore, we can compute higher-order corrections order by order.

In this article we only consider UV divergences, and do not treat IR divergences. Basically, there is no problem with IR divergences, because of the presence of the Einstein action and the cosmological constant, which play a role of diffeomorphism invariant IR regulators. Technically, we evaluate diagrams by replacing the propagator $1/k^4$ with $1/(k^2+z^2)^2$, with putting a small fictitious mass, z. Such a mass term is not gauge invariant. We take $z \to 0$ limit after extracting IR divergences.

The beta function for the renormalized coupling constant t_r is calculated to be

$$\beta_t = -\left(\frac{n_F}{40}+\frac{10}{3}\right)\frac{t_r^3}{(4\pi)^2} - \frac{7n_F}{72}\frac{e_r^2t_r^3}{(4\pi)^4} + O(t_r^5). \tag{1.2.50}$$

It shows that the model is asymptotically free with respect to the coupling t_r. For the renormalized coupling constant e_r, we obtain the beta function

$$\beta_e = \frac{4n_F}{3}\frac{e_r^3}{(4\pi)^2} + \left(4n_F - \frac{8}{9}\frac{n_F^2}{b_1}\right)\frac{e_r^5}{(4\pi)^4} + O(e_r^3 t_r^2). \tag{1.2.51}$$

Here, the term proportional to $1/b_1$ is a gravitational correction. We note that there is no order $t_r^2 e_r$ corrections, as similar to the fact that there is no order $e_r^2 t_r$ corrections in the beta function for the traceless field.

The residue of the counterterm for $\bar{G}_4$ can be calculated from the three-point functions of traceless field. For the simple and double pole terms, they are known to be [43]

$$b_1 = \frac{11n_F}{360} + \frac{40}{9} - \frac{n_F^2}{6}\frac{e_r^4}{(4\pi)^4} + O(t_r^2), \quad b_2 = \frac{2n_F^3}{9}\frac{e_r^6}{(4\pi)^6} + O(t_r^4). \tag{1.2.52}$$

There is no order e_r^2 corrections from the QED sector in b_1.

As the consistency check of the background-metric free theory, we demonstrate $Z_\phi = 1$ up to order e_r^6 for the QED coupling, and up to order t_r^2 for the traceless field coupling. The order e_r^2 correction is trivially finite. The diagrams up to order t_r^2 and e_r^4 are depicted in Figure 2.1 and Figure 2.2. Here, the wavy line and the dashed line with arrows denote the photon and the fermion propagators, respectively, and ϵ at the vertices denotes the factor $D - 4$. The subdiagram in Figure 2.2(5), appearing as a circle with s.e. inside, denotes the sum of two-loop diagrams for the ordinary photon self-energy. The vertices in Figure 2.1 are obtained from the induced action $b_1 S_1(\phi, \bar{g})$, and the vertices including the conformal field in Figure 2.2 come from the action (1.2.45). We can see that the sum of these diagrams in each order becomes finite. The finiteness at order e_r^6 is also demonstrated so that a simple pole produced from the sum of loop diagrams of this order does cancel with the direct counterterm for the kinetic term of the conformal field coming from the double-pole term b_2 in (1.2.52). In these computations, Hathrell's results [43] are included.

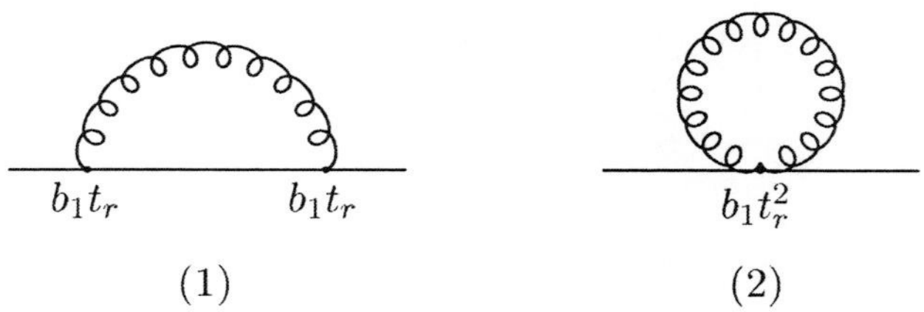

Figure 1.2.1: The order t_r^2 corrections to the kinetic term of the conformal field.

Furthermore, we can demonstrate that the order e_r^6 vertex corrections of the type $\phi F_{\mu\nu}^{r2}$ are renormalized with $Z_\phi = 1$, in which diagrams including the induced vertices of the types $\phi^2 \bar{\Delta}_4 \phi$ and $\phi^n F_{\mu\nu}^{r2}$ with $n = 1, 2$ contribute.

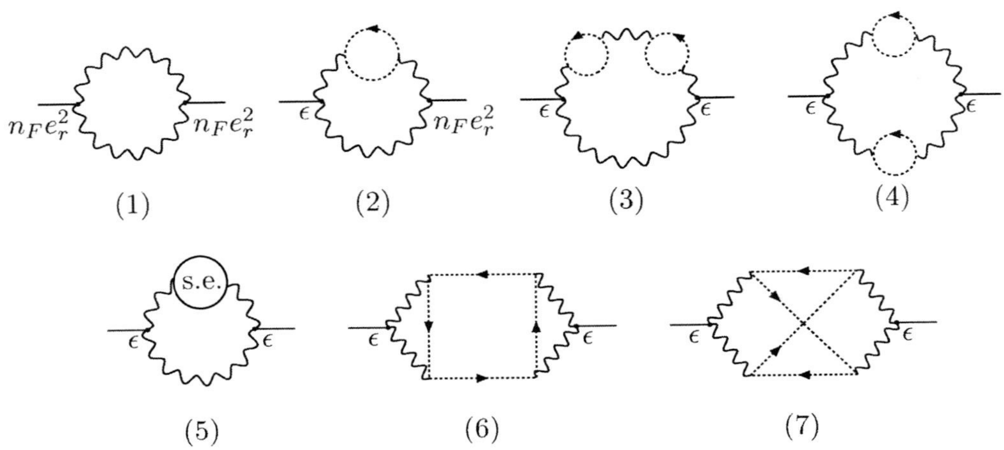

Figure 1.2.2: The order e_r^4 corrections to the kinetic term of the conformal field.

Effective Action We, now, can discuss the effective action given in Subsection 2.1. In this paragraph we go back to employ the Minkowski spacetime. According to the expansion of the running coupling constant,

$$t_r^2(\Delta_4) = \frac{2}{\beta_0 \log \frac{\Delta_4}{\Lambda_{\rm QG}^4}} = t_r^2 + 2\beta_0 t_r^4 \phi - \frac{\beta_0}{2} t_r^4 \log \frac{\bar{\Delta}_4}{\mu^4} + \cdots, \tag{1.2.53}$$

where $t_r = t_r(\mu)$, the effective action (1.2.9) is expanded in t_r for the Weyl sector as

$$\Gamma_{\rm W} = \int d^4x \left\{ -\frac{1}{t_r^2} \bar{C}_{\mu\nu\lambda\sigma}^2 + 2\beta_0 \phi \bar{C}_{\mu\nu\lambda\sigma}^2 - \frac{\beta_0}{2} \bar{C}^{\mu\nu\lambda\sigma} \log\left(\frac{\bar{\Delta}_4}{\mu^4}\right) \bar{C}_{\mu\nu\lambda\sigma} + \cdots \right\}, \tag{1.2.54}$$

and for the generalized Riegert action as

$$\begin{aligned} \Gamma_{\rm R} \;=\; & -\frac{b_1}{(4\pi)^2} \int d^4x \bigg\{ (1 - a_1 t_r^2 + \cdots) \left(2\phi \bar{\Delta}_4 \phi + \bar{E}_4 \phi + \frac{1}{18} \bar{R}^2 \right) \\ & -2a_1 \beta_0 t_r^4 \left(2\phi^2 \bar{\Delta}_4 \phi + \bar{E}_4 \phi^2 \right) \\ & + \frac{1}{2} a_1 \beta_0 t_r^4 \left[2\phi \bar{\Delta}_4 \log\left(\frac{\bar{\Delta}_4}{\mu^4}\right) \phi + \bar{E}_4 \log\left(\frac{\bar{\Delta}_4}{\mu^4}\right) \phi \right] + \cdots \bigg\}. \end{aligned} \tag{1.2.55}$$

These expressions are consistent with the perturbative computations of the effective actions in t_r. The logarithmic terms are finite contributions from loop diagrams which yield a simple pole.

The coefficient a_1 has the form, $a_1 = a_{\rm L} b_1 + a_{\rm NL} + a_{\rm NNL}/b_1$, where $a_{\rm L} = 1/12(4\pi)^2$ is the leading contribution in the large number limit of matter fields, obtained by the finite diagram shown in Figure 2.3. There are many contributions to the next leading term $a_{\rm NL}$ and the next to next leading term $a_{\rm NNL}$. The finite parts of diadrams (1) and (2) of Figure 2.1, for instance, give the positive contributions [19], 1/2 and 5/4, to the next leading coefficient $a_{\rm NL}$, respectively. In addition to these, there are trivially finite loop diagrams contributing to this term.

Figure 1.2.3: The diagram contributing to $a_{\rm L}$ in the coefficient a_1.

Finally, we compute the anomalous dimension for the composite field of the cosmological constant. The cosmological constant term is expanded as $\Lambda e^{D\phi} = Z_\Lambda \Lambda^r \sum_n (D\phi)^n/n!$, where Z_Λ is the renormalization factor and Λ^r is the renormalized cosmological constant. The diagrams contributing up to order $1/b_1^2$ in the large b_1 limit are depicted in Figure 2.4. They give

$$\gamma_\Lambda = \frac{\mu}{Z_\Lambda}\frac{dZ_\Lambda}{d\mu} = \frac{4}{b_1} + \frac{8}{b_1^2} + \cdots . \tag{1.2.56}$$

This result agrees with the exact value γ_0 in (1.2.31) computed using the conformal

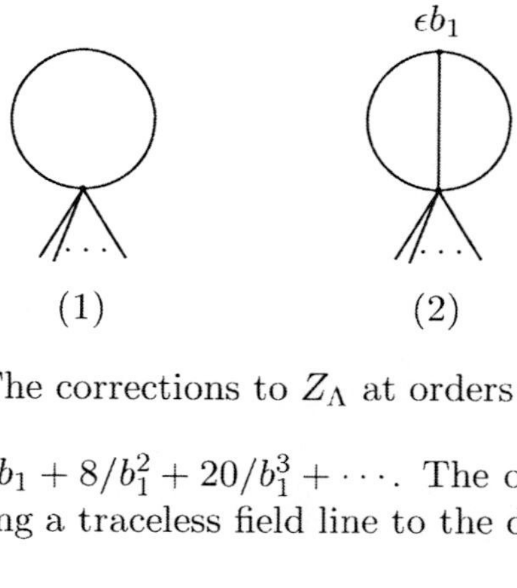

Figure 1.2.4: The corrections to Z_Λ at orders $1/b_1$ and $1/b_1^2$.

algebra, $\gamma_\Lambda = \gamma_0 - 4 = 4/b_1 + 8/b_1^2 + 20/b_1^3 + \cdots$. The order t_r^2 corrections will be given by diagrams attaching a traceless field line to the diagrams in Figure 2.4.

1.3 Numerical Quantum Gravity

In the last section a renormalizable model of quantum gravity is proposed and its property is examined extensively. It is constructed on the three basic requirements: (1) background metric independence, (2) finiteness, and (3) four spacetime dimensions. We have seen that the problem of non-renormalizabilty of Einstein quantum gravity could be overcome by introducing higher derivative terms in the action, such as the Weyl tensor term and the topological term. While the renormalization problem seemed to be avoided by such an action, it introduces a new problem of the unitarity. As one of possible resolutions a systematic method for constructing the physical space is proposed, but full understanding of the problem is still on the way.

Although it is widely believed that the Einstein quantum gravity is perturbatively non-renormalizable, there remains a chance to be quantized non-perturbatively. As one of efforts along such a direction we discuss here a method of quantization in the simplectic manifold by discretizing the four dimensional spacetime. Historically, discretization of spacetime has been proposed by Regge [44] in order to formulate general relativity without introducing a priori background metric. Because of background free nature of the theory by construction, it is readily reformulated for quantizing spacetime.

Simplicial Quantum Gravity In the simplicial quantum gravity, four dimensional spacetime is created by putting equilateral 4-simplices together to form a 4-dimensional piecewise linear (PL) manifold, regarding each simplex having flat geometry. Quantization is performed in the same sprit as the Feynman path integral: we calculate the partition function defined by

$$Z = \sum_{\mathcal{T}} \exp(-S_G - S_M), \tag{1.3.1}$$

where the sum is taken over all possible triangulations $\mathcal{T}$, *i.e.* combinations of 4-simplices put together to form a 4-dimensional manifold, and all possible topologies. This method is known as the dynamical triangulation [45, 46, 47, 48, 49, 50, 51]. Unlike two-dimension, the classification of topologies for 4-dimensional manifolds is not completely known, we shall restrict our discussion within the fixed spherical topology, S^4.

The action S $(= S_G + S_M)$ is required to be a real function in numerical simulations, and we regard the topological and the real number constraints on the action as the boundary conditions for simulating a closed universe of $R \times S^3$ topology with the Wick rotation along the R direction. The meaning of these conditions will be discussed in the context of creation and evolution of the universe, later. The total action is composed of the spacetime term S_G, and the matter term S_M. The local coordinate transformation invariance suggests that S_G is a function of global variables of the simplicial manifold such as numbers of i-simplices $N_i (i = 0, 1, .., 4)$. From the requirement for the existence of a proper thermodynamical limit it should also be a linear function of these variables. The Einstein-Hilbert action represented in the simplicial space satisfies these conditions [44, 52]:

$$\begin{aligned} \int d^4x\sqrt{g}R &\rightarrow 4\pi A_2 N_4 \left(\frac{N_2}{N_4} - \frac{5}{\pi} cos^{-1}\frac{1}{4} \right), \\ \int d^4x\sqrt{g} &\rightarrow A_4 N_4, \end{aligned} \tag{1.3.2}$$

where A_n $(= \sqrt{(n+1)/2^n} a^n/n!)$ is the 'volume' of an n-simplex with the edge length a.

These five variables are not independent because of the manifold conditions: (0) there should be no two d-simplices ($d = 1, 2, 3, 4$) having the same set of vertices. (1) Two 4-simplices are attached through one 3-simplex, and, inversely, any 3-simplex is shared by two 4-simplices. This condition gives

$$5N_4 = 2N_3. \tag{1.3.3}$$

(2) 4-simplices sharing a 0-simplex (vertex) form a 4-ball, which gives another condition;

$$-2N_1 + 3N_2 - 4N_3 + 5N_4 = 0. \tag{1.3.4}$$

Together with the Euler relation,

$$N_0 - N_1 + N_2 - N_3 + N_4 = \chi, \tag{1.3.5}$$

with χ being the Euler charactoristics, we have three relationships for five variables. Choosing N_2 and N_4 as the independent variables, other variables are written as

$$\begin{aligned} N_0 &= \chi + \frac{1}{2}N_2 - N_4, \\ N_1 &= \frac{3}{2}N_2 - \frac{5}{2}N_4, \\ N_3 &= \frac{5}{2}N_4. \end{aligned} \tag{1.3.6}$$

In terms of N_2 and N_4 the spacetime term of the action is written as

$$S_G = \kappa_0\chi - \kappa_2 N_2 + \kappa_4 N_4, \tag{1.3.7}$$

where the κ_0 term, which is a constant for the fixed spherical topology ($\chi = 2$), is added for completeness. Two parameters κ_2 and κ_4 are related to the constants in the Einstein-Hilbert action,

$$\int d^4x\sqrt{g}\left(-\frac{1}{16\pi G}R + \Lambda\right), \tag{1.3.8}$$

through

$$\begin{aligned} \kappa_2 &= \frac{1}{4G}A_2, \\ \kappa_4 &= \Lambda A_4 + \frac{5}{4\pi G}A_2 cos^{-1}\frac{1}{4}. \end{aligned} \tag{1.3.9}$$

Monte Carlo Quantization In the simplicial quantization the path integral of metric function is carried out by summing contributions over all possible combinations of triangulation, which may be called configurations. This method has been examined in the 2-dimensional case intensively [53], and known to be equivalent

Table 1.3.1: Changes of variables by each move.

P	(p,q)	ΔN_0	ΔN_1	ΔN_2	ΔN_3	ΔN_4
1	$(1,5)$	1	5	10	10	4
2	$(2,4)$	0	1	4	5	2
3	$(3,3)$	0	0	0	0	0
4	$(4,2)$	0	-1	-4	-5	-2
5	$(5,1)$	-1	-5	-10	-10	-4

to the matrix model and the Liouville theory, for example, through string susceptibilities for several matters and topologies [54]. The sums are performed by the Monte Carlo method creating a Markov chain by the so called (p,q)-moves. Any two configurations are connected by combinations of the following five elementary (p,q)-moves: $(1,5)$, $(2,4)$, $(3,3)$, $(4,2)$, and $(5,1)$-move. The parameters p and q represent numbers of 4-simplices annihilated and created by the (p,q)-move, respectively. Those six simplices, $p+q$, involved in each move constitute the surface of a smallest 5-ball, that is a 4-sphere, and thus all the moves conserve the topology. In Figure 3.1 we show those five moves, and in Table 3.1 we list changes of variables by each move.

When one starts from the smallest configuration with a fixed topology, the net increases of numbers of n-simplex, $\{N_n\} = N_n - N_n^{(0)}$ with the initial configuration $N_n^{(0)}$, are given by

$$\begin{aligned} \{N_0\} &= m_1 - m_5, \\ \{N_1\} &= 5(m_1 - m_5) + (m_2 - m_4), \\ \{N_2\} &= 10(m_1 - m_5) + 4(m_2 - m_4), \\ \{N_3\} &= 10(m_1 - m_5) + 5(m_2 - m_4), \\ \{N_4\} &= 4(m_1 - m_5) + 2(m_2 - m_4), \end{aligned} \tag{1.3.10}$$

for a configuration constructed by combinations of m_p moves for each (p,q)-move. Since $\{N_n\}$ cannot be negative for any possible configurations, it requires

$$\begin{aligned} m_1 - m_5 &= \frac{1}{2}\{N_2\} - \{N_4\} \geq 0, \\ m_2 - m_4 &= -\{N_2\} + \frac{5}{2}\{N_4\} \geq 0. \end{aligned} \tag{1.3.11}$$

This gives the constraint on the action. For example, starting with the surface of a 5-simplex made by six vertices as the minimum S^4 configuration, the action S_G is bounded as

$$\frac{1}{4G} A_2 N_4 \left(\frac{5}{\pi} cos^{-1}\frac{1}{4} - \frac{5}{2} \right) \leq S_G \leq \frac{1}{4G} A_2 N_4 \left(\frac{5}{\pi} cos^{-1}\frac{1}{4} - 2 \right), \tag{1.3.12}$$

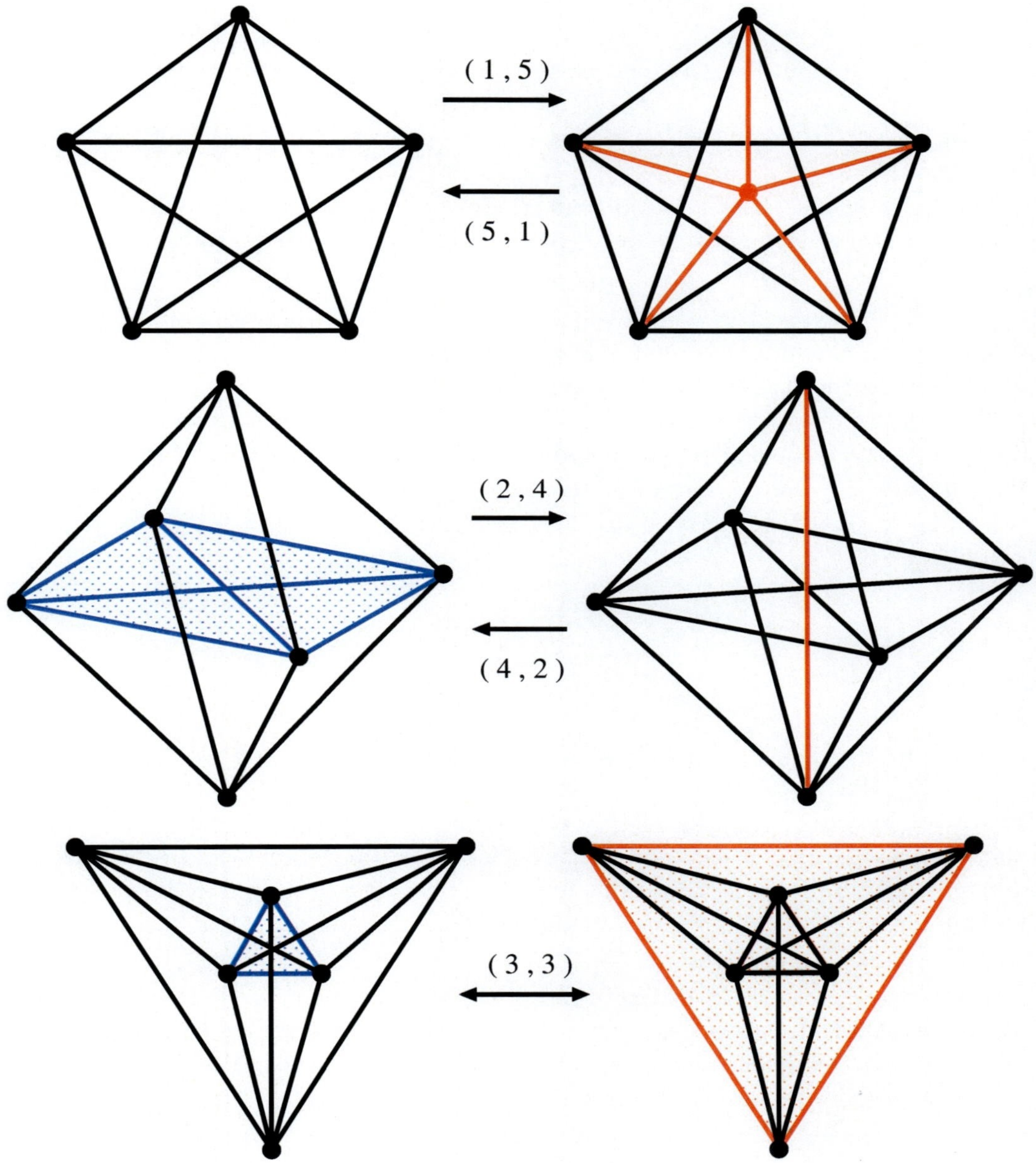

Figure 1.3.1: The five types of (p, q)-move.

for large N_4.

These moves except the (3,3)-move involve changes of N_4, and in such cases the grand canonical Monte Carlo method should be employed. To achieve the equilibrium distribution, $P_a \sim e^{-S_a}$, the detail balance condition for the transition between neighboring configurations a and b in the Markov process with the move probabilities $w_{a\to b}$ and $w_{b\to a}$,

$$\frac{1}{n_a} P_a w_{a\to b} = \frac{1}{n_b} P_b w_{b\to a}, \tag{1.3.13}$$

should be satisfied. Here, the additional factor, n_a, takes care of changes in number of states caused by the change of degrees of freedom. It is given by

$$n_a = n_a^{(1,5)} + n_a^{(2,4)} + n_a^{(3,3)} + n_a^{(4,2)} + n_a^{(5,1)}, \tag{1.3.14}$$

where $n_a^{(p,q)}$ is the number of possible (p,q)-moves satisfying the manifold conditions $(0) \sim (2)$ starting from a configuration a.

In order to obtain a non-singular configuration lying between two extreme manifolds, *i.e.* the branched polymer and the crumpled sphere, we add matter fields. We consider only those cases of adding N_X scalars $\{X_i^{(\alpha)}\}(\alpha = 1, \cdots, N_X)$ on vertices, and N_A vectors $\{A_{ij}^{(\beta)}\}(\beta = 1, \cdots, N_A)$ on links with $A_{ji}^{(\beta)} = -A_{ij}^{(\beta)}$. They are written as

$$\begin{aligned} S_X &= \frac{1}{2}\sum_{\alpha=1}^{N_X}\sum_{i,j} \mathcal{O}(l_{ij}) \left\{X_i^{(\alpha)} - X_j^{(\alpha)}\right\}^2, \\ S_A &= \frac{1}{2}\sum_{\beta=1}^{N_A}\sum_{i,j,k} \mathcal{O}(t_{ijk}) \left\{A_{ij}^{(\beta)} + A_{jk}^{(\beta)} + A_{ki}^{(\beta)}\right\}^2, \end{aligned} \tag{1.3.15}$$

where the factors $\mathcal{O}(l_{ij})$ and $\mathcal{O}(t_{ijk})$ are numbers of 4-simplices sharing the link l_{ij} and the triangle t_{ijk}, respectively. They guarantee the existence of proper thermodynamical limit. These expressions are also consistent with kinetic terms of fields written in terms of forms defined in the simplectic manifold [55].

Numerical Results In order to check the simulation procedures we have employed, we apply them to the 2-dimensional quantum gravity, where analytic solutions are available from the Liouville field theory and the matrix model. The grand canonical method makes it possible to obtain the number of states directly, if we give the factor $n_a^{(p,q)}$ in the detailed balance condition properly. For pure gravity the partition function is proportional to the number of sates, which is known to be $\exp(\lambda^c N_2)$ with $\lambda^c \sim \frac{1}{2}\log\frac{256}{27}$ [56], while the numerical simulation gives the correct value within four digits [54].

The partition function for the grand canonical simulation can be written as

$$Z_{SG}(\lambda) = \sum_{N_2} e^{-\lambda N_2} Z_{SG}(N_2). \tag{1.3.16}$$

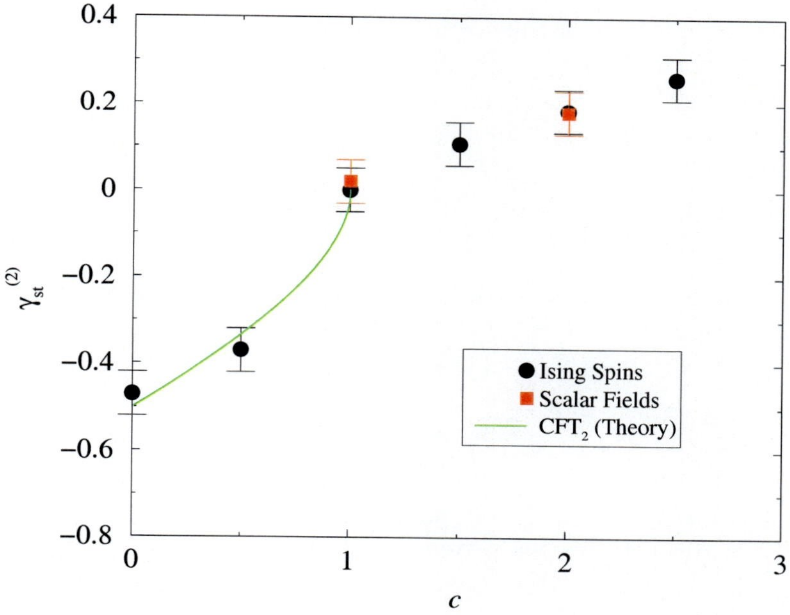

Figure 1.3.2: The string susceptibility $\gamma_{\rm st}^{(2)}$ versus the central charge c in the two-dimensioanl case.

By factorizing the exponential state number factor as $Z_{SG}(N_2) = z_{SG}(N_2)e^{\lambda^c N_2}$ the grand canonical partition function becomes

$$Z_{SG}(\lambda) = A_2 \sum_{N_2} e^{-\frac{\lambda-\lambda^c}{A_2}A_2N_2} \frac{1}{A_2} z_{SG}(N_2). \tag{1.3.17}$$

It has been known that there is the universal continuum limit: $N_2 \to \infty$, $a \to 0$ with a fixed area $A_2N_2 \to V_2$ and a 'cosmological' constant $(\lambda - \lambda^c)/A_2 \to \Lambda^{(2)}$, the partition function converges as

$$Z_{SG}(\lambda) \to \int_0^\infty dV_2 e^{-\Lambda^{(2)}V_2} z_C(V_2), \tag{1.3.18}$$

where $\frac{1}{A_2}z_{SG}(N_2)$ is assumed to converge to $z_C(V_2)$ in the limit.

The partition function has the logarithmic quantum correction, which is written for large N_2 as

$$Z_{SG}(N_2) \sim N_2^{\gamma_{\rm st}^{(2)}-3} e^{\lambda^c N_2}, \tag{1.3.19}$$

where $\gamma_{\rm st}^{(2)}$ corresponds the so-called string susceptibility. The string susceptibility in Liouville theory is obtained by the following way: the number of 2-simplices, N_2, is proportional to the dimensionless spacetime volume, $v_2 \sim V_2/a^2$, determined by the zero-mode dependence of the cosmological constant field, $v_2 = e^{\gamma_0 \phi_0}$, where γ_0 is given by (1.2.19) with $\Delta = 0$. Since the zero-mode dependence of the Liouville action (1.2.13) is given by $S_L(\phi_0) = b_L \chi \phi_0$, the partition function behaves as a function for the fixed volume v_2 like

$$z_C(v_2) \sim \int d\phi_0 e^{-b_L \chi \phi_0} \delta(e^{\gamma_0 \phi_0} - v_2) \sim v_2^{-1 - b_L \chi / \gamma_0}. \tag{1.3.20}$$

Comparing to (1.3.19) by identifying v_2 with N_2, we obtain the formula $\gamma_{\rm st}^{(2)} = 2 - b_L \chi / \gamma_0$, which is written as a function of the central charge c by

$$\gamma_{\rm st}^{(2)} = 2 + \frac{1}{12} \left\{ c - 25 - \sqrt{(25-c)(1-c)} \right\}, \tag{1.3.21}$$

for the spherical topology, $\chi = 2$. The numerical simulations have been done by the grand canonical method where the string susceptibility is extracted from logarithmic corrections of state numbers for systems with $n(=1,2)$ Ising spins for the central charge $c = n/2$ and $m(=1,2)$ scalar fields for $c = m$. They give good agreements with the theoretical predictions as shown in Figure 3.2 as far as $c \leq 1$. For $1 < c < 25$ the theoretical string susceptibility is predicted to be a complex number and it is considered as the reflection of unstable configurations, while the numerical simulation gives positive numbers, which indicates the instability of surface towards creation of baby universes [57].

In the four dimensional case the partition function is written as

$$Z_{SG}(\kappa_2, \kappa_4) = \sum_{N_4} e^{-\kappa_4 N_4} Z_{SG}(\kappa_2, N_4), \tag{1.3.22}$$

where

$$Z_{SG}(\kappa_2, N_4) = \sum_{\mathcal{T}(N_4)} \frac{1}{\mathcal{N}_{\mathcal{T}}} \int [\mathcal{D}X][\mathcal{D}A] e^{\kappa_2 N_2 - S_M(X,A)}. \tag{1.3.23}$$

Here, the sums are taken over all possible configurations with a fixed N_4, and $\mathcal{N}_{\mathcal{T}}$ is the symmetric factor in order to avoid over-counting due to discrete symmetries of a configuration $\mathcal{T}$. The number of states is bounded exponentially, and we factorize it as

$$Z_{SG}(\kappa_2, N_4) = z_{SG}(\kappa_2, N_4) e^{\kappa_4^c(\kappa_2) N_4}. \tag{1.3.24}$$

We show in Figure 3.3 κ_4^c and $\langle N_2 \rangle / N_4$ as functions of κ_2 for a simulation with one vector field.

By the similar manner as the 2-dimensional case the grand canonical partition function is written as

$$Z_{SG}(\kappa_2, \kappa_4) = A_4 \sum_{N_4} e^{-\frac{\kappa_4 - \kappa_4^c}{A_4} V_4} \frac{1}{A_4} z_{SG}(\kappa_2, N_4), \tag{1.3.25}$$

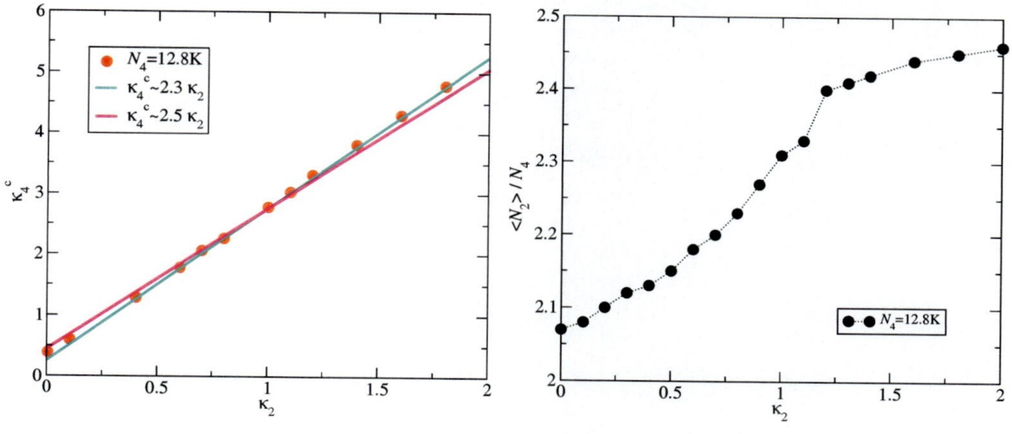

Figure 1.3.3: κ_4^c and $\langle N_2\rangle/N_4$ versus κ_2.

which converges to

$$Z_{SG}(\kappa_2, \kappa_4)) \to \int_0^\infty dV_4 e^{-\Lambda V_4} z_C(G, \Lambda), \tag{1.3.26}$$

in the continuum limit: $N_4 \to \infty$, $a \to 0$ with a fixed volume, $A_4 N_4 \to V_4$, a gravitational constant, $A_2/4\kappa_2 \to G$, and a cosmological constant $(\kappa_4 - \kappa_4^c)/A_4 \to \Lambda$

If the simplicial quantum gravity has the universal continuum limit, we expect that the system makes a second order phase transition at $\kappa_2 = \kappa_2^c$. In such a case it could be observed by measuring the N_i-susceptibility

$$\chi_i = \langle (N_i - \langle N_i \rangle)^2 \rangle. \tag{1.3.27}$$

It has been known that for pure gravity the transition is of first order, and in order to obtain the universal continuum limit we need to put matter fields, at least one vector field [58]. We have made simulations with $N_{\rm X}$ scalar matters and $N_{\rm A}$ vector matters and observe the nature of phase transitions through the finite size scaling of the N_2-susceptibility. Figure 3.4 shows the phase diagram as a function of κ_2 and $N_{\rm X} + 62N_{\rm A}$, which form is expected from the conformal gravity. Change of order of the phase transition is also expected from the change of sign in the string susceptibility defined as the same manner as the 2-dimensional case in the scaling limit:

$$Z(\kappa_2, N_4) \sim N_4^{\gamma_{\rm st}^{(4)} - 3} e^{\kappa_4^c N_4}. \tag{1.3.28}$$

Here, the susceptibility $\gamma_{\rm st}^{(4)}$ can be determined from the volume-dependence of the patition function derived by the Euclidean continuum theory. Since the zero-mode dependence of the Riegert action (1.2.20) is given by $S_1(\phi_0) = 2b_1\chi\phi_0$, the

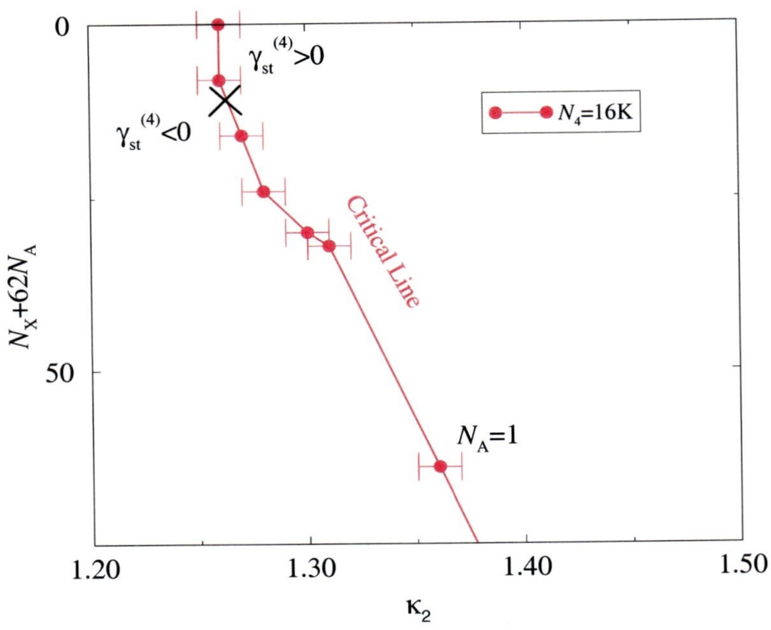

Figure 1.3.4: The phase transition line of the four dimensional case with matter fields (N_X,N_A). Place for the change of sign of $\gamma_{\rm st}^{(4)}$ is indicated by $\times$.

partition function behaves as a function for the fixed volume $V_4 \sim a^4 v_4$ like

$$Z(\kappa_2, v_4) \sim \int d\phi_0 \exp\left\{-2b_1\chi\phi_0 + \kappa_2 e^{\gamma_2\phi_0}\mathcal{R}_2\right\}\delta(e^{\gamma_0\phi_0} - v_4)$$
$$\sim v_4^{-1-2b_1\chi/\gamma_0} f(\kappa_2 v_4^{\gamma_2/\gamma_0}). \tag{1.3.29}$$

Thus, we obtain $\gamma_{\rm st}^{(4)} = 2 - 2b_1\chi/\gamma_0$. At the lowest of the perturbation for the coupling t, γ_0 are given by (1.2.31), and hence,

$$\gamma_{\rm st}^{(4)} = 2 - \frac{1}{2}\left(b_1 + \sqrt{b_1^2 - 4b_1}\right) \tag{1.3.30}$$

for the spherical topology. We show in Figure 3.5 values of b_1's calculated through eq.(1.3.30) using the measured $\gamma_{\rm st}^{(4)}$'s. It is remarkable that the non-perturbative calculation shows the small difference from $b_1 = (N_{\rm X} + 62N_{\rm A})/360 + 769/180 = 0.00278(N_{\rm X} + 62N_{\rm A}) + 4.27$, which seems to justify the perturbative treatment of traceless modes in the background-metric independent quantum gravity.

1.4 Cosmology

In 2003, the WMAP team [3] released the first year result together with the the-

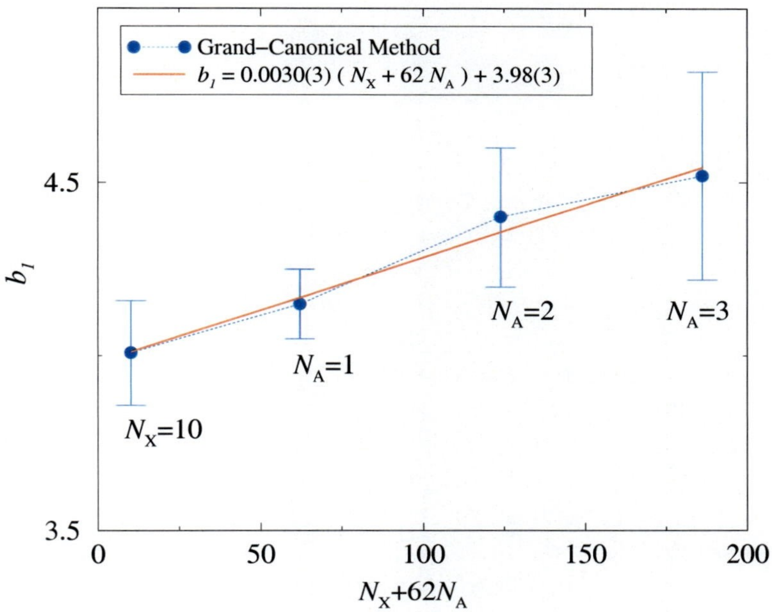

Figure 1.3.5: Comparison of central charges between numerical simulations and conformal gravity predictions.

oretical analysis using the ΛCDM model, which showed surprisingly good fit except low multipole components of the angular power spectra corresponding to the long distance correlation. Although the deviation from the theoretical prediction is normally explained as the cosmic variance [59], there are various alternative ideas such as the appearance of quantum nature of spacetime. We shall take the physical view point that it is a reflection of the scale of quantum gravity [6], which has appeared in the very early universe when it made a sharp transition from the quantum spacetime with the full conformal invariance to the Friedmann spacetime of the classical Einstein gravity.

We shall give an evolutional scenario including the inflation driven by quantum gravity without any artificial fields in addition to the full metric field. Based on this scenario we make fit the angular power spectrum to the observed data of WMAP within low multipole components. We also present a method to check the theory numerically in terms of the simplicial quantum gravity.

Quantum Gravity Scenario of Inflation Let us first illustrate the quantum gravity scenario of the inflationary universe. There are three mass scales involved in the model, namely the Planck mass M_P, the dynamical scale Λ_{QG}, and the cosmological constant Λ. We set their ordering as [4]

$$M_{\rm P} \gg \Lambda_{\rm QG} \gg \Lambda^{1/4}. \tag{1.4.1}$$

The inverse of the energy scale is regarded as the typical size of the universe.

In the very early epoch when the space extends much smaller than the Planck scale, typical energy is higher than $M_{\rm P}$. Corrections of order $\alpha_{\rm G}$ in the asymptotic free regime can be neglected and the dynamics is governed by the $\rm CFT_4$ with the full conformal invariance. Then, there exists no classical spacetime and the universe is filled with quantum fluctuations of the conformal field.

Evolution of the universe is governed by the equation of motion derived from the effective action (1.2.9). As the universe expands typical energy gets lowered to $M_{\rm P}$, and the Einstein action gradually becomes effective, so that the conformal symmetry starts to be broken to develop the classical solution. At the $\Lambda_{\rm QG}$ scale, this symmetry is completely broken and the universe transforms to the classical solution dominant spacetime.

Since the equation of motion is rather complicated containing non-linear as well as higher order derivative terms, we simplify it under the following physical consideration: the running coupling constant $\alpha_G = \{2\pi\beta_0 \log(\Delta_4/\Lambda_{\rm QG}^4)\}^{-1}$ is an operator which acts on the field ϕ, and it may fluctuate very much from a smooth background. When the scale of local fluctuation of the order $1/M_{\rm P}$ is very small compared to the size of the system ($\sim 1/\Lambda_{\rm QG}$) we approximate the running coupling operator by its average known as the mean-field approximation: $\alpha_G \to \langle\alpha_G\rangle$. The equation of motion obtained from the effective action (1.2.9) for the homogeneous solution $\phi = \phi(\eta)$ is then written as

$$-\frac{4}{(4\pi)^2} b_1 B(\tau) \partial_\eta^4 \phi + 6 M_{\rm P}^2 e^{2\phi} \left\{ \partial_\eta^2 \phi + (\partial_\eta \phi)^2 \right\} = 0, \tag{1.4.2}$$

where $B(\tau) = 1/(1 + 4\pi a_1 \langle\alpha_{\rm G}\rangle)$. Instead of calculating the averaged running coupling we make a bold guess as $\langle\alpha_{\rm G}\rangle = \{2\pi\beta_0 \log(1/\tau^4 \Lambda_{\rm QG}^4)\}^{-1}$ for $\tau < \tau_\Lambda = 1/\Lambda_{\rm QG}$, from the dimensional consideration, where τ is the proper time defined by $d\tau = a d\eta$ with $a = e^\phi$. For $\tau \geq \tau_\Lambda$, the factor B vanishes. The conservation equation for the energy density, under this approximation scheme, becomes

$$\frac{b_1}{(4\pi)^2} \partial_\eta B(\tau) \left\{ 4 \partial_\eta^3 \phi \partial_\eta \phi - 2 \partial_\eta^2 \phi \partial_\eta^2 \phi \right\} + e^{4\phi} \left\{ \partial_\eta \rho + 4 \partial_\eta \phi \rho \right\} = 0, \tag{1.4.3}$$

where ρ is the matter energy density, assuming it to be a perfect fluid.

In order to solve the equation analytically, we simply replace the function $B(\tau)$ by the step function $B(\tau) = 1 - \theta(\tau - \tau_\Lambda)$. Introducing the variable $H = \dot{a}(\tau)/a(\tau)$,

---

[4] The condition $M_{\rm P} \gg \Lambda_{\rm QG}$ implies that not only the spacetime singularity but also the horizon of an elementary black hole with the Planck mass disappear above the energy byond the Planck mass.

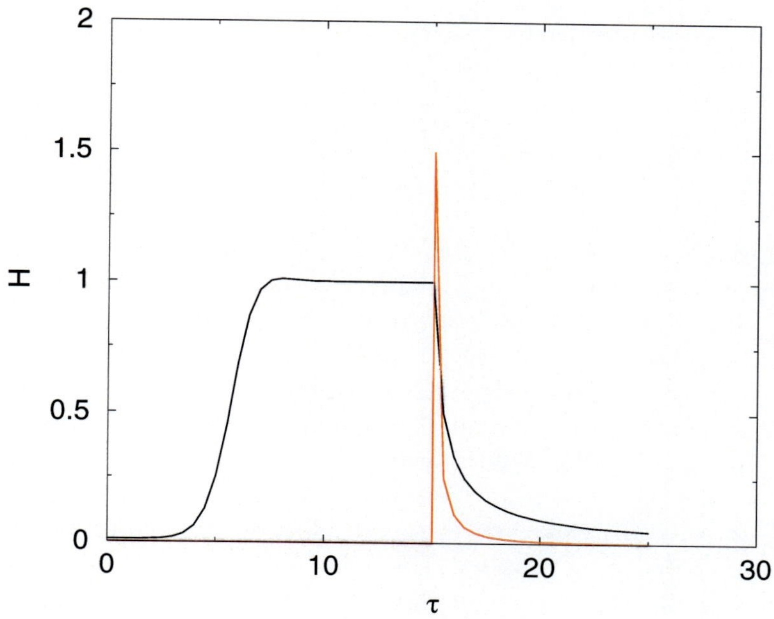

Figure 1.4.1: The time evolution of H and ρ. $H_{\mathbf{D}}$ is normalized to 1. $H(0)$ is taken to be 0.01(black), and $\rho_{\mathbf{D}} = 1.5$.

the equation of motion becomes

$$\begin{cases} b_1(\dddot{H} + 7H\ddot{H} + 4\dot{H}^2 + 18H^2\dot{H} + 6H^4) = 24\pi^2 M_{\rm P}^2(\dot{H} + 2H^2) & \text{if } \tau < \tau_\Lambda \\ \dot{H} + 2H^2 = 0 & \text{otherwise,} \end{cases} \tag{1.4.4}$$

and the conservation equation is

$$\dot{\rho} + 4H\rho = \frac{b_1}{8\pi^2}\delta(\tau - \tau_\Lambda)(2H\ddot{H} - \dot{H}^2 + 6H^2\dot{H} + 3H^4). \tag{1.4.5}$$

For $\tau < \tau_\Lambda$, it has two stationary solutions: a *stable* inflationary solution with $H = H_{\mathbf{D}}$ and $\rho = 0$,

$$a(\tau) \propto e^{H_{\mathbf{D}}\tau}, \qquad H_{\mathbf{D}} = \sqrt{8\pi^2/b_1}M_{\rm P}, \tag{1.4.6}$$

and an unstable solution with $H = 0$. Starting with the unstable solution, the solution shifts to the stable one (1.4.6) by an infinitely small perturbation $H \approx \epsilon$ (Figure 4.1). Quantum fluctuations of the conformal field still keep the long range correlation over the horizon scale on the de Sitter background (1.4.6).

The equations of motion indicate that at $\tau = \tau_\Lambda$, the spacetime shifts to the Friedmann universe with initial conditions for the Hubble constant $H_{\mathbf{D}}$, and the matter density $\rho_{\mathbf{D}} = 24\pi^2 M_{\mathrm{P}}^4/b_1 = 3H_{\mathbf{D}}^2/8\pi G$. For $\tau \geq \tau_\Lambda$ we obtain

$$\begin{aligned} H(\tau) &= \frac{H_{\mathbf{D}}}{2H_{\mathbf{D}}(\tau - \tau_\Lambda) + 1}, \\ \rho(\tau) &= \frac{\rho_{\mathbf{D}}}{\{2H_{\mathbf{D}}(\tau - \tau_\Lambda) + 1\}^2}. \end{aligned} \tag{1.4.7}$$

This solution suggests that the long-range quantum fluctuation of the conformal field covering over the whole quantum universe makes a transition to matter fluctuations at τ_Λ, and the expansion slows down from the inflating conformal spacetime to the Friedmann spacetime. At the transition point, the scalar curvature tends to vanish discontinuously. The entropy is then generated to develop the thermal universe, which is regarded as the big bang in our scenario. The primordial fluctuation we extract from the CMB observation is expected to be the reflection of quantum fluctuation of the conformal field developed right before this transition.

In order to solve the equations, we have adopted the step function for the time dependence of the running coupling constant B for simplicity. In reality, the transition process would be rather complicated: until the universe is thermalized, it evolves under the influence of rapidly growing traceless-field. Adopting the step function means neglecting such a dynamical process. We also simplified the thermalization process, assuming the universe reached to the equilibrium soon after the inflation. In order to make our big bang scenario complete, we need an appropriate field theoretical analysis including the non-linear effects of the traceless mode to describe the transition dynamically.

Since the inflation starts at the Planck scale $\tau_{\mathrm{P}} = 1/H_{\mathbf{D}}$ and ends at $\tau_\Lambda = 1/\Lambda_{\mathrm{QG}}$, the number of e-foldings of inflation is estimated as

$$\mathcal{N}_e = \log\frac{a(\tau_\Lambda)}{a(\tau_{\mathrm{P}})} = H_{\mathbf{D}}(\tau_\Lambda - \tau_{\mathrm{P}}) \simeq \frac{H_{\mathbf{D}}}{\Lambda_{\mathrm{QG}}}. \tag{1.4.8}$$

Once the value of Λ_{QG} is chosen to fit the observed data, we can check whether the model can give the right number of e-foldings large enough to solve the flatness and horizon problems.

CMB Anisotropies The observed CMB anisotropies are understood as a reflection of the gravitational potential on the last scattering surface(lss), when photons decouple to the recombined matter. The temperature fluctuation follows the relationship known as the Sachs-Wolfe effect [60], $\delta T/T = \Phi(\mathbf{x}_{\mathrm{lss}})/3$, where $\Phi(\mathbf{x}_{\mathrm{lss}})$ is the gravitational potential on the lss, $\mathbf{x}_{\mathrm{lss}}$, in the comoving frame.

Observationaly the angular power spectrum is obtained from the two point correlation function of $\delta T/T = \sum_{l,m} a_{lm} Y_{lm}$ taken over the ensemble of sub-universes. The ensemble of sub-universes is considered to be Gaussian with $\langle a_{lm}\rangle = 0$, and $\langle a_{lm} a^*_{l'm'}\rangle = C_l \delta_{ll'}\delta_{mm'}$. The two point correlation function is then expressed in terms of the angular power spectrum $C_l = \langle |a_{lm}|^2\rangle$.

In order to relate the observed anisotropies to the quantum fluctuation, we separate the transition process into two steps: the step at the recombination time $\tau_{\rm rec}$ and the step at the dynamical transition τ_Λ. At $\tau_{\rm rec}$ the potential fluctuation Φ is related to the density contrast, $\delta_\rho = \delta\rho/\rho$, through the Poisson equation, $\vec{\nabla}^2\Phi = 4\pi G\rho\delta_\rho$, and the Friedmann equation, $H^2 = 8\pi G\rho/3$, resulting $\delta_\rho(\mathbf{k}) = -(2/3)(k/m_{\rm rec})^2\Phi(\mathbf{k})$, where $\mathbf{k}$ is the comoving wave number and $m_{\rm rec} = aH$ evaluated at $\tau_{\rm rec}$. At τ_Λ we assume the density fluctuation reflects faithfully the scalar curvature fluctuation, which is the unique observable with the general covariance:

$$\left.\frac{\delta R}{R}\right|_{\tau_\Lambda^-} \sim \left.\frac{\delta\rho}{\rho}\right|_{\tau_\Lambda^+}, \tag{1.4.9}$$

where $\tau_\Lambda^{-(+)}$ is the time just before(after) τ_Λ. The fluctuation patterns of large scale with super-horizon separations will be preserved until $\tau_{\rm rec}$ because of the lack of dynamical communication.

Therefore, the observed two point correlation function is related to the quantum mechanical correlation of the scalar curvature contrast operator, $\delta_R = (\delta R/R)_{op}$, as

$$c_2(\theta) = \left\langle \frac{\delta T}{T}(\mathbf{n})\frac{\delta T}{T}(\mathbf{n}')\right\rangle \sim \int \frac{d^3\mathbf{k}}{(2\pi)^3}\frac{1}{4}\left(\frac{m_{\rm rec}}{k}\right)^4 \langle\langle \delta_R(\mathbf{k})\delta_R(-\mathbf{k})\rangle\rangle e^{i\mathbf{k}\cdot(\mathbf{n}-\mathbf{n}')x_{\rm lss}}, \tag{1.4.10}$$

where $\langle\langle\cdots\rangle\rangle$ indicates the quantum mechanical expectation value, and $x_{\rm lss} = |\mathbf{x}_{\rm lss}|$. Expanding exponential functions in terms of the spherical harmonic functions, and integrating over angles we obtain the angular power spectrum as

$$C_l \sim \frac{1}{2\pi}\int \frac{dk}{k}\frac{m_{\rm rec}^4}{k}\langle\langle \delta_R(\mathbf{k})\delta_R(-\mathbf{k})\rangle\rangle j_l^2(kx_{\rm lss}), \tag{1.4.11}$$

where $j_l(x)$ is the spherical Bessel function.

According to the CFT_4 the two point correlation function of the scalar curvature contrast is given by

$$\langle\langle \delta_R(\tau_\Lambda,\mathbf{r})\delta_R(\tau_\Lambda,\mathbf{r}')\rangle\rangle \sim \left(H_{\mathbf{D}}|\mathbf{r}-\mathbf{r}'|\right)^{-2\Delta_R}, \tag{1.4.12}$$

where $|\mathbf{r}-\mathbf{r}'| = a(\tau_\Lambda)|\mathbf{x}-\mathbf{x}'|$ is the physical distance on the hypersurface at τ_Λ, and Δ_R is the scaling dimension of δ_R.

The scaling dimension can be obtained from the conformal properties of operators in CFT_4 discussed in Section 2.2. Any physical operator should be conformal invariant, and in this sense the dynamical fields, such as ϕ, $h^\mu{}_\nu$, A_μ, are not physical operators. A physical operator with the conformal dimension Δ is written as a composite field operator by $O_\Delta = \sqrt{-\hat{g}}e^{\gamma\phi}\mathcal{O}_{\Delta_0}$, where γ is the conformal charge, and Δ_0 is the naive dimension of the operator $\mathcal{O}_{\Delta_0}(\partial\phi, h, A, \cdots)$. Within the lowest order of α_G, the conformal charge is calculated in CFT_4 to be $\gamma = 2b_1\{1-\sqrt{1-(4-\Delta_0)/b_1}\}$ (1.2.31). For example, the cosmological constant and the scalar curvature operators, which have naive dimension 0 and 2, respectively, are written as $\sqrt{-\hat{g}}e^{\gamma_0\phi}$ and $\sqrt{-\hat{g}}e^{\gamma_2\phi}\mathcal{R}_2(\partial\phi)$. The curvature contrast operator is then given by $\delta_R = e^{\gamma_2\phi}\mathcal{R}_2/12H_{\mathbf{D}}^2$, where the denominator is the curvature of the de Sitter background (1.4.6).

The scaling dimension of the scalar curvature Δ_R is determined from the scale transformation properties of operators: the operator with the scaling dimension Δ is considered to transform as $d^4xO_\Delta \to \omega^{4-\Delta}d^4xO_\Delta$ under the constant Weyl rescaling defined such that the cosmological constant operator transforms as $\Delta = \Delta_0 = 0$. The Weyl rescaling is equivalent to the constant shift $\phi \to \phi + (4/\gamma_0)\log\omega$. By this shift the operator with the scaling dimension Δ, d^4xO_Δ, changes to $\omega^{4\gamma/\gamma_0}d^4xO_\Delta$. Thus, we obtain the relation $4 - \Delta = 4\gamma/\gamma_0$, which gives $\Delta_R = 4 - 4\gamma_2/\gamma_0$ for the scalar curvature operator. The difference $\Delta - \Delta_0$ is the anomalous dimension of the operator.

The Fourier component of scalar curvatures correlation function is written as

$$\frac{d^3\mathbf{p}}{H_{\mathbf{D}}^3}\left(\frac{|\mathbf{p}|}{H_{\mathbf{D}}}\right)^{2\Delta_R-3} = \frac{d^3\mathbf{k}}{m_\lambda^3}\left(\frac{|\mathbf{k}|}{m_\lambda}\right)^{2\Delta_R-3}, \tag{1.4.13}$$

where $\mathbf{p} = \mathbf{k}/a(\tau_\Lambda)$ is the physical wave vector at τ_Λ, $\mathbf{k}$ is the comoving wave vector, and the constant m_λ is defined by $m_\lambda = a(\tau_\Lambda)H_{\mathbf{D}}$. The exponent $n = 2\Delta_R - 3$ [61] is known as the spectral index. The deviation from the Harrison-Zel'dovich spectrum [62], $n - 1 = 2/b_1 + 4/b_1^2 + o\left(1/b_1^3\right)$, is then the contribution from the conformal field.

In general, the conformal charges depend on the traceless field, and thus the scaling dimension should have a correction of the traceless-field coupling. Hence, the index of the correlation should be replaced by

$$\bar{n} = n + u\alpha_{\mathrm{G}}, \tag{1.4.14}$$

where u is assumed to be a positive constant.[5]

The two point correlation function involves integration over the momentum, which runs in the region where the coupling with the traceless field is not negligible. In order to take into account the non-perturbative effect we replace the effective coupling constant α_{G} by the running coupling (1.2.8). We then obtain the angular power spectrum for large angles to be

$$C_l = \int_\lambda^\infty \frac{dk}{k} j_l^2(kx_{\mathrm{lss}})P(k) \tag{1.4.15}$$

with

$$P(k) = A\left(\frac{k}{m_\lambda}\right)^{n-1+\frac{v}{\log(k^2/\lambda^2)}}, \tag{1.4.16}$$

[5]This constant is calculable, but is not given here, because renormalizations of composite operators with conformal charges are rather complicated. In practice, the precise value does not matter much in the following discussion.

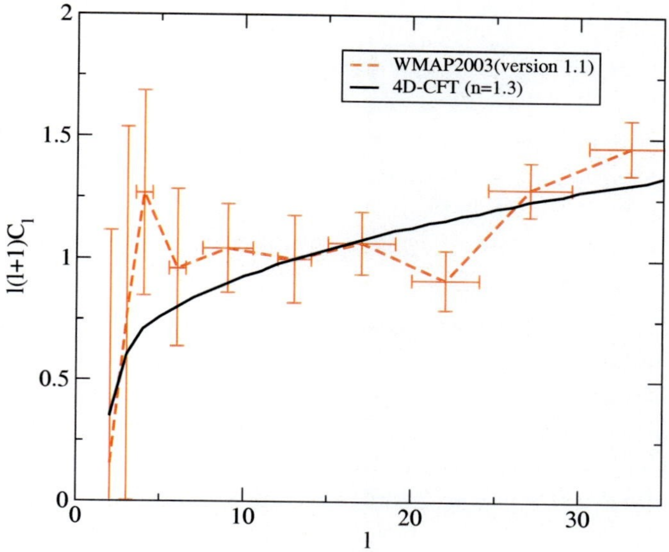

Figure 1.4.2: The CMB spectra for $n = 1.3$ together with the WMAP data (red).

where A is a dimensionless normalization constant and $v = u/4\pi\beta_0$. The parameter λ is the comoving dynamical scale at the time τ_Λ defined by $\lambda = a(\tau_\Lambda)\Lambda_{\rm QG}$.

Spectrum Let us determine parameters by comparing low multipole components of the angular power spectra of WMAP observation. We regard the sharp fall off at the $l = 2$ and 3 multipole components as an appearance of the dynamical scale of quantum gravity. This determines a value for the comoving dynamical scale to be about $\lambda = 3/x_{\rm lss}$. If we take $x_{\rm lss} = 14000$ Mpc, we obtain $\lambda = 0.0002\ {\rm Mpc}^{-1}$. The number of e-foldings, $\mathcal{N}_e \simeq H_{\rm D}/\Lambda_{\rm QG} = m_\lambda/\lambda$, can be taken an arbitrarily large value depending on the value m_λ. For example, we can choose $m_\lambda = 0.02\ {\rm Mpc}^{-1}$ so that $\mathcal{N}_e = 100$ which is large enough to solve the flatness problem. Since $H_{\rm D}$ is order of 10^{19} GeV, we have $\Lambda_{\rm QG} \sim 10^{17}$ GeV.

The spectral index n of the Standard Model ($n = 1.41$ obtained from $N_A = 12$, $N_W = 45$) predicts large blue spectrum. From the observation it looks favorable to make n a little smaller by adding extra matter and gauge fields such as the GUT or SUSY models suggest. The deviations of multi-pole components $4 \sim 8$ would be improved by taking into account the integrated Sachs-Wolfe(ISW) effect [63]. However, the conclusive statement should be postponed until the overall analysis based on our scenario is completed. The CMB spectra, $l(l+1)C_l$, are calculated up to $l = 40$ for $n = 1.3$ as shown in Figure 4.2 together with the WMAP data. The normalization constant A is chosen appropriately, so that $l = 13$ multipole components coincide with the observed value, and v is taken to be 0.001.

Here, we consider the angular power spectrum only in the super-horizon region where the quantum nature of gravity shows up significantly. In order to achieve the overall fit to the data with our primordial spectrum, more detail consideration on various physical process is necessary [63, 64]. For example, we have assumed that the unique decoupling time τ_Λ for entire momentum range. However, if there is a time lag in the phase transition, the short scale delay will change m_λ to be an increasing function of the comoving wave number k.

Simulation of the simplicial universe Since the simplicial quantum gravity is aimed to formulate quantum fluctuation of spacetime dynamics without a priori metric, it is a necessary but difficult task to embed the coordinate system in such a universe for measuring the geometrical structure. Although it is not so obvious to imagine a fixed coordinate frame in fluctuating quantized spacetime, we expect that an appropriate classical background metric will emerge at the epoch when the dynamical scale in the last chapter becomes physically significant. Since a dynamical mechanism for the scenario is not yet well-understood, we seek alternative method in the followings based on the analogy of 2-dimensional quantum gravity. This method will also make it possible to analyze the anisotropies of CMB without referring to the classical background because of two-dimensional nature of the light source known on the lss.

The method is based on the observation [65] that regarding a 2-dimensional spacetime plane as a homogeneous conducting sheet the resistance is invariant under local scale transformation. The resistance between two sides of a rectangular segment of the length a and the width b is given by

$$R = r\frac{a}{b}, \tag{1.4.17}$$

where r is the resistivity constant of the sheet, which is obviously invariant under the scale change $a \to a\eta$, and $b \to b\eta$. By picking up four points $\{z_i\}$ on a complex plane as the conducting surface , measurement of the resistance between two points z_1 and z_2 can be carried out by applying the current input I at z_3, and the sink current $-I$ at z_4:

$$V(z_1) - V(z_2) = R(z_1, z_2; z_3, z_4)I. \tag{1.4.18}$$

In the case of the surface with S^2 topology it can be projected on the infinite flat sheet, and we obtain

$$R = -\frac{r}{2\pi} \log |[z_1, z_2; z_3, z_4]|, \tag{1.4.19}$$

where

$$[z_1, z_2; z_3, z_4] = \frac{z_1 - z_3}{z_2 - z_3}\frac{z_2 - z_4}{z_1 - z_4} \tag{1.4.20}$$

is known as the anharmonic ratio.

In order to embed a coordinate system in a triangulated surface we regard its dual graph as a trivalent network of registers with each having unit resistance. Since the harmonic ratio is invariant under a $SL(2, C)$ transformation,

$$z \longrightarrow \frac{az+b}{cz+d} \quad (ad - bc = 1), \tag{1.4.21}$$

we can fix three points among $\{z_i\}$ at $z_2 = 1$, $z_3 = 0$, and $z_4 = \infty$ without changing the resistance, by choosing parameters a, b, c and d, appropriately. In practice, four points are not enough to determine z_1 and r, we add one more measuring point at z_5.

After fixing $\{z_i\}$ for all the dual lattice point corresponding to a triangle, we calculate the density distribution

$$\rho(z) = \sum_i \langle \delta(z - z_i) \rangle, \tag{1.4.22}$$

with appropriate smoothing which we indicate by $\langle \cdots \rangle$. We can check our procedure by comparing the density-density correlation function between the numerical result and the theoretical prediction which is known to be [66]

$$\langle \rho(z)\rho(z') \rangle \sim |z - z'|^{-4}, \tag{1.4.23}$$

where the exponent is expected to be independent of the central charge. We confirm this within the statistical error.

Now, we try to apply the simplicial quantum gravity for the evolution of universe, especially to analyses of the CMB anisotropies. We assume the following simplicial scenario of creation and evolution of the universe: at the beginning one 4-simplex appeared through quantum tunneling. Making it as the seed another simplex is created attaching to form the complex of 4-ball topology with the probability proportional to the surface area, *i.e.* numbers of 3-simplices of the surface with the S^3 topology, which corresponds to the space dimension of the universe. The complex will grow to a bigger complex by attaching 4-simplices over and over, or it may sometimes shrink because of annihilation of 4-simplices. Since one 4-simplex has five faces (3-simplices) it can attach through at most five faces of the mother complex. We restrict the construction within simply connected manifolds, such as a 4-ball or a 4-sphere for simplicity. In principle, it may grow infinitely large when the parameter κ_4 is at the critical value κ_4^c. The simplicial universe in general has the topology $R \times S^3$, and we choose the growth rate, or in other word scale along the time direction R, in such a way that space and time directions have the similar lattice structure in average to form the local spacetime homogeneity.

To construct the last scattering surface we consider a large S^4 universe instead of a 4-ball as a representative of a generic spacetime. As for the first step we dig a hollow (an empty 4-ball) in the S^4 spacetime complex whose boundary is a 3-sphere. The radius of the hollow is measured in terms of the geodesic distance which corresponds to the time when the universe made the phase transition from the conformal to the Friedmann spacetime. The second step is to dig again a 3-ball in the 3-sphere to create a S^2 surface as the last scattering surface. The density-density correlation function can be obtained in the same manner as the two dimensional case, by noting that the density of triangles of the S^2 surface should be considered as the square root of the density in the 4-dimensional universe. Using the correlation function we can calculate the angular power spectra. At the moment we do not have data with large enough size, which makes it possible to analyze the long distance correlation with a statistically meaningful accuracy. Our preliminary result of the correlation function suggests a possibility to determine the angular power spectra numerically, and check the cause of sharp fall off.

1.5 Conclusion

In this article we review two non-perturbative methods for four-dimensional quantum gravity capable of describing spacetime dynamics beyond the Planck scale: the renormalizable quantum gravity and the simplicial quantum gravity. The models were constructed on the three basic conditions: background-metric independence, finiteness and four spacetime dimensions. In the continuum theory, these conditions are represented by the diffeomorphism invariance, renormalizablility and integrability. In the discretized theory, the finiteness is apparent, and the background-metric independence is realized by taking the sum over all possible triangulations of four-dimensional simplicial maniforld.

Spacetime sigularities are prohibited quantum mechanically, which was shown by the asymptotically free property of the traceless tensor mode in the metric field. The asymptotic freedom also implies that there should be a dynamical scale of gravity, Λ_{QG}, which separates quantum spacetime with the conformal invariance from the classical Einstein spacetime. We have suggested that at this scale a kind of phase transition occurs, and the correlation becomes short-range of order of $1/\Lambda_{QG}$. Field fluctuations freeze to localized objects, and they eventually decay into the classical matter driving the universe into the big bang phase.

We have shown an inflationary scenario of the universe induced by quantum gravity and derived the CMB angular power spectrum at large angles from the two point correlation function. The sharp damping at low multipole components is obtained reflecting the dynamical scale of quantum gravity. Since the primordial spectrum is produced by CFT_4, we expect that fluctuations of the tensor mode are relatively small in comparison to those of the conformal mode, and multi-point correlation functions of non-Gaussian type can exist.

The strong repulsive effect in quantum gravity causes the inflation, and also erase the singularity inside a black hole by balancing the pressure of the collapsing matter. However, if the collapsing goes too far and exceedes the balance, the black hole may explode something like as mini-inflation.

Finally, as mentioned in Introduction the goal of quantum gravity is to break the wall of Planck scale. The wall of Planck scale reminds us the wall of sound speed. When an airplane speeds up and approaches to the sound speed, it faces violent vibrations due to the sound made by the airplane itself and sometimes breaks the airplane into pieces. People of old days thought that the sound speed is the unpassable wall. However, we know now once we pass the wall with durable body, a peaceful space without sounds spreads about us. Similary, we might think that the Planck scale is the wall that we can never pass. However, once we go beyond the Planck scale, there is no singularity, but a harmonious space of conformal symmetry.

References

[1] A. Penzias and R. Wilson, *Astrophys. J.* **142** (1965) 419.

[2] C. Bennett et al., *Astrophys. J.* **464** (1996) L1.

[3] C. Bennett et al., *Astrophys. J. Suppl.* **148** (2003) 1.

[4] A. Guth, *Phys. Rev.* **D23** (1981) 347; K. Sato, *Mon. Not. R. Astron. Soc.* **195** (1981) 467; A. Linde, *Phys. Lett.* **B108** (1982) 389; A. Albrecht and P. Steinhardt, *Phys. Rev. Lett.* **48** (1982) 1220.

[5] A. Starobinsky, *Phys. Lett.* **B91** (1980) 99; A. Vilenkin, *Phys. Rev.* **D32** (1985) 2511.

[6] K. Hamada and T. Yukawa, *Mod. Phys. Lett.* **A20** (2005) 509 [astro-ph/0401070].

[7] A. Einstein, *Annalen der Phys.* **49** (1916) 769.

[8] B. DeWitt, in Relativity, *Groups and Topology*, eds. B. DeWitt and C. DeWitt (Gordon and Breach, New York, 1964).

[9] G. 't Hooft and M. Veltman, *Ann. Inst. Henri Poincare* **XX** (1974) 69; M. Veltman, in *Methods in Field Theory*, Les Houches 1975.

[10] M. Goroff and A. Sagnotti, *Phys. Lett.* **160** (1985) 81.

[11] K. Stelle, *Phys. Rev.* **D16** (1977) 953.

[12] E. Tomboulis, *Phys. Lett.* **70B** (1977) 361.

[13] E. Fradkin and A. Tseytlin, *Nucl. Phys.* **B201** (1982) 469.

[14] R. Riegert, *Phys. Lett.* **134B** (1984) 56.

[15] I. Antoniadis and E. Mottola, *Phys. Rev.* **D45** (1992) 2013.

[16] I. Antoniadis, P. Mazur and E. Mottola, *Nucl. Phys.* **B388** (1992) 627.

[17] I. Antoniadis, P. Mazur and E. Mottola, *Phys. Rev.* **D55** (1997) 4756; 4770.

[18] K. Hamada and F. Sugino, *Nucl. Phys.* **B553** (1999) 283 [hep-th/9810095].

[19] K. Hamada, *Prog. Theor. Phys.* **103** (2000) 1237 [hep-th/9912098].

[20] K. Hamada, *Prog. Theor. Phys.* **108** (2002) 399 [hep-th/0203250].

[21] K. Hamada and S. Horata, *Prog. Theor. Phys.* **110** (2003) 1169 [hep-th/0307008].

[22] K. Hamada, *Int. J. Mod. Phys.* **A20** (2005) 5353 [hep-th/0402136].

[23] A. Polyakov, *Phys. Lett.* **103B** (1981) 207; *Mod. Phys. Lett.* **A2** (1987) 893.

[24] V. Knizhnik, A. Polyakov and A. Zamolodchikov, *Mod. Phys. Lett.* **A3** (1988) 819.

[25] J. Distler and H. Kawai, *Nucl. Phys.* **B321** (1989) 509; F. David, *Mod. Phys. Lett.* **A3** (1988) 1651.

[26] N. Seiberg, *Prog. Theor. Phys. Suppl.* **102** (1990) 319.

[27] J. Wess and B. Zumino, *Phys. Lett.* **37B** (1971) 95.

[28] D. Capper and M. Duff, *Nuovo Cimento* **23A** (1974) 173.

[29] S. Deser, M. Duff and C. Isham, *Nucl. Phys.* **B111** (1976) 45.

[30] M. Duff, *Nucl. Phys.* **B125** (1977) 334.

[31] L. Bonora, P. Cotta-Ramusino and C. Reina, *Phys. Lett.* **B126** (1983) 305.

[32] T. Lee and G. Wick, *Nucl. Phys.* **B9** (1969) 209; *Phys. Rev.* **D2** (1970) 1033; *Phys. Rev.* **D3** (1971) 1046.

[33] N. Nakanishi, *Prog. Theor. Phys. Suppl.* **51** (1972) 1.

[34] A. Belavin, A. Polyakov and A. Zamolodchikov, *Nucl. Phys.* **B241** (1984) 333.

[35] B. Lian and G. Zuckermann, *Phys. Lett.* **B266** (1991) 21.

[36] P. Bouwknegt, J. McCarthy and K. Pilch, *Commun. Math. Phys.* **145** (1992) 541.

[37] E. Witten, *Nucl. Phys.* **B373** (1992) 187.

[38] K. Hamada, *Phys. Lett.* **B324** (1994) 141.

[39] A. Pais and G. Uhlenbeck, *Phys. Rev.* **79** (1950) 145; R. Riegert, *Phys. Lett.* **A105**(1984) 110.

[40] D. Varshalovich, A. Moskalev and V. Khersonskii, *Quanyum Theory of Angular Momentum* (World Scientific, Singapore, 1988).

[41] M. Rubin and C. Ordóñez, *J. Math. Phys.* **25** (1984) 2888.

[42] G. 't Hooft and M. Veltman, *Nucl. Phys.* **B44** (1972) 189.

[43] S. Hathrell, *Ann. Phys.* **142** (1982) 34.

[44] T. Regge, *Nuovo Cim.* **19** (1961) 45.

[45] D. Weingarten, *Phys. Lett.* **B90** (1980) 285, *Nucl. Phys.* **B210** [FS6] (1982) 229.

[46] F. David, *Nucl. Phys.* **B257** (1985) 45.

[47] V. Kazakov, *Phys. Lett.* **150B** (1985) 282.

[48] J. Ambjørn, B. Durhuus and J. Fröhlich, *Nucl. Phys.* **B257** (1985) 433.

[49] D. Boulatov, V. Kazakov, I. Kostov and A. Migdal, *Nucl. Phys.* **B275** (1986) 641.

[50] M. Agishtein and A. Migdal, *Nucl. Phys.* **B385** (1992) 395.

[51] J. Ambjørn and J. Jurkiewicz, *Phys. Lett.* **B278** (1992) 42.

[52] R. Friedberg and T. D. Lee, *Nucl. Phys.* **B242** (1984) 145.

[53] D. A. Johnston, *Nucl. Phys. B (Proc. Suppl.)* **53** (1997) 43; G. Thorleifsson, *Nucl. Phys. B (Proc. Suppl.)* **73** (1998) 133; R. Loll, *Nucl. Phys. B (Proc. Suppl.)* **94** (2001) 96.

[54] S. Oda, N. Tsuda and T. Yukawa, *Nucl. Phys. B (Proc. Suppl.)* **63** (1998), 733. S. Oda, N. Tsuda and T. Yukawa, *Prog. Theor. Phys.* **99** (1998) 875.

[55] C. Itzykson and J. M. Drouffe, *Statistical Field Theory* (Cambridge University Press, Cambridge, 1989).

[56] E. Brezin, C. Itzykson, G. Parisi and J. B. Zuber, *Comm. Math. Phys.* **59** (1978) 35.

[57] F. David, *Nucl. Phys.* **B368** (1992) 671.

[58] H. Egawa, A. Fujitu, S. Horata, N. Tsuda and T. Yukawa, *Nucl. Phys. Proc. Suppl.* **73** (1999) 795 [hep-lat/9809028]; S. Horata, H. Egawa, N. Tsuda and T. Yukawa, *Nucl. Phys. B (Proc. Suppl.)* **83-84** (2000) 751 [hep-lat/9908048]; *Prog. Theor. Phys.* **106** (2001) 1037 [hep-lat/0004021]; S. Horata, H. Egawa and T. Yukawa, *Nucl. Phys. B (Proc. Suppl.)* **94** (2001) 701 [hep-lat/0010050];

S. Horata, H. Egawa and T. Yukawa, *Nucl. Phys. Proc. Suppl.* **106** (2002) 971 [hep-lat/0110042]; S. Horata, H. Egawa and T. Yukawa, *Nucl. Phys. Proc. Suppl.* **119** (2003) 921 [hep-lat/0209004].

[59] L. Abbott and M. Wise, *Astrophys. J.* **282** (1984) L47; R. Scaramella and N. Vittorio, *Astrophys. J.* **353** (1990) 372.

[60] R. Sachs and A. Wolfe, *Astrophys. J.* **147** (1967) 73.

[61] I. Antoniadis, P. Mazur and E. Mottola, *Phys. Rev. Lett.* **79** (1997) 14.

[62] E. Harrison, *Phys. Rev.* **D1** (1970) 2726; Ya. B. Zel'dovich, *Mon. Not. R. Astron. Soc.* **160** (1972) P1.

[63] W. Hu and N. Sugiyama, *Phys. Rev.* **D51** (1995) 2599.

[64] A. Liddle and D. Lyth, *Cosmological Inflation and Large-Scale Structure*, Cambridge Univ. Press, 2000.

[65] H. Kawai, N. Tsuda and T. Yukawa, *Phys. Lett.* **B351** (1995) 162; H. Kawai, N. Tsuda and T. Yukawa, *Nucl. Phys. B (Proc. Suppl.)* **53** (1997) 777.

[66] A. Zamolodchikov and Al. Zamolodchikov, *Nucl. Phys.* **B429** (1994) 375.

In: Focus on Quantum Gravity Research
Editor: David C. Moore, pp. 49-87

ISBN 1-59454-660-6

Chapter 2

Testing Lorentz Invariance Violation in Quantum Gravity Theories

Héctor Vucetich[†]
[†]Observatorio Astronómico, Universidad Nacional de La Plata,
Paseo del Bosque S/N, (900) La Plata, Argentina
E-mail: vucetich@fcaglp.unlp.edu.ar

Much research has been done in the latter years on the subject of Lorentz violation induced by Quantum Gravity effects. On the theoretical side it has been shown that both Loop Quantum Gravity and String Theory predict that Lorentz violation can be induced at an energy near to the Planck scale. On the other hand, most of the experimental results in the latter years, have confirmed that the laws of physics are Lorentz invariant at low energy with very high accuracy.

The inclusion of one- and two-loop contributions from a Lorentz violating Lagrangian dramatically change the above picture: the loop momenta run into the Planck scale and above and from the "divergent" terms finite Lorentz violating contributions of order one arise. These can be suppressed through suitable counterterms in the Lagrangian, originating a strong fine tuning problem.

A brief discussion of these issues and their possible influence in future research follows.

2.1 Introduction

For a long time, the search of experimental clues on the nature of quantum gravity, was dismissed as impractical, based on the simplistic argument that those effects should appear only at energies of the order of Planck scale ($E_P = [1.2 \times 10^{19}]GeV$), far beyond present day experimental possibilities. But recently there was a revolutionary change in this attitude, originated in references [1, 2, 3, 4, 5]. Here, a spontaneous breakdown of Lorentz symmetry at the Planck scale motivated in string theory was proposed, breaking CPT symmetry and consequently of Lorentz invariance. Later, an extension of the standard model was developed by the same group, including all possible CPT and Lorentz violating interactions [6, 7], which has been used to develop sensitive tests of Lorentz symmetry.

A different approach [8, 9, 10, 11, 12] was based on the possibility that Quantum Gravity effects would modify the dispersion relations for particle propagation, such as photons. These modifications in turn would change the propagation velocity of photons, introducing delays for particles of different energies which could be detected if these particles would travel cosmological distances.

Such modifications of the dispersion relations have been found in the two most popular approaches to Quantum Gravity: Loop Quantum Gravity [13, 14, 15, 16, 17] and String theory [18, 19, 20, 21]. These theories predict corrections to the dispersion relations which depend on energy in the form $(E\ell_P)^n$, with $\ell_P = [1.6 \times 10^{-33}]cm$ the Planck length scale.

String theory has suggested another form of Lorentz violation: noncommutative field theory [22, 23], where it is assumed that the coordinate themselves are functions from the differential variety to a noncommutative algebra. This was originally found in [22] where it was shown that the low energy behavior of a bosonic string propagating in a Neveu-Schwartz condensate can be represented by a noncommutative field theory.[1] Noncommutativity, of course, induces Lorentz violation and it constitutes one of the most interesting sources of it.

More recently, a new approach to the testing of Quantum Gravity effects was introduced [25, 26], based on the fact that if the dispersion relations of photons in the vacuum are modified in the form $v(E) \neq c$, this implies a breakdown of Lorentz invariance since such a statement can be valid at most in a single reference frame. In this way a privileged reference frame (the *New Aether*) is introduced in the theory where a particular simple form of the equations of motion is valid. Since Lorentz invariance must be broken, the motion of the laboratory with respect to the privileged frame should be detected in suitable experiments. Moreover, modern cosmology suggests a single candidate for the New Aether, namely, the reference system at rest with respect to the Cosmic Microwave Background. In this way, the possibility of accurate tests of Quantum Gravity induced Lorentz breakdown opens.

The rest of the paper is organized as follows: Section 2.2 is devoted to a brief discussion of the main differences between Lorentz abiding and Lorentz violating theories. Section 2.3 discusses several of the test theories that have been used to design and interpret tests Lorentz invariance. In section 2.4 the origin of Lorentz invariance violation in Quantum Gravity is discussed, centering on the predictions of

[1] A didactic introduction to noncommutative field theory can be found in [24].

Loop Quantum Gravity, and in Section 2.4 some of the tests carried on the theory at the tree-level are discussed. The dramatic consequences of the inclusion of radiative correction effects are discussed in section 2.6. Finally, in section 2.7 we state our conclusions.

2.2 Issues of Lorentz violation

Lorentz invariant theories have well defined properties, such as the nonexistence of privileged reference frames. Many of these properties are lost if Lorentz invariance does not hold and new phenomena may arise in these conditions. In this section we discuss some of these issues since they may be used to check the validity of Lorentz symmetry.

2.2.1 Privileged reference systems

A characteristic fact of Lorentz violating theories is that most of them predict the existence of *privileged reference systems*, where the equations of motion take their simplest form. This is akin to the old notion of "luminiferous aether", before the formulation of special relativity one century ago, and we shall call sometimes these privileged reference systems "the new aether".

That Lorentz violating theories should generically predict the existence of privileged reference systems can be easily inferred if propagating particles have general functions of energy as dispersion relations. Consider for instance a photon: if its dispersion relation does not have the Lorentz covariant form $\omega = ck$, but propagate with an energy dependent velocity $v(E)$, such statement can be at best valid in one specific inertial frame. This selects a preferred frame of reference, where a particular form of the equations of motion is valid, and one should then be able to detect the laboratory velocity with respect to that frame [25]. It is this fact that opens the possibility of detecting tiny violations of Lorentz symmetry.

Fortunately, we have today, in contrast with the situation at the end of the 19th century, a rather unique choice for that "preferred inertial frame": the frame where the Cosmic Microwave Background (CMB) looks isotropic. Our velocity $\boldsymbol{w}$ with respect to that frame has already been determined to be $w/c \approx 1.23 \times 10^{-3}$ by the measurement of the dipole term in the CMB by COBE, for instance [27]. We shall usually refer tests of Lorentz invariance to the CMB reference system.

2.2.2 "Particle" and "Observer" Lorentz transformations

As it is ell known, there are two possible formulations of Lorentz-Poincaré transformations: *passive* or *observer* transformations, where coordinates are transformed in the form

$$x' = \Lambda^{-1}(x - a) \tag{2.2.1}$$

and *active* or *particle* transformations, where fields and states are transformed

$$U(\Lambda, a)\phi(x)U^\dagger(\Lambda, a) = \phi(\Lambda x + a) \qquad U(\Lambda, a)|0\rangle = |0\rangle \tag{2.2.2}$$

These two forms of the Lorentz transformation are equivalent in Lorentz-invariant theories, but this will not be so in Lorentz-violating theories. In particular, particle transformations may be limited to a given subset of Lorentz transformations, while observer transformations, in principle, are unrestricted. These will be interpreted as the transformations related to the laboratory (L) frame, and these are important in the design and interpretation of experiments.

2.2.3 Discrete Lorentz transformations

Among the testable forms of Lorentz violations, the breakdown of CPT symmetry is paramount, since it signals the breakdown of Lorentz symmetry altogether. Indeed, the *CPT Theorem*, *i.e.*, the validity of CPT symmetry, can be proved from very weak assumptions in axiomatic field theory [28, 29]. The converse proposition has been proved recently [30].

The consequences of CPT invariance are well known: masses, magnetic moments and charge of particles and antiparticles must be equal in absolute value, as well as cross sections and decay rates. Any of these properties may be used to check the validity of CPT invariance and some of the most sensitive tests of Lorentz Invariance are based on them.

2.3 Test theories for Lorentz Invariance Violation

Testing symmetries of nature such as Lorentz invariance is better made through the development of *test theories*: theories that suitably generalize the symmetry being tested through the introduction of a set of well defined parameters C_i. These are chosen in such a way that the symmetry being tested is recovered for particular values of the parameters C_i^0, while the other values represent a breakdown of the symmetry. Let us examine a few of the proposed test theories of Lorentz invariance.

2.3.1 The Robertson model

The simplest of the proposed test theories are the Robertson model [31] and its generalization by Mansouri and Sexl [32, 33, 34]. The connection between both references has been discussed in reference [35]. These are purely kinematic test theories, where the parameters are introduced through suitable generalizations of the Lorentz transformation.

Consider the privileged reference system S_0 chosen in such a way that Maxwell's equations are there valid. Let us introduce the metric, valid in this rest frame

$$ds^2 = dx_0^2 - (dx^2 + dy^2 + dz^2) \tag{2.3.3}$$

This quantity is important, since it has the following properties

1. A time interval at a given point $\boldsymbol{x}$ is expressed as

$$dt = \frac{ds}{c} \qquad (d\boldsymbol{r} = 0) \tag{2.3.4}$$

2. The distance between two points $\boldsymbol{x}_1, \boldsymbol{x}_2$ is expressed as

$$dl = +\sqrt{-ds^2} \qquad (dt = 0) \tag{2.3.5}$$

3. Light rays are along the normal to the spheres

$$ds^2 = 0 \tag{2.3.6}$$

Now consider an inertial reference system S', moving with respect to S with velocity $\boldsymbol{w}$. The Robertson transformation between both reference systems, with a suitable choice of axis, is defined as

$$\begin{aligned} x &= Ax' + Bx'_0 \\ x_0 &= B'x'_0 + A'x' \\ y &= Cy' \\ z &= Cz' \end{aligned} \tag{2.3.7}$$

with A, B, C arbitrary functions of the relative velocity w. These are, of course, "observer" transformations.

Imposing Einstein's synchronization on S' the conditions

$$A' = \frac{w}{c}A \qquad\qquad B' = \frac{w}{c}B \tag{2.3.8}$$

are obtained. Finally, the metric in S' results

$$ds^2 = g_0^2 dx_0'^2 - g_1^2 dx'^2 - g_2^2(dy'^2 + dz'^2) \tag{2.3.9}$$

where

$$\begin{aligned} g_0(w) &= \sqrt{1 - \frac{w^2}{c^2}}B \\ g_1(w) &= \sqrt{1 - \frac{w^2}{c^2}}A \\ g_2(w) &= C \end{aligned} \tag{2.3.10}$$

Thus, these tests theories are equivalent to the proposal of a metric of the form

$$ds^2 = g_0^2 dx_0^2 - g_1^2 dx^2 - g_2^2(dy^2 + dz^2) \tag{2.3.11}$$

for space-time, where $g_i(V)$ are arbitrary functions of the velocity of the local reference system with respect to a privileged inertial frame. Lorentz symmetry is recovered for the particular case $g_i(V) = 1$. For low velocity experiments, it is usual to parametrize these functions in terms of the *Mansouri-Sexl parameters* [32]

$$g_0(V) = 1 + \left(\frac{1}{2} - \alpha\right)\frac{V^2}{c^2} + \dots \tag{2.3.12}$$

$$\frac{g_1}{g_0} = 1 + (\alpha - \beta)\frac{V^2}{c^2} + \dots \tag{2.3.13}$$

$$\frac{g_2}{g_1} = 1 + (\delta - \beta)\frac{V^2}{c^2} + \dots \tag{2.3.14}$$

which have the values

$$\alpha = \frac{1}{2} \qquad \beta = \frac{1}{2} \qquad \delta = 0 \tag{2.3.15}$$

if Lorentz invariance is valid.

Robertson [31] showed that the classical experiments of Michelson-Morley [36], Kennedy-Thorndike [37] and Yves-Stillwell [38] restricted the functions g_i to a small neighborhood of 1. We shall discuss modern limits on g_i later on.

2.3.2 The Dispersion Relations models

A set of simple tests theories for testing Lorentz violation has been proposed by Amelino-Camelia [39] (See also [40]). These are kinematic models, but applied to the dynamical dispersion relations they introduce a bridge between kinematics an dynamics.

The dispersion relation for a low-energy particle (*i.e.* small with respect to the Planck scale $E_P = [1.2 \times 10^{19}]GeV$) in a Lorentz violating theory, can be written in the form

$$m^2 \simeq E^2 - \boldsymbol{p}^2 + \eta \boldsymbol{p}^2 \left(\frac{E}{E_P}\right)^n + O\left(\frac{E}{E_P}\right)^{n+3} \tag{2.3.16}$$

where η and n are free parameters. The exponent n is characteristic of the magnitude of the effects expected. On the other hand, for Lorentz violating effect generated in quantum gravity, one expects $\eta \sim O(1)$. The case $n = 1$ is specially interesting, since it arises in many contexts such as low energy limits of Loop Quantum Gravity.It should be noted, however, that a photon dispersion relation of the form (2.3.16) with n odd is forbidden by causality. The detection of such a correction would signal not only the breakdown of Lorentz invariance but of causality relations.

To close the test theory, two prescriptions must be added to the above dispersion relation

1. The law of energy-momentum conservation is valid

$$E_1 + E_2 = E_1' + E_2' \qquad \boldsymbol{p}_1 + \boldsymbol{p}_2 = \boldsymbol{p}_1' + \boldsymbol{p}_2' \tag{2.3.17}$$

2. The velocity of the particle is computed from the Lorentzian expression

$$\boldsymbol{v} = \frac{dE}{d\boldsymbol{p}} \tag{2.3.18}$$

These simple prescriptions suffice for analyzing many Lorentz-violating phenomena, specially those related to threshold modifications by Quantum Gravity effects.

The set of equations (2.3.16), (2.3.17) and (2.3.18) are the basis of *threshold analysis*: the very powerful tool to discuss many astrophysical phenomena in the presence of Lorentz violation [41, 42, 39, 11].

A variation of the above dispersion relation is obtained assuming that different polarization states of the particle have different propagation velocities

$$m^2 = E^2 - \boldsymbol{p}^2 \pm \theta_i \boldsymbol{p}^2 \left(\frac{E}{E_P}\right) \tag{2.3.19}$$

where the sign depends on the helicity of the particle. This type of test theories should be analyzed in the context of low energy field theories.

2.3.3 The $TH\epsilon\mu$ model

A *dynamical* test theory for local Lorentz invariance was introduced in reference [43] as a test theory for the validity of Einstein's Equivalence Principle. In this model, the breakdown of Local Lorentz Invariance comes from the structure of the Lagrangian of the system.

Let us consider a system of classical particles interacting through the electromagnetic field in a background spherically symmetric gravitational field U. The action of the system is

$$\begin{aligned} S_{TH\epsilon\mu} = &- \sum_a m_{0a} \int \sqrt{T - H v_a^2} dt \\ &+ \sum_a e_a \int A_\mu(x_a^\nu) v_a^\mu dt \\ &+ \frac{1}{8\pi} \int \left(\epsilon \boldsymbol{E}^2 - \frac{\boldsymbol{B}^2}{\mu}\right) \end{aligned} \tag{2.3.20}$$

where $\hbar = c = 1$, m_a, e_a are the mass and charge of the a-th particle, x_a^μ, v_a^μ its world-line and velocity and $\boldsymbol{E}, \boldsymbol{B}$ the electric and magnetic fields. T, H, ϵ, μ are arbitrary functions of the spherically symmetrical gravitational field $U = GM/r$.

To see how a Lorentz invariance violation is generated in this test theory, let us pass to a freely falling reference frame. Consider a given point in space-time $x_0^\mu = (t_0, \boldsymbol{r}_0)$, and a local reference frame with origin in x_0^μ. If $\boldsymbol{g} = \boldsymbol{\nabla} U$ is the local acceleration of gravity, the transformation equations to the freely falling reference frame, correct to first order in the small quantities $\boldsymbol{g}t$ and $\boldsymbol{g} \cdot \boldsymbol{x}$ are

$$\hat{t} = \sqrt{T_0}t\left(1 + \frac{T_0'}{2T_0}\boldsymbol{g}\cdot\boldsymbol{x}\right) \tag{2.3.21}$$

$$\hat{x} = \sqrt{H_0}\left[\boldsymbol{x} + \frac{T_0'}{4H_0}\boldsymbol{g}t^2 + \frac{H_0'}{4H_0}(2\boldsymbol{x}(\boldsymbol{g}\cdot\boldsymbol{x}) - \boldsymbol{g}x^2)\right] \tag{2.3.22}$$

The transformed action, keeping only the lowest order terms, is

$$\begin{aligned}\hat{S}_{TH\epsilon\mu} = &-\sum_a m_{0a}\int\sqrt{1-\hat{v}^2}d\hat{t} \\ &+\sum_a e_a\int\hat{A}_\mu\hat{v}_a^\mu d\hat{t} \\ &+\frac{1}{8\pi}\epsilon_0\sqrt{\frac{T_0}{H_0}}\int\left(\hat{E}^2 - \frac{H_0}{T_0\epsilon_0\mu_0}\hat{B}^2\right)d^4\hat{x}\end{aligned} \tag{2.3.23}$$

where the "hatted" quantities are referred to the local reference system. The first two terms in (2.3.23) are locally Lorentz invariant, but the third one breaks the symmetry unless

$$\frac{H_0}{T_0\epsilon_0\mu_0} = 1 \qquad \epsilon_0\sqrt{\frac{T_0}{H_0}} = 1 \tag{2.3.24}$$

which are the conditions for the validity of Local Lorentz Invariance. The first one implies that the limiting speed for massive particles is equal to the propagation speed for electromagnetic signals

$$c_L^2 = \frac{T_0}{H_0} = \frac{1}{\epsilon_0\mu_0} = c_{em}^2 \tag{2.3.25}$$

The second one, which is related to the validity of the Weak Equivalence Principle, can be recast as position invariance of the electric charge.

The $TH\epsilon\mu$ model has been generalized to the Standard Model [44] where a set of conditions similar to (2.3.24) should hold for the validity of Lorentz symmetry.

2.3.4 The Kostelecký Model

The most general treatment of Lorentz Invariance violations within the Standard Model is the model developed by Kostelecký and coworkers [6, 7, 45, 46], which we shall simply call the Kostelecký Model (KM).

The KM Lagrangian density is based on a careful inclusion of all terms of dimension 4 that break Lorentz invariance for the particles in the Standard Model. Thus, the Dirac particle modified Lagrangian takes the form

$$\mathcal{L}_D = \frac{1}{2}i\bar{\psi}\Gamma^\mu\partial_\mu\psi - \bar{\psi}M\psi \tag{2.3.26}$$

where the Γ and M operators are

$$\Gamma^\mu = \gamma^\mu + c^{\mu\nu}\gamma_\nu + d^{\mu\nu}\gamma_5\gamma_\nu + e^\mu + if^\mu\gamma_5 + \frac{1}{2}ig^{\lambda\nu\mu}\sigma_{\lambda\nu} \tag{2.3.27}$$

$$M = m + a_\mu\gamma^\mu + b_\mu\gamma_5\gamma^\mu + \frac{1}{2}H^{\mu\nu}\sigma_{\mu\nu} \tag{2.3.28}$$

These operators have well defined transformation properties under CPT symmetry. In particular, the a_μ and b_μ terms are CPT-odd, making the masses of particle and antiparticle different.

There are corresponding modifications for the gauge field and Higgs Lagrangians. For the particular case of the photon Lagrangian, these modifications take the form

$$\mathcal{L}_{em} = -\frac{1}{4}(k_F)_{\kappa\lambda\mu\nu}F^{\kappa\lambda}F^{\mu\nu} + \frac{1}{2}(k_{AF})^\kappa\epsilon_{\kappa\lambda\mu\nu}A^\lambda F^{\mu\nu} \tag{2.3.29}$$

where the first term is CPT-even while the second is CPT-odd.

The CPT-even tensor k_F has 19 independent components, which can be grouped into a set of 10 P-odd components (responsible for vacuum birefringence phenomena) and a set of 9 P-even components, which describe breakdown of Lorentz boost invariance [47, 48, 49].

KM is an extremely general test theory for Lorentz violation: it includes all dimension 4- terms and thus it is "closed" under Lorentz transformations. The physical properties of the model have been extensively studied [6, 50, 51, 52, 46, 53, 54] and it has been successfully used to design and analyze [55, 56, 57, 58, 59, 60, 61, 62] experiments on Lorentz violation.

The nonrelativistic limit of the generalized Dirac operator is specially interesting since same of the most accurate test of Lorentz invariance have been carried in that régime. Although the Lorentz violating terms are small and can be treated perturbatively, it is not straightforward to carry out the nonrelativistic limit because several of the small terms have nontrivial time derivatives. A generalization of the Foldy-Wouthuysen transformation developed n reference [50] can be used to obtain it. Neglecting terms of order $(P/m)^3$ the following Hamiltonian is obtained

$$
\begin{aligned}
H_{NR} =& m + \frac{p^2}{2m} \\
&+ [a_0 + m(c_{00} + e_0)] + [-b_j + \frac{1}{2}\epsilon_{jkl}H_{kl} + m(d_{j0} - \epsilon_{jkl}g_{kl0})]\sigma^j \\
&+ [-a_j + m(c_{j0} + c_{0j} + e_j)]\frac{p^j}{m} \\
&+ [b_0\delta_{kj} - \epsilon_{jkl}H_{l0} - m(d_kj + d_{00}\delta_{kj} + \frac{1}{2}\epsilon_{klm}g_{mlj} - \epsilon_{jkm}g_{m00})]\frac{p^j}{m}\sigma^k \\
&- m(c_{jk} + \frac{1}{2}c_{00})\frac{p^j p^k}{m^2} \\
&+ \{[m(d_{j0} + d_{0j}) - \frac{1}{2}(b_j + md_j + \frac{1}{2}m\epsilon_{jmn}g_{mn0} + \frac{1}{2}\epsilon_{jmn}H_{mn})]\delta_{kl} \\
&+ \frac{1}{2}(b_l + \frac{1}{2}m\epsilon_{lmn}g_{mn0})\delta_{jk} - m\epsilon_{jlm}(g_{m0k} + g_{ml0})\}\frac{p^j p^k}{m^2}\sigma^l
\end{aligned}
\tag{2.3.30}
$$

The first line of (2.3.30) corresponds to the usual NR Hamiltonian for a free particle. The rest of the terms describe Lorentz violating phenomena of different types. We shall discuss some of these phenomena in the following pages.

2.3.5 Other models of Lorentz violation

As a last example of test theories for Lorentz violation, let us mention the Myers-Pospelov model [63, 64]. In this model, terms of dimension five are added to the Lagrangian of the standard model satisfying the following criteria

1. The terms must be quadratic in the same field.
2. They should have one more derivative than the kinetic energy term.
3. They must respect gauge invariance.
4. They must be Lorentz invariant except for the appearance of a unit vector $n \cdot n = 1$ parameterizing a privileged direction in spacetime.
5. The terms should not reduce to lower dimension terms through the equations of motion.
6. They should not reduce to a total derivative.

The simplest correction satisfying the above criteria are

$$\delta\mathcal{L}_S = i\kappa\ell_P\phi(n\cdot\partial)^3\phi \tag{2.3.31}$$

$$\delta\mathcal{L}_\gamma = \xi\ell_P n^\mu F_{\mu\alpha}(n\cdot\partial)n_\nu\tilde{F}^{\nu\alpha} \tag{2.3.32}$$

$$\delta\mathcal{L}_f = \ell_P\tilde{\psi}\left(\eta_1 \not{n} + \eta_2 \not{n}\gamma_5\right)(n\cdot\partial)^2\psi \tag{2.3.33}$$

In these equations ξ, η_1, η_2 are free parameters and

$$\tilde{F}_{\mu\nu} = \frac{1}{2}\epsilon_{\mu\nu\alpha\beta}F^{\alpha\beta} \tag{2.3.34}$$

is the dual of the electromagnetic field tensor.

These terms are CPT-odd and thus induce a host of Lorentz breaking phenomena. The model can be applied to the tree level, but radiative corrections will generate terms of dimension 2 or 3 badly breaking Lorentz invariance [65, 66]. These issues will be treated later on.

Last but not least I would like to mention *double special relativity* as a possible test theory, not of Lorentz violation but of new physics near the Planck scale [67, 68]. Rather than a violation of Lorentz invariance, double special relativity is a nonlinear realization of the Lorentz group, that exhibits both a limiting velocity (the speed of light c) and a limiting energy (he Planck energy E_P). This beautiful theory falls beyond the scope of this article, which is concerned with "true" violations of Lorentz invariance. For a discussion on double special relativity, see for instance [69].

2.4 Quantum Gravity as a source of Lorentz violation

Quantum Gravity has been proposed as a source of Lorentz violations [1, 8, 13, 70, 71] by many authors. The rationale behind the proposal is that quantum gravity effects will make space-time granular at a scale near the Planck scale ℓ_P. This in turn will change the dispersion relations of elementary particles, breaking in general Lorentz invariance.

As it has been mentions, such modifications have been found within the two currently most popular approaches to quantum gravity, namely *Loop Quantum Gravity* and *String theory*. Let us discuss briefly how these phenomena arise, taking Loop Quantum Gravity as an example. The analysis of the resulting Lorentz violations can be carried out in the framework of the test theory developed by Kostelecký and coworkers.

2.4.1 A reminder of Loop Quantum Gravity

Loop Quantum Gravity is a theory of canonical quantization of the gravitational field. References [72, 73, 74] are introductions in different levels, while [75] is a critical review on the subject.

The details of the theory are complicated, so in this note we shall only give some qualitative description.

As it is well-known, gravitation has gauge invariance under diffeomorphisms and thus physical states of the gravitational field must satisfy the *momentum constraint* and the *Hamiltonian constraint*. The strategy of Loop Quantum Gravity is to find a suitable way of dealing with these constraints.

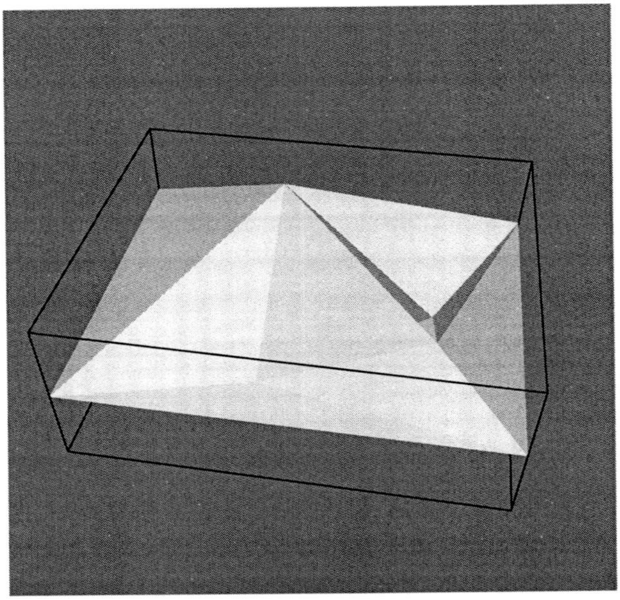

Figure 2.4.1: A discretized state of the gravitational field.

1. In order to perform the quantization, a suitable set of gravitational canonical variables must be introduced. These are cunningly contrived modifications of the Ashtekar variables [76, 77].

2. A regularization is introduced in the theory by breaking space-time in a lattice-like structure, at a scale of the order of ℓ_P (Figure 2.4.1). With this regularization, all physical quantities became well defined, although gauge dependent.

3. In terms of these variables,a suitable auxiliary Hilbert space $\mathcal{H}_{\rm aux}$ is built [72]. However, because of the invariance of the theory under diffeomorphisms, not all of the states in $\mathcal{H}_{\rm aux}$ represent physical states of the gravitational field.

4. The physical states are chosen so that the Wilson loops built from the canonical variables are well defined. This is the crucial step, from which the theory can be developed. This representation, originally introduced for the electromagnetic field [78, 79, 80] allows the definition of well defined physical quantities.

5. In principle, the regularization introduced in step 2 can be eliminated shrinking the scale of the lattice to zero. However, it can be shown that the volume element operator acts like a regulator, yielding well-defined quantities in this limit.

Although the technical details are very complex, the above steps have been carried to success (see, however, Ref. [75]). Among the most interesting results obtained, let us mention

1. Geometrical quantities, such as areas and volumes, are quantized s multiples of ℓ_P^2 (or ℓ_P^3), even in the limit of the auxiliary lattice size going to zero [72, 73].
2. Other field theories can be included in the scheme. The lattice-like discretization provides a suitable regularization of the field quantities [72].
3. Solutions for highly symmetrical states, such as spherically symmetrical or cosmological ones, can be found by solving additional constraints. For instance, a beautiful solution for a Robertson-Walker universe has been found, free of singularities (For a review, see [81, 82]).

2.4.2 Lorentz Invariance Violation in Loop Quantum Gravity

Let us examine the origin of Lorentz invariance violation in Loop Quantum Gravity with an example: the lowest order corrections to the Maxwell Hamiltonian [13]. The latter can be written in the form

$$H_{\rm M} = \frac{1}{2\alpha}\int d^3x \frac{g_{ab}}{g}\left(\tilde{E}^a\tilde{E}^b + \tilde{B}^a\tilde{B}^b\right) \tag{2.4.35}$$

where the tildes denote that the fields are tensor densities in the canonical framework, g is the determinant of the metric and α is the fine structure constant. We wish to average (2.4.35) over a semiclassical state $|g, \mathcal{L}\rangle$. Indexes a, b are purely spatial.

There is currently no clear characterization in Loop Quantum Gravity of a semiclassical state. Following [13] we shall use a *weave*: this is a state characterized by a semiclassical length scale $\mathcal{L}$ such that

$$\ell_P \ll \mathcal{L} \ll \lambda \tag{2.4.36}$$

Intuitively, a weave can be thought as a big carpet, of size $\mathcal{L}$, weaved of tiny pieces of scale ℓ_P (Fig. 2.4.2). The characteristic size $\mathcal{L}$ should be smaller than any "macroscopic" characteristic length λ such that spacetime can be considered continuous. Thus, weaves are states where whose scale $\mathcal{L}$ marks the transition from quantum (discrete) spacetime to classical one.

In the following, we shall assume that the weave state $|g, \mathcal{L}\rangle$ correctly describes the *privileged* reference frame S.

Consider the averages of the Hamiltonian (2.4.35) in the weave state $|g, \mathcal{L}\rangle$

$$H_{\rm ef} = \langle g, \mathcal{L}\,| H_{\rm M} |\, g, \mathcal{L}\rangle \tag{2.4.37}$$

The average is performed for fixed values of the electromagnetic field operators $\boldsymbol{E}, \boldsymbol{B}$ which, for all practical purposes, can be considered classical fields. That is,

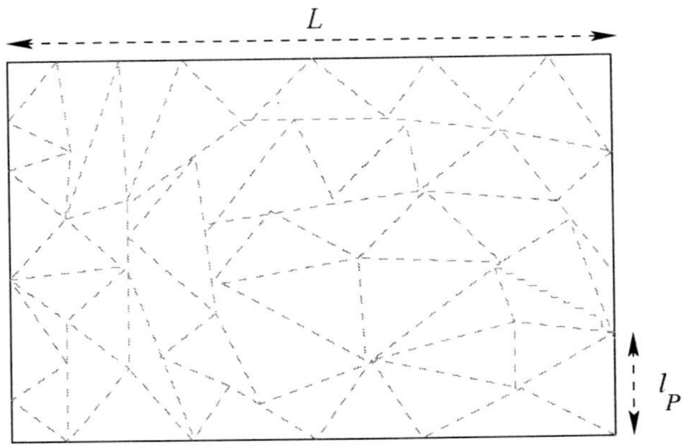

Figure 2.4.2: A semiclassical weave state

the average is made on the fluctuations of the quantum gravitational field. We shall also assume that the classical geometry in the privileged reference frame is flat euclidean and so we expect

$$\langle g, \mathcal{L} \left| g_{ab} g^{-1} \right| g, \mathcal{L} \rangle = \delta_{ab} + O\left(\frac{\ell_P}{\mathcal{L}}\right) \tag{2.4.38a}$$

$$\langle g, \mathcal{L} \left| A_{ia} \right| g, \mathcal{L} \rangle = 0 + \frac{1}{\mathcal{L}}\left(\frac{\ell_P}{\mathcal{L}}\right)^{\Upsilon} \tag{2.4.38b}$$

where A is the Ashtekar variable operator and Υ is a real number parameter.

Let us now introduce the Thiemann factorization of the metric operator

$$g_{ab} g^{-1} = \hat{w}_a \hat{w}_b \tag{2.4.39}$$

where the operators $\hat{w}$ are finite and have nonzero contributions only at the vertexes v_i. Introducing a point splitting regulator $\Delta(x, y) \to \delta(x - y)$ when the triangulation scale goes to zero, the Maxwell Hamiltonian can be written as a sum over the vertexes in a weave

$$H_{\text{ef}} = \frac{1}{2\alpha} \sum_{v_i, v_j} \langle g, \mathcal{L} \left| \hat{w}(v_i) \hat{w}(v_j) \right| g, \mathcal{L} \rangle \left[E^a(v_i) E^b(v_j) + B^a(v_i) B^b(v_j) \right] \tag{2.4.40}$$

The electromagnetic field, being quasiclassical, is slowly varying across the weave and can be expanded in a Taylor series about some point x_0

$$E^a(v_i) = E^a(x_0) + (v_i - x_0)_c \partial^c E^a(x_0) \tag{2.4.41}$$

and after a simple calculation the low energy Maxwell Hamiltonian takes the form

$$H_{\text{ef}} = \frac{1}{2\alpha} \int d^3x \left[\left(\boldsymbol{E}^2 + \boldsymbol{B}^2 \right) + 2 t_{abc} \left(E^a \partial^c E^b + B^a \partial^c B^b \right) \right] \tag{2.4.42}$$

where the 3-tensor

$$t_{abc} = \frac{1}{2} \sum_{v_i, v_j} \langle g, \mathcal{L} \,| \hat{w}(v_i) \hat{w}(v_j) |\, g, \mathcal{L} \rangle \, (v_i - x_0)_c. \tag{2.4.43}$$

Now, let us consider the electromagnetic field in the *privileged* reference frame S. We expect that the electromagnetic field will behave isotropically in this particular reference frame and so we shall write the averages as isotropic tensors in S: that is, tensors built from the metric tensor δ_{ab} or the Levi-Civita tensor ϵ_{abc}. Thus the 3-tensor t_{abc} must have the form

$$t_{abc} = \theta_{GP} \ell_P \epsilon_{abc} \tag{2.4.44}$$

the last factor coming from the fact that a nonzero value of t_{abc} can be expected only at scales near the Planck length, where the distributions of vertexes is very anisotropic.

As a final result, we get the *Gambini-Pullin Hamiltonian*

$$H_{GP} = \frac{1}{2\alpha} \int d^3x \left[\left(\boldsymbol{E}^2 + \boldsymbol{B}^2 \right) + 2\theta_{GP} \ell_P \left(\boldsymbol{E} \cdot \boldsymbol{\nabla} \times \boldsymbol{E} + \boldsymbol{B} \cdot \boldsymbol{\nabla} \times \boldsymbol{B} \right) \right]. \tag{2.4.45}$$

The second term is parity odd and originates birefringence in vacuum propagation. This happens because we have implicitly assumed that weaves can break parity, although in the privileged system rotational invariance is preserved.

The above informal method was developed in a much more rigorous way in reference [15], where a detailed exposition of the used techniques is given, and higher order corrections in the small parameter $\ell_P/\mathcal{L}$ were found. These higher order corrections depend on the Υ and $\mathcal{L}$ parameters and have a rather complex structure.

The method was also applied to the Dirac Hamiltonian [14, 16]. These latter corrections are particularly important, since they can be submitted to strict experimental tests.

The modified Dirac equation, in the privileged reference frame, takes the form

$$\left[i\gamma^\mu \partial_\mu + \Theta_1 m \ell_P i \boldsymbol{\gamma} \cdot \boldsymbol{\nabla} - K/2 \gamma_5 \gamma^0 - m \left(\alpha - i \Theta_2 \ell_P \boldsymbol{\Sigma} \cdot \boldsymbol{\nabla} \right) \right] \Psi = 0 \tag{2.4.46}$$

where the spin operator is $\Sigma^k = (i/2) \epsilon_{klm} \gamma^l \gamma^m$. Here

$$K = m \Theta_4 m \ell_P, \qquad \alpha = \left(1 + \Theta_3 m \ell_P \right), \tag{2.4.47}$$

and $\Theta_1, \Theta_2, \Theta_3, \Theta_4$ are constants of order one. Besides we have assumed $\mathcal{L} = 1/m$, where m is typically the particle mass.

The term $m\,(1 + \Theta_3\, m\ell_P)$ can be interpreted as a renormalization of the mass whose physical value is taken to be $M = m\,(1 + \Theta_3\, m\ell_P)$. The other terms, however cannot be eliminated by a simple transformation, since they break discrete symmetries. Indeed, the first two correction terms are CPT odd, while the last one is CP odd. Thus, they represent new physical effects induced by Quantum Gravity.

2.4.3 Transforming into the Laboratory System

Although many phenomena can be conveniently analyzed in the Privileged reference system S, many others are manifest in the Laboratory reference system S', moving with four velocity W^μ with respect to S. This transformation can be accomplished in a very elegant way using the Lagrangian density, which should be a scalar density [25]. As an example, consider transforming (2.4.46) to the lab system.

We shall write (2.4.46) in a formally covariant form using W^μ and projectors constructed from it. In this way, we find

$$\begin{aligned}\mathcal{L} =& \frac{1}{2}\bar{\Psi} i\gamma^\mu \partial_\mu \Psi - \frac{1}{2}\bar{\Psi} M \Psi \\ &+ \frac{1}{2} i(\Theta_1 M\ell_P)\bar{\Psi}\gamma_\mu \left(g^{\mu\nu} - W^\mu W^\nu\right)\partial_\nu \Psi \\ &+ \frac{1}{4}(\Theta_2 M\ell_P)\bar{\Psi}\epsilon_{\mu\nu\alpha\beta} W^\mu \gamma^\nu \gamma^\alpha \partial^\beta \Psi \\ &- \frac{1}{4}(\Theta_4 M\ell_P) M W^\mu \bar{\Psi}\gamma_5\gamma_\mu \Psi\end{aligned} \tag{2.4.48}$$

In this way, the Dirac particle Lagrangian takes the form of the Kostelecký Lagrangian, with

$$a_\mu = H_{\mu\nu} = d_{\mu\nu} = e_\mu = f_\mu = 0 \tag{2.4.49}$$

$$c_{\mu\nu} = \Theta_1 M\ell_P \left(g^{\mu\nu} - W^\mu W^\nu\right) \tag{2.4.50}$$

$$g_{\alpha\beta\gamma} = \Theta_2 M\ell_P \epsilon_{\mu\alpha\beta\gamma} W^\mu \tag{2.4.51}$$

$$b_\mu = \frac{1}{2}\Theta_4 M^2 \ell_P W^\mu \tag{2.4.52}$$

and from these expressions the low energy Hamiltonian for the fermion can be found in the form

$$\begin{aligned}\tilde{H} =& Mc^2(1 + \Theta_1\, M\ell_P\, (\boldsymbol{w}/c)^2) \\ &+ \left(1 + 2\,\Theta_1 M\ell_P\left(1 + \frac{5}{6}(\boldsymbol{w}/c)^2\right)\right)\left(\frac{p^2}{2M} + g\,\mu\,\boldsymbol{s}\cdot\boldsymbol{B}\right) + \\ &+ \left(\Theta_2 + \frac{1}{2}\Theta_4\right) M\ell_P \left[\left(2Mc^2 - \frac{2p^2}{3M}\right)\boldsymbol{s}\cdot\frac{\boldsymbol{w}}{c} + \frac{1}{M}\boldsymbol{s}\cdot Q_P\cdot\frac{\boldsymbol{w}}{c}\right] \\ &+ \Theta_1 M\ell_P \left[\frac{\boldsymbol{w}\cdot Q_P\cdot\boldsymbol{w}}{Mc^2}\right],\end{aligned} \tag{2.4.53}$$

where g is the standard gyromagnetic factor and Q_P represents the momentum quadrupole tensor, with components $Q_{Pij} = \langle p_i p_j - 1/3 p^2 \delta_{ij}\rangle$.

The first two lines of (2.4.53) represent the Schrödinger-Pauli Hamiltonian, with corrections to the rest and inertial mass, coming from the motion with respect to the "New Aether". The third and fourth lines represent an "Aether wind" blowing on the spin and and an anisotropy of the inertial mass of the particle, which could be tested in Hughes-Drever experiments.

2.4.4 Other models of Lorentz violation by Quantum Gravity

A similar treatment can be applied to other models of Lorentz violation generated by Quantum Gravity. As an example let us briefly discuss the so called Liouville approach to Non-critical String Theory. In this scheme our universe is identified with a 4 dimensional D-brane, which will naturally contains a type of topological defect called a D-particle. These will interact with ordinary particles, represented by strings in the corresponding mode, by elastic scattering, which would produce a recoil of the D-particle. This recoil will produce a local disturbance of the space-time geometry which will in turn affect the propagation of the particle [18, 19, 20].

We take this part of the analysis directly from Ref. [83], starting with their modified Dirac equation

$$\left[\gamma^\mu(i\partial_\mu - eA_\mu) - v^i\gamma^0(i\partial_i - eA_i) - m\right]\Psi \;=\; 0. \qquad (2.4.54)$$

Here γ^μ are the standard flat-space gamma matrices and v^i is the D-particle recoil velocity in the CMB frame, where it is initially at rest. The above modified Dirac equation preserves gauge invariance due to the standard minimal coupling. In the following we set $A_\mu = 0$. We will be concerned with a perturbative expansion in the small parameter v^i.

The above equation can be interpreted as describing the motion of a fermion in an effective metric

$$p^\mu G_{\mu\nu} p^\nu = E^2 - 2E\boldsymbol{v}\cdot\boldsymbol{p} - \boldsymbol{p}^2, \quad p^\mu = (E,\boldsymbol{p}), \quad |\boldsymbol{v}| \approx \frac{|\,\boldsymbol{p}\,|}{M}. \qquad (2.4.55)$$

The analysis of the above equation requires some care to define correctly the D-particle recoil velocity in the nonrelativistic limit [26]. Passing to the laboratory frame the following form of the equation is found

$$\left[i\gamma^\mu\partial_\mu - i\gamma^\rho\,(W_\rho V^\nu)\,\partial_\nu - m\right]\Psi = 0, \qquad (2.4.56)$$

which can be put in the standard form of [50]. The identification of the corresponding Kostelecký parameters leads to

$$a_\mu \;=\; b_\mu = 0 = H_{\mu\nu}, \qquad g_{\lambda\mu\nu} = 0 = d_{\mu\nu}, \qquad c_{\mu\nu} = -W_\mu V_\nu. \qquad (2.4.57)$$

Let us note that $c_{\mu\nu}$ is not symmetrical in this case.

Non commutative field theory is another important model of Lorentz violating field theory. We shall discuss issues connected with these theories in Section 2.6.

2.5 Testing Lorentz Invariance Violation at the tree level

The results collected in Section 2.4 can be tested with sensitive well defined experiments if one neglects the effects of radiative corrections. In this section we shall examine these tests at the "tree level", leaving a discussion of radiative corrections for the next one.

There is a host of phenomena which can be used to detect Lorentz Invariance violation. Among them, let us mention

1. Breakdown of local rotational symmetry ("Aether wind effects").
2. Breakdown of Lorentz Boost Invariance ("Kennedy-Thorndike experiments").
3. Occurrence of "forbidden processes" (such as photon decay in vacuum).
4. Dispersive processes in vacuum (such as energy dependent velocities or birefringence).
5. Breakdown of discrete symmetries, principally violation of the CPT theorem.

There are two main group of observations with high enough sensitivity to probe the Quantum Gravity regime: Cosmological or astrophysical tests of Lorentz invariance ("Threshold analysis type"), based on the modification of the dispersion relations of particles, and Clock comparison experiments ("Hughes-Drever type"). The sensitivity requirements are indeed very strict; for instance, for terms linear in the Planck length ℓ_P, one should have an accuracy better than

$$\ell_P m \sim 10^{-19} \tag{2.5.58}$$

for a mass of the order of the nucleon mass.

In the following we shall consider some of the most interesting tests of Lorentz invariance. Our selection is motivated by their effectiveness in testing Quantum Gravity induced Lorentz violation. More complete covering of the phenomenology can be found in references [59, 84, 40, 60, 69, 62, 85, 86, 87]

2.5.1 Kinematic tests of Lorentz invariance

These are the classical Michelson-Morley [36] and Kennedy-Thorndike [37] experiments, which test isotropy and boost independence of the local velocity of light. Using the Robertson metric (2.3.11) it is easy to show that the light velocity is given by

$$\frac{c(\theta, V)}{c_0} = 1 + (\alpha + \beta - 1)\frac{V^2}{c^2} + \left(\beta - \delta - \frac{1}{2}\right)\frac{V^2}{c^2} \tag{2.5.59}$$

Te last term is a local anisotropy in the velocity of light, while the second one represents a boost dependence. Several modern versions of both experiments have been carried in the latter years [88, 89, 90, 49], with strict limits on the Mansouri-Sexl parameters

$$|\alpha + \beta - 1| < 6.9 \times 10^{-7} \tag{2.5.60a}$$

$$\left|\delta - \beta + \frac{1}{2}\right| < 4.5 \times 10^{-9} \tag{2.5.60b}$$

The result of the new version Michelson-Morley experiment [49] were also interpreted also in terms of the Kostelecký k_F parameter tensor. It tests only the P-even part of the tensor and it imposes limits of

$$|k_{Fe^-}| < 10^{-14} \qquad |k_{Fo^-}| < 10^{-10} \tag{2.5.61}$$

where the limits refer to the P-even and P-odd decomposition of k_F [48, 49].

α is an important parameter (or rather, the function $g_0(V)$ is) since its value is related to the dispersion relation for the particle. Indeed, it can be shown [35] that the energy of a moving particle is related to its velocity in the form

$$E = \frac{mc^2}{g_0(V)\sqrt{1 - \frac{V^2}{c^2}}} \tag{2.5.62}$$

which provides an interesting connection between the Robertson-Mansouri-Sexl test theory and some dispersion relation ones. But (2.5.62) defines also the transverse Döppler effect, from which α can be measured. The most accurate value obtained is [91]

$$\left|\alpha + \frac{1}{2}\right| < 8 \times 10^{-7} \tag{2.5.63}$$

In spite of these impressive results, the accuracy of the above experiments is still below the requirement for testing Quantum Gravity at the tree level.

2.5.2 "Forbidden" processes

If the limiting velocity of a high energy particle is different from the light velocity, several processes forbidden in a Lorentz invariant context became possible [92, 93]. As an example, consider the decay of a photon in an electron-positron pair, a process which is usually forbidden by energy-momentum conservation (Fig. 2.5.3)

$$\gamma \to e^+ + e^- \tag{2.5.64}$$

But if the speed of photons is greater than the limiting speed of electrons this is no more so. Indeed, energy-momentum conservation implies

$$q = p_+ + p_- \tag{2.5.65}$$

Let us examine this process in the context of the $TH\epsilon\mu$ formalism 2.3.23. There is a threshold for this reaction: the photon energy should be enough to produce an electron-positron pair at rest, so that

$$E_\gamma > \frac{2m}{\sqrt{\left(\frac{c_\gamma}{c_e}\right)^2 - 1}} = E_0 \tag{2.5.66}$$

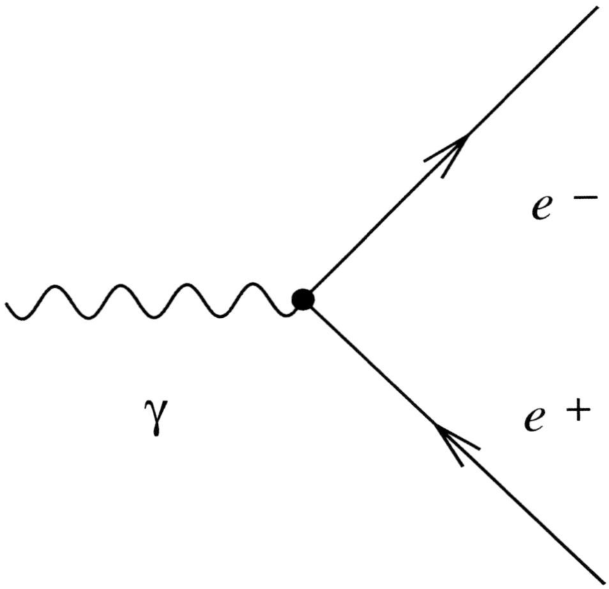

Figure 2.5.3: The forbidden process $\gamma \to e^+ + e^-$.

Any photon with energy $E > E_0$ will rapidly decay into electron-positron pairs so that they should never arrive to earth. However, photons of energy $E_\gamma > [20] TeV$ have been observed so that an upper limit is obtained [92]

$$\frac{c_\gamma}{c_e} - 1 < 1.5 \times 10^{-15} \tag{2.5.67}$$

On the other hand, if the limiting speed is greater than the photon speed, the energy of charged particles is limited by vacuum Cerenkov process

$$p \to p + \gamma \tag{2.5.68}$$

whose threshold is

$$E_0' = \frac{m_p}{\sqrt{\left(1 - \frac{c_\gamma}{c_e}\right)^2}} \tag{2.5.69}$$

From the fact that the upper limit of cosmic ray spectrum is $E_u \sim [10^{20}] eV$ a lower bound of

$$\frac{c_\gamma}{c_e} - 1 > -5 \times 10^{-23} \tag{2.5.70}$$

was obtained [92].

Of course, the same phenomena can be analyzed in the context of any dispersion relation model, once the effective limit- and light- speeds can be determined. For instance, much more stringent limits have been obtained from the γ-ray spectrum of the Crab nebula [94]. These limits have been obtained within the Myers-Pospelov test theory.

2.5.3 Propagation phenomena: delays and polarization

The propagation of high energy particles, specially photons, has been one of the most important tests of Lorentz Invariance since the seminal paper of Amelino-Camelia *et. al.* [8]. These are based on the simple fact that particles of different energy, with dispersion relations of the form (2.3.16), will have different velocities and so different travel times from the same sources.

For a pair of photons with energies E_1, E_2 and dispersion relation of the form (2.3.16) with $n = 1$ the difference in arrival time would be

$$\frac{\Delta t}{T} \simeq \eta \frac{\Delta E}{E_P} \tag{2.5.71}$$

where $T \simeq D/c$ is the travel time of the photon. For a photon energy difference of $[1] GeV$ and a travel time of $[10^{17}] s$ a time delay of the order of a second results, depending on the value of η. A typical bound from TeV flares of Makarian 421 is [95]

$$\frac{E_P}{\eta} \geq [4 \times 10^{16}] GeV \tag{2.5.72}$$

implying

$$\eta > 300 \tag{2.5.73}$$

This limit is still below the accuracy requirements (2.5.58) for testing Quantum gravity.

On the other hand, very interesting results have been found studying the polarization behavior of the traveling radiation. As an example, consider the modified Maxwell equations derived from the Hamiltonian (2.4.45) [13, 96]

$$\dot{\boldsymbol{E}} = -\boldsymbol{\nabla} \times \boldsymbol{B} + 2\theta_{GP}\ell_P \nabla^2 \boldsymbol{B} \tag{2.5.74a}$$

$$\dot{\boldsymbol{B}} = \boldsymbol{\nabla} \times \boldsymbol{E} - 2\theta_{GP}\ell_P \nabla^2 \boldsymbol{B} \tag{2.5.74b}$$

These modified Maxwell equations (written in the privileged reference system) have a solution [13, 96]

$$\boldsymbol{E}_{\pm} = (\hat{\mathbf{e}}_1 \pm \hat{\mathbf{e}}_2) e^{i(\Omega_{\pm} t - \boldsymbol{k} \cdot \boldsymbol{x})} \tag{2.5.75a}$$

and a similar expression for $\boldsymbol{B}$. Here $\hat{\mathbf{e}}_{1,2}$ denote the polarization unit vectors and

$$\Omega_{\pm} = k \left(1 \mp 2\theta_{GP}\ell_P k\right) \tag{2.5.75b}$$

is the dispersion relation for photon propagation in vacuum.

The latter equation predicts birefringence of the vacuum. Indeed, for a wave packet traveling along the x axis, formed with a superposition of plane waves with central frequency Ω_0 (2.5.75a), one finds

$$\boldsymbol{E} \simeq e^{i\Omega_0(t-x)} \left[\mathcal{A}(x - v_+ t) e^{-i\theta_{GP}\ell_P \Omega_0^2 z} \hat{\boldsymbol{e}_+} + \mathcal{A}(x - v_- t) e^{i\theta_{GP}\ell_P \Omega_0^2 z} \hat{\boldsymbol{e}_-} \right] \tag{2.5.76}$$

with $A(x)$ the initial wave packet, $\hat{\boldsymbol{e}}_\pm = \hat{\boldsymbol{e}}_1 \pm \hat{\boldsymbol{e}}_2$ and the group velocities are given by

$$v_\pm = 1 \mp 4\theta_{GP}\ell_P k \tag{2.5.77}$$

For short enough distances, so that

$$8\theta_{GP}\ell_P \Omega_0 \delta\Omega \ll 1 \tag{2.5.78}$$

where $\delta\Omega$ is the frequency width of the wave packet, equation (2.5.76) can be recast in the form

$$\boldsymbol{E} \sim B(t - x) \left[\cos\left(\theta_{GP}\ell_P \Omega_0^2 z\right) \hat{\boldsymbol{e}_1} + \sin\left(\theta_{GP}\ell_P \Omega_0^2 z\right) \hat{\boldsymbol{e}_2} \right] \tag{2.5.79}$$

showing explicitly a rotation of the polarization vector.

From the fact that the optical polarization spectrum of many distant galaxies is flat, the following bound was found for the Gambini-Pullin parameter [96]

$$|\theta_{GP}| < 10^{-4} \tag{2.5.80}$$

This is a very strict bound on a parameter directly connected to Quantum Gravity.

Even more strict bounds were obtained for the P-odd part of the k_F tensor in the Lorentz violating extension of the Standard Model [47, 48]. From an analysis of a carefully selected set of cosmologically distant sources, whose polarization had been measured. From a careful analysis of these sources, a limit

$$|k_F^\alpha| < 10^{-31} \tag{2.5.81}$$

was found.

The bounds (2.5.80) and (2.5.81) are in a certain sense complementary, since the Lorentz violating extension of the Standard Model does not include terms of dimension 5, such as those in the Gambini-Pullin Hamiltonian.

2.5.4 Laboratory experiments

There has been a great number of laboratory experiments designed to detect tiny violations of Lorentz invariance. Among them, the Hughes-Drever-like experiments [97, 98] designed to detect tiny anisotropies in the laboratory reference system, are the most sensitive ones.

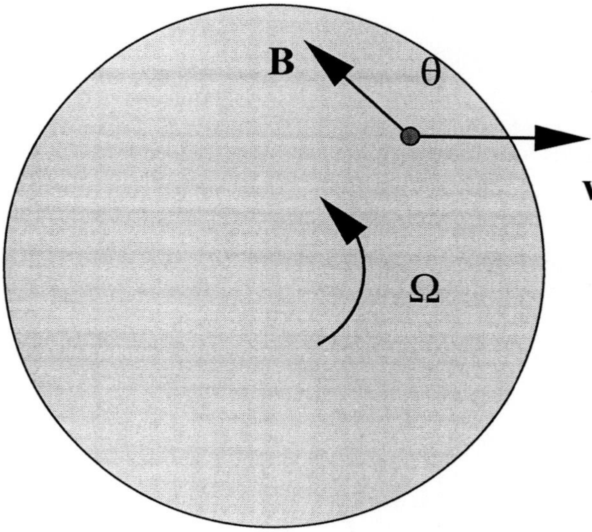

Figure 2.5.4: Schematic view of a Hughes-Drever experiment

In a Hughes-Drever experiment the spin of an atomic or nuclear system is oriented in a magnetic field in a reference system rotating with angular velocity $\boldsymbol{\Omega}$ and moving with velocity $\boldsymbol{w}$ with respect of the privileged one (fig. 2.5.4). The motion with respect to the privileged system generates local anisotropies that in turn generate a signal of frequency $n\Omega$.

In the latter years many new versions of the original Hughes-Drever experiment have been devised, with ever increasing accuracy [99, 100, 101, 102, 103, 104, 105]. As an example, let us consider bounds on the Dirac Hamiltonian parameters (2.4.53) from Loop Quantum Gravity [25].

As we have already said, the last line in (2.4.53) represent a coupling of the spin to the velocity with respect to the privileged frame $\boldsymbol{w}$. Keeping only the dominant terms and passing to the usual units we find the dominant perturbation

$$\delta H_S = \left(\Theta_1 + \frac{1}{2}\Theta_4\right) M\ell_P(2Mc^2)\left[1 + O\left(\frac{p^2}{2Mc^2}\right)\right] \boldsymbol{s}\cdot\frac{\boldsymbol{w}}{c} \tag{2.5.82}$$

The third line in (2.4.53) represents an anisotropy of the inertial mass, to test which the Hughes-Drever experiments were designed. With the approximation

$$Q_P \simeq -\frac{5}{3}\langle p^2\rangle\frac{Q}{R^2} \tag{2.5.83}$$

for the momentum quadrupole moment, with Q the electric quadrupole moment and R the nuclear radius, we obtain

$$\delta H_Q = -\Theta_1 M\ell_P\frac{5}{3}\left\langle\frac{p^2}{2M}\right\rangle\frac{Q}{R^2}\left(\frac{w}{c}\right)^2 P_2(\cos\theta) \tag{2.5.84}$$

for the quadrupole moment perturbation, where θ is the angle between the quantization axis and $\boldsymbol{w}$, $\langle p^2/2M\rangle \sim [40]MeV$ for the kinetic energy of the last shell of a typical heavy nucleus. from the results of references [101, 104] we find [25]

$$\left|\Theta_1 + \frac{1}{2}\Theta_4\right| < 2 \times 10^{-9} \qquad |\Theta_1| < 3 \times 10^{-5} \tag{2.5.85}$$

This set of experiments can be interpreted in different test theories. For instance, the above results have been used to set bounds for the Myers-Pospelov parameters [63].

Besides, the parameter $c - 1$ in the $TH\epsilon\mu$ model has been constrained from the results of references [99, 100, 101]

$$|c - 1| < 10^{-21} \tag{2.5.86}$$

the result being less sensitive than (2.5.85) because of several suppressing nuclear factors.

Much more interesting is the possibility of constraining the mass of the D-particle from the above results. Indeed, from equation (2.4.56) one obtains the low-energy perturbation Hamiltonian

$$H'_Q = -4\frac{m_N}{M_D}\frac{\mathbf{w}\cdot Q_P\cdot\mathbf{w}}{Mc^2} \tag{2.5.87}$$

and comparison with (2.4.53) shows that

$$M_D = \frac{4}{\Theta_1 \ell_P} > 1.2 \times 10^5 M_P \tag{2.5.88}$$

This is the strongest bound found so far for the mass of the D-particle. The corresponding recoil velocity

$$v < 2 \times 10^{-27} c \tag{2.5.89}$$

is extremely small, even by the standards of everyday experience (e.g. the speed of a crawling snail is $10^{-11}c$), that it seems quite unlikely that it will be detected in more direct experiments, such as time delays.

2.5.5 The GZK cutoff

The spectrum of high energy cosmic rays should show a cutoff at energies about $[50]EeV$, called *the GZK cutoff* [106, 107]. The origin of this cutoff comes from several processes that eat up the energy of primary cosmic rays, such as inverse Compton effect by the Cosmic Microwave Background photons. A modern evaluation of the ultrahigh energy spectrum of cosmic rays can be found in [108] and a complete review of the problem in [109].

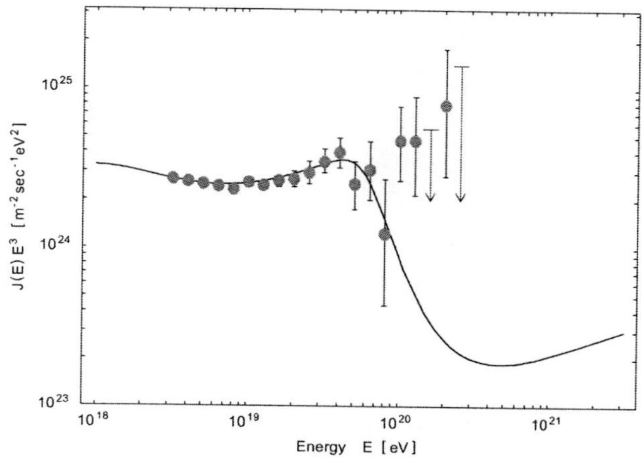

Figure 2.5.5: The GZK cutoff and the AGASA High energy data.

In spite of the theoretical prediction, quite a few cosmic rays of energies greater than the GZK cutoff have been found in several cosmic ray observatories, mainly AGASA [110, 111]. This is usually called *the GZK anomaly.*

Alfaro and Palma [112, 41] have shown that an explanation of the GZK anomaly can be found within the Loop Quantum gravity model, and besides, that this can be used to test quadratic terms in the Planck scale.

To see how this can be done, observe that the dispersion relations for particles in the Loop Quantum Gravity of section 2.4.2 take the form

$$E_{\pm}^2 \simeq \left[1 + 2\kappa_\alpha \left(\frac{\ell_P}{\mathcal{L}}\right)^2\right] p^2 + \kappa_\eta \ell_P^2 p^4 \pm \kappa_\lambda \frac{\ell_P}{2\mathcal{L}^2} p + m^2 \tag{2.5.90a}$$

$$\omega_{\pm} \simeq k \left[1 + \kappa_\gamma \left(\frac{\ell_P}{\mathcal{L}}\right)^{2+2\Upsilon} - \theta_3 (\ell_P k)^2 \pm \theta_{GP} \ell_P k\right] \tag{2.5.90b}$$

for massive fermions and photons respectively. In these equations the κ_i and θ parameters are numbers of order unity. The explicit dependence with the $\mathcal{L}$ and Υ parameters has been left explicit.

With these equations, a careful analysis of the main reactions in the presence of Lorentz violations was carried with the result that the GZK anomaly could be explained. A fit to the data is possible assuming that $\mathcal{L}$ is an universal constant with a value

$$[2.6 \times 10^{-18}]eV^{-1} \lesssim \mathcal{L} \lesssim [1.6 \times 10^{-17}]eV^{-1} \tag{2.5.91}$$

The good fit shown in Figure 2.5.6 shows the importance of threshold analysis for the test of even quadratic order in the Planck scale.

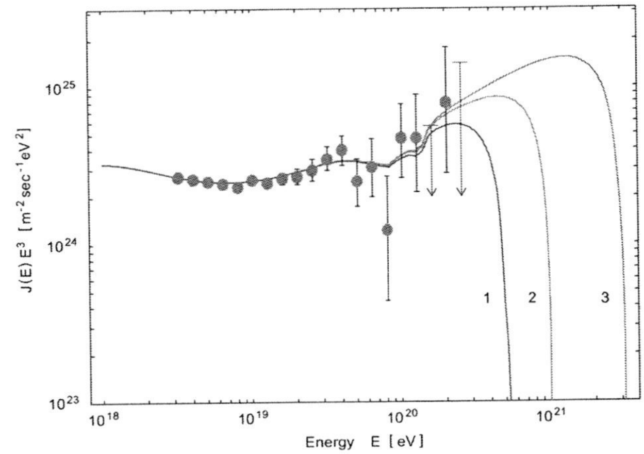

Figure 2.5.6: Lorentz Noninvariant fit of the GZK anomaly

2.5.6 Other tests of Lorentz invariance

There are many other tests of Lorentz invariance, but few of them satisfy our accuracy requirement (2.5.58). The most important of them are the laboratory test for CPT symmetry [113, 114, 115].

These experiments test the CPT symmetry of an electron (proton) in a Penning trap. Although potentially they should be very sensitive, they have not yet our required accuracy (2.5.58). Their impressive result

$$\frac{|\Delta a|\bar{\omega}_c}{2m_e c^2} = (3 \pm 12) \times 10^{-22} \tag{2.5.92}$$

translates into the weak constraint [84]

$$\left|\Theta_2 + \frac{1}{2}\Theta_4\right| \lesssim 1 \tag{2.5.93}$$

Another important group of experiments are made with spin-polarized matter [116, 84]. In these experiments a torsion pendulum with a total spin $S \sim 10^{22}$ can be used to extract a signal coupling spin and velocity with respect to a privileged frame. Their result translates into the mild bound

$$\left|\Theta_2 + \frac{1}{2}\Theta_4\right| \lesssim 0.002 \tag{2.5.94}$$

Although the increasing accuracy of these experiments will improve very much the above bound, they are still far from the results of Hughes-Drever experiments. Their main interest is that the parameters Θ_i are estimated from electron instead of proton, yielding independent tests of Lorentz invariance.

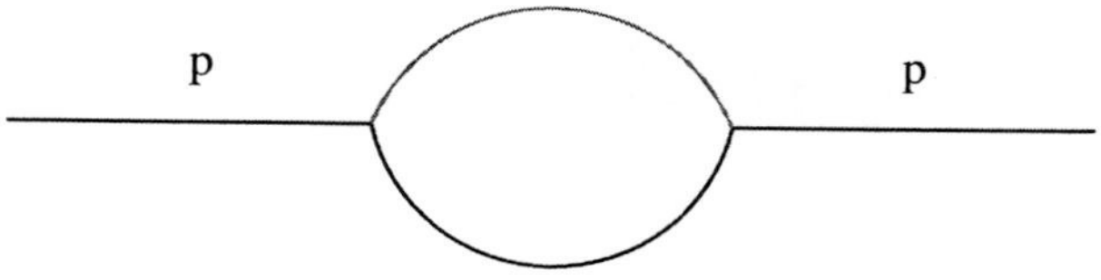

Figure 2.6.7: A typical self-energy diagram

2.6 The influence of radiative corrections

Radiative corrections, whether at one-loop or higher order, open the possibility of testing energies as high as Planck's in nowadays existing laboratory experiments. Indeed, consider the diagram in Figure 2.6.7: the loop variable k runs over all energies, probing thus near Planck scale.

Although quite a lot of work has been done on radiative corrections in Lorentz non-invariant theories [52, 117, 118, 65] only quite recently it was shown that they impose very stringent constraints on Lorentz violation phenomena [66]. In the context of renormalization theory, this originates a new fine tuning problem.

The main idea is that Planck scale physics should cutoff the range of the loop variable at a given scale $\Lambda \sim E_P$. If this cutoff is anisotropic, this anisotropy will be "dredged up" to a low energy scale. This would include corrections to the dispersion relations of the order of the coupling constants of the standard model, rather than $O(\ell_P p)$.

In the following sections, we shall discuss first a couple of examples and then discuss the general case.

2.6.1 A couple of examples

Let us consider first the electron self-energy, introducing a cutoff function $C(k^2)$ to model Planck scale effects

$$\Sigma^{(2)}(p,\Lambda,\xi) = \frac{e^2}{4\pi i}\int d^4k\gamma^\mu \frac{\gamma\cdot(p-k)+m}{(p-k)^2-m^2}\gamma_\mu \frac{1}{k^2+i\epsilon} C\left(k^2+\xi(W\cdot k)^2\right) \quad (2.6.95)$$

where ξ parametrizes an anisotropy in the laboratory system. Consider the particular case

$$C(k^2) = \frac{\Lambda^2}{\Lambda^2-k^2} \quad (2.6.96)$$

with $\Lambda \sim E_P$ and $\xi \lesssim 1$. Then we can expand the self energy in powers of ξ and keep the lowest order term

$$\Sigma^{(2)}(p,\Lambda,\xi) = \Sigma^{(2)}(p,\Lambda,0) + \delta\Sigma \quad (2.6.97)$$

with

$$\delta\Sigma(p,\Lambda,\xi) = -i\alpha \int d^4k\gamma^\mu \frac{\gamma\cdot(p-k)+m}{(p-k)^2-m^2}\gamma_\mu \frac{1}{k^2+i\epsilon}\frac{\Lambda^2}{\Lambda^2-k^2}\frac{\xi(W\cdot k)^2}{\Lambda^2-k^2} \tag{2.6.98}$$

After the usual procedure of combining denominators and shifting the loop variable one is lead to the on-shell expression

$$\delta\Sigma(\not p = m,\Lambda,\xi) = 12\alpha\xi \int_0^1 zdz \int_0^1 y^2dy \int d^4q \frac{P(q)\Lambda^2}{[q^2-m^2(1-y)^2-\Lambda^2xy]} \tag{2.6.99}$$

with

$$P(q) = \frac{1}{4}\left[(4+2y)mW^2 - 4(\gamma\cdot W)(W\cdot p)(1-y)\right]q^2 + O(q^0) \tag{2.6.100}$$

To compute this integral one usually performs a Wick rotation, but it is not clear that this can be done in the highly fractal structure of spacetime at the Planck scale. Indeed, one expects that the physical cutoff induced by discreteness will be in real spacetime. But this can be done in this particular model with the result

$$\delta\Sigma(\not p = m,\Lambda,\xi) = \pi^2\alpha\xi \int_0^1 zdz \int_0^1 y^2dy\tilde{P}(y)\frac{\Lambda^2}{m^2(1-y)^2+\Lambda^2xy} \tag{2.6.101}$$

with

$$\tilde{P}(y) = [(4+2y)mW^2 - 2(\gamma\cdot W)(W\cdot p)(1-y)] \tag{2.6.102}$$

The integral is finite in the limit $\Lambda\to\infty$ and we find thus a correction term in the nonrelativistic limit

$$\delta\Sigma \sim \xi\alpha m(\boldsymbol{s}\cdot\boldsymbol{w}) \tag{2.6.103}$$

that contradicts experiment unless $\xi \ll 10^{-28}$.

As a second example, let us consider a non-commutative version of QED, with

$$[x^\mu, x^\nu] = i\theta^{\mu\nu} \tag{2.6.104}$$

with $\theta^{\mu\nu}$ a c-number.

This case has been analyzed in reference [119] with the result

$$\delta\mathcal{L}_{\text{ef}} = \begin{cases} \frac{3}{4}m\Lambda^2\alpha^2\bar{\psi}\theta^{\mu\nu}\sigma_{\mu\nu}\psi, & \theta\Lambda^2 \ll 1 \\ 2\pi m\alpha^2\bar{\psi}\frac{\theta^{\mu\nu}\sigma_{\mu\nu}}{\frac{1}{2}\operatorname{Tr}\theta^2}\psi, & \theta\Lambda^2 \gg 1 \end{cases} \tag{2.6.105}$$

where θ is a typical scale of non-commutativity, such as $\theta = \sqrt{\theta^{\mu\nu}\theta_{\mu\nu}}$.

The second result predicts an anisotropy of order α^2, independent of both the scale of non-commutativity and cutoff and thus is in strong disagreement with experiment, while the first can only be consistent in the strange situation where the cutoff Λ is much smaller than the interesting non-commutative scale. This means,

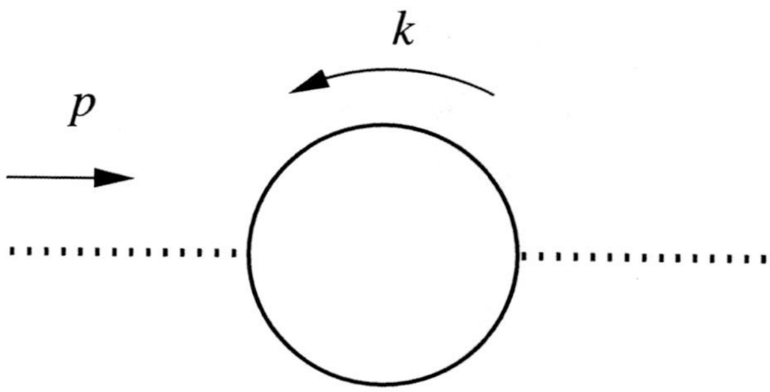

Figure 2.6.8: Lowest order self-energy graph in a Yukawa theory.

of course that the non-commutative field theory ceases to be valid at energy scales much smaller than the non-commutativity scale.

For our last example, consider the second order self energy of a scalar particle in a Yukawa theory $\Pi^{(2)}(p)$ (Figure 2.6.8). A measure of Lorentz violating in this graph is given by the quantity [66]

$$\xi = \left. \frac{\partial^2 \Pi^{(2)}(p)}{\partial p_0^2} + \frac{\partial^2 \Pi^{(2)}(p)}{\partial p_1^2} \right|_{p=0} \tag{2.6.106}$$

Without Planck scale modifications one would obtain

$$\xi = -\frac{ig^2}{\pi^4} \int d^4k \frac{\left(k_0^2 + k_1^2\right)\left(k^2 + 3M^2\right)}{\left(k^2 - M^2 + i\epsilon\right)^4} \tag{2.6.107}$$

This integral could be shown to be zero by a Wick rotation, after a suitable regularization to avoid the logarithmic divergence. However, if Planck scale physics introduces a cutoff, things are different. For instance, assume that the fermion propagator has a cutoff in the form

$$S'(k) = f(\boldsymbol{k}) \frac{\gamma \cdot k + M}{k^2 - M^2} \qquad f(0) = 1 \qquad f(\infty) = 0 \tag{2.6.108}$$

then the integral can be computed and the corresponding Lorentz violation is

$$\xi = \frac{g^2}{6\pi^2} \left[1 + \int_0^{\infty} dx x f'(x)^2\right] \tag{2.6.109}$$

This shows again that a Lorentz violation at the Planck sale is translated into a Lorentz violating term at low energy, of the order of the coupling constant squared.

2.6.2 A general argument

Let us now show that the above examples are really parts of a general property of radiative corrections. To fix our ideas, consider the self energy of a scalar particle in a Yukawa theory. The dispersion relation for the particle is obtained by solving

$$E^2 - \boldsymbol{p}^2 - m^2 - \Pi(E, \boldsymbol{p}) = 0 \tag{2.6.110}$$

with Π the sum of all self-energy graphs, to which we have added any small Lorentz-violating corrections that stem from tree-level theory.

Now we reason as follows: without a cutoff the graphs have divergences from large momenta. In the Lagrangian defining the theory the divergences correspond to terms of dimension 4 (or less) that obey the symmetries of the microscopic theory. In a Lorentz invariant case these divergences are removed by renormalization of the parameters of the theory. But Planck scale physics can both cutoff the divergences and modify the expressions that define the graph in the neighborhood of the Planck energy. The same power counting that determines the divergences also determines the natural size of the contributions from Planck scale momenta: the dominant contributions correspond to operators of dimension 4 (or less) in the Lagrangian. If the microscopic theory violates Lorentz invariance at the Planck scale, then generically we get Lorentz violating terms at low energy, without any suppression by powers of E/E_P.

Of course, the above Lorentz violating terms can be removed by explicitly including Lorentz violating counterterms of dimension 4 in the Lagrangian which are fine-tuned to give the observed low-energy Lorentz invariance. Such fine-tuning is unacceptable [120, 121, 122] in a fundamental theory.

In the models considered in Section 2.3 and 2.5 Lorentz violating corrections to the dispersion relations, suppressed by some power of E/E_P, were found by considering the propagation of free particles in a granular spacetime background. The above reasoning shows that there are effects that are only suppressed by two powers of the standard model coupling constants. These effects change the effective value of c in the usual dispersion relation

$$E^2 = c^2p^2 + m^2c^4 \tag{2.6.111}$$

The well-known structure of the standard model shows that (2.6.111) predicts differences up to 10% between different particles, in violent disagreement with experiment that shows that the fractional differences in c is below 10^{-20}.

The above result is a consequence of well-known properties of quantum field theories, of which the standard model is an example. Indeed the actual technical result is related to more or less explicit statements that can be found in several references [123, 6, 93, 7, 118, 63] although never fully recognized except n the recent [64]. For a criticism of these results see [65, 66].

2.7 Conclusion

In spite of the difficult nature of the research, a group of delicate experiments has been devised to test Lorentz invariance violations induced by Quantum Gravity. Up to now, no clear signal of Lorentz violation has been found (See, however, Section 2.5.5).

Tree-level tests of Quantum Gravity modifications include attempts to measure the parameters of the effective Lagrangian's describing these modifications. Several of the above mentioned experiments are accurate enough to measure the parameters of corrected dispersion relations with high precision. Even better results can be found from Hughes-Drever experiments. Indeed, the bounds on the parameters are so small that they call into question the full scheme of Lorentz invariance violations induced by Quantum Gravity, at least with respect to terms linear in the Planck length [25, 26].

Even more strict results stem from the analysis of radiative corrections. We have found that the loop variables can dredge up Lorentz violations at the Planck scale up to the low-energy scale. This would violently contradict experiment, unless extremely fine tuning counterterms are introduced in the Lorentz violating Lagrangian. This shows that Lorentz invariance should be added to the well-known list of fine-tuning problems; namely, the cosmological constant, the Higgs bare mass and mass hierarchies.

The implications of the above argument for both experiment and theory are quite profound. First, the present unsuccessful searches for Lorentz violations suffice by many orders of magnitude. Of course, since it is correct science to question and test accepted principles, tests of Lorentz invariance are worthwhile. However, many searches that were started by estimates of specific orders of magnitude may be misled and should be revised.

As to the theory, the critical task concerns any proposal in which Lorentz symmetry is substantially broken at the Planck scale: to find and implement a mechanism to give automatic Lorentz invariance at low energy despite a violation at Planck scale. We assume here that the treatment involves real time, not an analytic continuation to imaginary time, as is common in quantum field theory in flat spacetime (See, however, Section 2.6.1). One mechanism is to have a custodial symmetry that is sufficient to prohibit Lorentz-violating symmetry without being itself the full Lorentz group. Such a symmetry does not appear to be known and the Coleman-Mandula theorem [124] suggests it does not exists.

It should be noted that there is not necessarily a conflict between discreteness and the absence of a preferred frame. For instance in reference [125] a Lorentz invariant macroscopic space is constructed by the use of a random causal set of points. On the other hand, the authors of [126] argue that the existence of a minimum length does not imply local Lorentz invariance violation, anymore than the discreteness of angular momentum eigenvalues signal a violation of rotational invariance.

An optimistic point of view should be stressed: a branch of theoretical physics long considered to suffer from detachment from experimental guidance is now in the opposite situation. The subtle interplay of cosmology, atomic and nuclear physics is shedding light on such a recondite subject as quantum gravity.

Our results show that Lorentz invariance still imposes stringent requirements on the mathematical theories hat attempt to describe experimental results.

Acknowledgements

The author would like to thank NovaScience for its invitation to write this paper. This work has been partially sponsored by the Project 42026-F of CONACYT, México and Grant 11/G071 Of Universidad de La Plata, Argentina.

References

[1] Kostelecky, V. A., and Samuel, S., *Phys. Rev.*, **D39**, 683 (1989).

[2] Kostelecky, V. A., and Samuel, S., *Phys. Rev. Lett.*, **63**, 224 (1989).

[3] Kostelecky, V. A., and Samuel, S., *Phys. Rev.*, **D40**, 1886–1903 (1989).

[4] Kostelecky, V. A., and Potting, R., *Nucl. Phys.*, **B359**, 545–570 (1991).

[5] Kostelecky, V. A., and Potting, R., *Phys. Lett.*, **B381**, 89–96 (1996).

[6] Colladay, D., and Kostelecky, V. A., *Phys. Rev.*, **D55**, 6760–6774 (1997).

[7] Colladay, D., and Kostelecky, V. A., *Phys. Rev.*, **D58**, 116002 (1998).

[8] Amelino-Camelia, G., Ellis, J. R., Mavromatos, N. E., Nanopoulos, D. V., and Sarkar, S., *Nature*, **393**, 763–765 (1998).

[9] Amelino-Camelia, G., *Nature*, **410**, 1065–1069 (2001).

[10] Ahluwalia, D. V., *Nature*, **398**, 199 (1999).

[11] Amelino-Camelia, G., Introduction to quantum-gravity phenomenology (2004), gr-qc/0412136.

[12] Sorkin, R. D., "Space-time and causal sets," in *Procedings of Silarg VII*, edited by J. C. d'Olivo et al., World Scientific, Singapore, 1991, p. 150.

[13] Gambini, R., and Pullin, J., *Phys. Rev.*, **D59**, 124021 (1999).

[14] Alfaro, J., Morales-Tecotl, H. A., and Urrutia, L. F., *Phys. Rev. Lett.*, **84**, 2318–2321 (2000).

[15] Alfaro, J., Morales-Tecotl, H. A., and Urrutia, L. F., *Phys. Rev.*, **D65**, 103509 (2002).

[16] Alfaro, J., Morales-Tecotl, H. A., and Urrutia, L. F., *Phys. Rev.*, **D66**, 124006 (2002).

[17] Urrutia, L. F., Flat space modified particle dynamics induced by loop quantum gravity (2004), hep-ph/0402271.

[18] Ellis, J. R., Mavromatos, N. E., and Nanopoulos, D. V., *Gen. Rel. Grav.*, **32**, 127–144 (2000).

[19] Ellis, J. R., Mavromatos, N. E., Nanopoulos, D. V., and Volkov, G., *Gen. Rel. Grav.*, **32**, 1777–1798 (2000).

[20] Ellis, J. R., Farakos, K., Mavromatos, N. E., Mitsou, V. A., and Nanopoulos, D. V., *Astrophys. J.*, **535**, 139–151 (2000).

[21] Ellis, J. R., Mavromatos, N. E., and Nanopoulos, D. V., *Phys. Rev.*, **D63**, 024024 (2001).

[22] Seiberg, N., and Witten, E., *JHEP*, **09**, 032 (1999).

[23] Douglas, M. R., and Nekrasov, N. A., *Rev. Mod. Phys.*, **73**, 977–1029 (2001).

[24] Grandi, N. E., *Teorías de gauge no conmutativas: Chern-Simmons y Born-Infeld*, Ph.D. thesis, U. N. L. P. (2002), (In spanish).

[25] Sudarsky, D., Urrutia, L., and Vucetich, H., *Phys. Rev. Lett.*, **89**, 231301 (2002).

[26] Sudarsky, D., Urrutia, L., and Vucetich, H., *Phys. Rev.*, **D68**, 024010 (2003).

[27] Lineweaver, C. H., The CMB dipole: The most recent measurement and some history (1996), astro-ph/9609034.

[28] Streater, R. F., and Wightman, A. S., *PCT, Spin & Statistics, and all that*, Benjamin, New York, 1968.

[29] Weinberg, S., *The quantum theory of fields: Foundations*, vol. I, Cambridge U. P., 1995.

[30] Greenberg, O. W., *Phys. Rev. Lett.*, **89**, 231602 (2002).

[31] Roberson, H. P., *Rev. Mod. Phys.*, **21**, 378 (1949).

[32] Mansouri, R., and Sexl, R. U., *Gen. Rel. and Grav.*, **8**, 497 (1977).

[33] Mansouri, R., and Sexl, R. U., *Gen. Rel. and Grav.*, **8**, 515 (1977).

[34] Mansouri, R., and Sexl, R. U., *Gen. Rel. and Grav.*, **8**, 809 (1977).

[35] MacArthur, D. W., *Phys. Rev. A*, **33**, 1 (1986).

[36] Michelson, A. A., and Morley, E. W., *Am. J. Sci.*, **34**, 333 (1887).

[37] Kennedy, R. J., and Thorndike, E. M., *Phys. Rev.*, **42**, 400 (1932).

[38] Yves, H. E., and Stillwell, G. R., *J. Opt. Soc. Am.*, **28**, 215 (1938).

[39] Amelino-Camelia, G., *New J. Phys.*, **6**, 188 (2004).

[40] Konopka, T. J., and Major, S. A., *New J. Phys.*, **4**, 57 (2002).

[41] Alfaro, J., and Palma, G., *Phys. Rev.*, **D67**, 083003 (2003).

[42] Lehnert, R., *Phys. Rev.*, **D68**, 085003 (2003).

[43] Lightman, A. P., and Lee, D. L., *Phys. Rev. D*, **8**, 364 (1973).

[44] Horvath, J., Logiúdice, E. A., Riveros, C., and Vucetich, H., *Phys. Rev. D*, **38**, 1754 (1988).

[45] Kostelecky, V. A., Lorentz- and CPT-violating extension of the standard model (1999), `hep-ph/9912528`.

[46] Kostelecky, V. A., *Phys. Rev.*, **D69**, 105009 (2004).

[47] Kostelecky, V. A., and Mewes, M., *Phys. Rev. Lett.*, **87**, 251304 (2001).

[48] Kostelecky, V. A., and Mewes, M., *Phys. Rev.*, **D66**, 056005 (2002).

[49] Müller, H., Herrmann, S., Braxmaier, C., Schiller, S., and Peters, A., *Phys. Rev. Lett.*, **91**, 020401 (2003).

[50] Kostelecky, V. A., and Lane, C. D., *J. Math. Phys.*, **40**, 6245–6253 (1999).

[51] Kostelecky, V. A., and Lehnert, R., *Phys. Rev.*, **D63**, 065008 (2001).

[52] Kostelecky, V. A., Lane, C. D., and Pickering, A. G. M., *Phys. Rev.*, **D65**, 056006 (2002).

[53] Bailey, Q. G., and Kostelecky, V. A., *Phys. Rev.*, **D70**, 076006 (2004).

[54] Montemayor, R., and Urrutia, L. F., Radiation in Lorentz violating electrodynamics (2004), `hep-ph/0412023`.

[55] Bluhm, R., Kostelecky, V. A., and Russell, N., *Phys. Rev. Lett.*, **79**, 1432–1435 (1997).

[56] Bluhm, R., Kostelecky, V. A., and Russell, N., *Phys. Rev. Lett.*, **82**, 2254–2257 (1999).

[57] Bluhm, R., and Kostelecky, V. A., *Phys. Rev. Lett.*, **84**, 1381–1384 (2000).

[58] Bluhm, R., Kostelecky, V. A., Lane, C. D., and Russell, N., *Phys. Rev. Lett.*, **88**, 090801 (2002).

[59] Bluhm, R., Electromagnetic tests of Lorentz and CPT symmetry (2001), `hep-ph/0112318`.

[60] Bluhm, R., Kostelecky, V. A., Lane, C. D., and Russell, N., *Phys. Rev.*, **D68**, 125008 (2003).

[61] Bluhm, R., *Nucl. Instrum. Meth.*, **B221**, 6–11 (2004).

[62] Bluhm, R., QED tests of Lorentz symmetry (2004), `hep-ph/0411149`.

[63] Myers, R. C., and Pospelov, M., *Phys. Rev. Lett.*, **90**, 211601 (2003).

[64] Myers, R. C., and Pospelov, M., Experimental challenges for quantum gravity (2004), gr-qc/0402028.

[65] Perez, A., and Sudarsky, D., *Phys. Rev. Lett*, **91**, 179101 (2003).

[66] Collins, J., Perez, A., Sudarsky, D., Urrutia, L., and Vucetich, H., *Phys. Rev. Lett.*, **93**, 191301 (2004).

[67] Magueijo, J., and Smolin, L., *Phys. Rev. Lett.*, **88**, 190403 (2002).

[68] Amelino-Camelia, G., *Nature*, **418**, 34–35 (2002).

[69] Amelino-Camelia, G., Kowalski-Glikman, J., Mandanici, G., and Procaccini, A., Phenomenology of doubly special relativity (2003), `gr-qc/0312124`.

[70] Carmona, J. M., Cortes, J. L., Gamboa, J., and Mendez, F., *Phys. Lett.*, **B565**, 222–228 (2003).

[71] Groot Nibbelink, S., and Pospelov, M., Lorentz violation in supersymmetric field theories (2004), `hep-ph/0404271`.

[72] Thiemann, T., *Lect. Notes Phys.*, **631**, 41–135 (2003).

[73] Perez, A., Introduction to loop quantum gravity and spin foams (2004), `gr-qc/0409061`.

[74] Smolin, L., An invitation to loop quantum gravity (2004), `hep-th/0408048`.

[75] Nicolai, H., Peeters, K., and Zamaklar, M., Loop quantum gravity: an outside view (2005), `hep-th/0501114`.

[76] Ashtekar, A., *Phys. Rev. Lett.*, **57**, 2244–2247 (1986).

[77] Ashtekar, A., *Phys. Rev.*, **D36**, 1587–1602 (1987).

[78] Gambini, R., and Trias, A., *Phys. Rev.*, **D22**, 1380 (1980).

[79] Gambini, R., and Trias, A., *Phys. Rev.*, **D23**, 553 (1981).

[80] di Bartolo, C., Nori, F., Gambini, R., and Trias, A., *Nuovo Cim. Lett.*, **38**, 497 (1983).

[81] Bojowald, M., and Morales-Tecotl, H. A., *Lect. Notes Phys.*, **646**, 421–462 (2004).

[82] Bojowald, M., *Pramana*, **63**, 765–776 (2004).

[83] Ellis, J. R., Gravanis, E., Mavromatos, N. E., and Nanopoulos, D. V., Impact of low-energy constraints on lorentz violation (2002), `gr-qc/0209108`.

[84] Bluhm, R., Probing the planck scale in low-energy atomic physics (2001), `hep-ph/0111323`.

[85] Stecker, F. W., High energy astrophysics tests of Lorentz invariance violation (2004), `astro-ph/0409731`.

[86] Jacobson, T., Liberati, S., and Mattingly, D., Astrophysical bounds on Planck suppressed Lorentz violation (2004), `hep-ph/0407370`.

[87] Jacobson, T., Liberati, S., and Mattingly, D., Quantum gravity phenomenology and Lorentz violation (2004), `gr-qc/0404067`.

[88] Brillet, A., and Hall, J. L., *Phys. Rev. Lett.*, **42**, 549 (1979).

[89] Hils, D., and Hall, J. L., *Phys. Rev. Lett.*, **64**, 1697 (1990).

[90] Braxmaier, C., Müller, H., Pradl, O., Mlynck, J., and Peters, A., *Phys. Rev. Lett.*, **88**, 010401 (2002).

[91] Grieser, R., et al., *Appl. Phys.*, **B59**, 127–133 (1994).

[92] Coleman, S. R., and Glashow, S. L., *Phys. Lett.*, **B405**, 249–252 (1997).

[93] Coleman, S. R., and Glashow, S. L., *Phys. Rev.*, **D59**, 116008 (1999).

[94] Jacobson, T., Liberati, S., and Mattingly, D., *Nature*, **424**, 1019–1021 (2003).

[95] Biller, S. D., et al., *Phys. Rev. Lett.*, **83**, 2108–2111 (1999).

[96] Gleiser, R. J., and Kozameh, C. N., *Phys. Rev.*, **D64**, 083007 (2001).

[97] Hughes, V. W., Robinson, H. G., and López, V. B., *Phys. Rev. Lett.*, **4**, 342 (1960).

[98] Drever, R. W. P., *Phil. Mag.*, **6**, 683 (1961).

[99] Prestage, J. D., Bollinger, J. J., Itano, W. M., and Wineland, D. J., *Phys. Rev. Lett.*, **54**, 2387 (1985).

[100] Lamoreaux, S. K., Jacobs, J. P., Heckel, B. R., Raab, F. J., and Fortson, E. N., *Phys. Rev. D*, **39**, 1082 (1989).

[101] Chupp, T. E., Hoare, R. J., Loveman, R. A., Oteiza, E. R., Richardson, J. M., Wangshul, M. E., and Thompson, A. K., *Phys. Rev. Lett.*, **63**, 1541 (1989).

[102] Phillips, D. F., Humphrey, M. A., Mattison, E. M., Stoner, R. E., Vessot, R. F. C., and Walsworth, R. L., *Phys. Rev. D*, **63**, 111101 (2001).

[103] Berglund, C. J., Hunter, L. R., D. Krause, J., Prigge, E. O., Ronfeldt, M. S., and Lamoreaux, S. K., *Phys. Rev. Lett.*, **75**, 1879 (1995).

[104] Bear, D., Stoner, R. E., Walsworth, R. L., Kostelecky, V. A., and Lane, C. D., *Phys. Rev. Lett.*, **85**, 5038–5041 (2000).

[105] Cane, F., et al., *Phys. Rev. Lett.*, **93**, 230801 (2004).

[106] Greisen, K., *Phys. Rev. Lett.*, **16**, 748–750 (1966).

[107] Zatsepin, G. T., and Kuzmin, V. A., *JETP Lett.*, **4**, 78–80 (1966).

[108] Scully, S. T., and Stecker, F. W., *Astropart. Phys.*, **16**, 271–276 (2002).

[109] Stecker, F. W., Astrophysics at the highest energy frontiers (2002), `astro-ph/0208507`.

[110] Takeda, M., et al., *Phys. Rev. Lett.*, **81**, 1163–1166 (1998).

[111] Takeda, M., et al., *Astrophys. J.*, **522**, 225–237 (1999).

[112] Alfaro, J., and Palma, G., *Phys. Rev.*, **D65**, 103516 (2002).

[113] Dehmelt, H., Mittleman, R., van Dyck, J., R. S., and Schwinberg, P., *Phys. Rev. Lett.*, **83**, 4694–4696 (1999).

[114] Mittleman, R. K., Ioannou, I. I., Dehmelt, H. G., and Russell, N., *Phys. Rev. Lett.*, **83**, 2116–2119 (1999).

[115] Gabrielse, G., et al., *Phys. Rev. Lett.*, **82**, 3198–3201 (1999).

[116] Heckel, B. R., Adelberger, E. G., Gundlach, J. H., Harris, M. G., and Swanson, H. E., Torsion balance test of spin coupled forces, prepared for International Conference on Orbis Scientiae 1999: Quantum Gravity, Generalized Theory of Gravitation and Superstring Theory Based Unification (28th Conference on High-Energy Physics and Cosmology Since 1964), Coral Gables, Florida, 16-19 Dec 1999.

[117] Carroll, S. M., Harvey, J. A., Kostelecky, V. A., Lane, C. D., and Okamoto, T., *Phys. Rev. Lett.*, **87**, 141601 (2001).

[118] Burgess, C. P., Cline, J., Filotas, E., Matias, J., and Moore, G. D., *JHEP*, **03**, 043 (2002).

[119] Anisimov, A., Banks, T., Dine, M., and Graesser, M., *Phys. Rev.*, **D65**, 085032 (2002).

[120] Weinberg, S., *Phys. Rev. D*, **8**, 4482 (1973).

[121] Susskind, L., *Phys. Rev.*, **D20**, 2619–2625 (1979).

[122] Weinberg, S., *Rev. Mod. Phys.*, **61**, 1–23 (1989).

[123] Kostelecky, V. A., and Potting, R., *Phys. Rev.*, **D51**, 3923–3935 (1995).

[124] Coleman, S. R., and Mandula, J., *Phys. Rev.*, **159**, 1251–1256 (1967).

[125] Dowker, F., Henson, J., and Sorkin, R. D., *Mod. Phys. Lett.*, **A19**, 1829–1840 (2004).

[126] Rovelli, C., and Speziale, S., *Phys. Rev.*, **D67**, 064019 (2003).

In: Focus on Quantum Gravity Research
Editor: David C. Moore, pp. 89-120
ISBN 1-59454-660-6

Chapter 3

Gravitons as Super-Strong Interacting Particles and Low-Energy Quantum Gravity

Michael A. Ivanov
Physics Dept.,
Belarus State University of Informatics and Radioelectronics,
6 P. Brovka Street, BY 220027, Minsk, Republic of Belarus
E-mail: ivanovma@gw.bsuir.unibel.by

It is shown by the author that if gravitons are super-strong interacting particles and the low-temperature graviton background exists, the basic cosmological conjecture about the Dopplerian nature of redshifts may be false. In this case, a full magnitude of cosmological redshift would be caused by interactions of photons with gravitons. A new dimensional constant which characterizes one act of interaction is introduced and estimated. Non-forehead collisions with gravitons will lead to a very specific additional relaxation of any photonic flux. It gives a possibility of another interpretation of supernovae 1a data - without any kinematics. Of course, all of these facts may implicate a necessity to change the standard cosmological paradigm. Some features of a new paradigm are discussed here, too.

A quantum mechanism of classical gravity based on an existence of this sea of gravitons is described for the Newtonian limit. This mechanism needs graviton pairing and "an atomic structure" of matter for working it, and leads to the time asymmetry. If the considered quantum mechanism of classical gravity is realized in the nature, then an existence of black holes contradicts to Einstein's equivalence principle. It is shown that in this approach the two fundamental constants - Hubble's and Newton's ones - should be connected between themselves. The theoretical value of the Hubble constant is computed. In this approach, every massive body would be decelerated due to collisions with gravitons that may be connected with the Pioneer 10 anomaly. It is shown that the predicted and observed values of deceleration are in good agreement. Some unsolved problems are discussed, so as possibilities to verify some conjectures in laser-based experiments.

3.1 Introduction

By a full coincidence of the forms of Coulomb's and Newton's laws, which describe an interaction of electric charges and a gravitational interaction of bodies, we see a dramatic difference in developing the pictures of these interactions on a quantum level. Constructed with pillars on multiple experiments QED is one of the most exact physical theories and an archetype for imitation by creation of new models. While the attempts to quantize the remarkable in its logical beauty theory of general relativity, which describes gravitation on a classical level so fully and delicately, (see review [1]) not only have not a hit until today but gave a specific side psychological effect - there exists conceptualization that quantum gravity may be described only by some sophisticated theory. An opinion is commonly accepted, too, that quantum gravity should manifest itself only on the Planck scale of energies, i.e. it is a high-energy phenomenon. The value of the Planck energy $\sim 10^{19}$ GeV has been got from dimensional reasonings. Still one wide-spread opinion is that we know a mechanism of gravity (bodies are exchanging with gravitons of spin 2) but cannot correctly describe it. If an apple from the legend about Newton's afflatus can imagine all these complications, it would hesitate to fall to the ground so artlessly as it is accepted among the apples.

Perhaps, physicists would be able to refuse easier the preconceived stereotypes which balk - as it seems to me - to go ahead in understanding quantum gravity if experiments or observations would give more essential meat for reasonings. But in this area, at least up to recent years, there was observed nothing that may serve if not Ariadne's clew but such a simple physical contradiction that an aspiration to overcome it would advantage introduction of new ideas and revision of the "inviolable".

In a few last years, the situation has been abruptly changed. I enumerate those discoveries and observations which may force, in my opinion, the ice to break up.

1. In 1998, Anderson's team reported about the discovery of anomalous acceleration of NASA's probes Pioneer 10/11 [12]; this effect is not embedded in a frame of the general relativity, and its magnitude is somehow equal to $\sim Hc$, where H is the Hubble constant, c is the light velocity.

2. In the same 1998, two teams of astrophysicists, which were collecting supernovae 1a data with the aim to specificate parameters of cosmological expansion, reported about dimming remote supernovae [13, 14]; the one would be explained

on a basis of the Doppler effect if at present epoch the universe expands with acceleration. This explanation needs an introduction of some "dark energy" which is unknown from any laboratory experiments.

3. In January 2002, Nesvizhevsky's team reported about discovery of quantum states of ultra-cold neutrons in the Earth's gravitational field [15]. Observed energies of levels (it means that and their differences too) in full agreement with quantum-mechanical calculations turned out to be equal to $\sim 10^{-12}$ eV. The formula for energy levels had been found still by Bohr and Sommerfeld. If transitions between these levels are accompanied with irradiation of gravitons then energies of irradiated gravitons should have the same order - but it is of 40 orders lesser than the Planck energy by which one waits quantum manifestations of gravity.

The first of these discoveries obliges to muse about the borders of applicability of the general relativity, the third - about that quantum gravity would be a high-energy phenomenon. It seems that the second discovery is far from quantum gravity but it obliges us to look at the traditional interpretation of the nature of cosmological redshift critically. An introduction into consideration of an alternative model of redshifts [4] which is based on a conjecture about an existence of the graviton background gives us odds to see in the effect of dimming supernovae an additional manifestation of low-energy quantum gravity. Under the definite conditions, an effective temperature of the background may be the same one as a temperature of the cosmic microwave background, with an average graviton energy of the order $\sim 10^{-3}$ eV.

In this chapter, the main results of author's research in this approach are described. Starting from a statistical model of the graviton background with low temperature, it is shown - under the very important condition that gravitons are super-strong interacting particles - that if a redshift would be a quantum gravitational effect then one can get from its magnitude an estimate of a new dimensional constant characterizing a single act of interaction in this model. It is possible to calculate theoretically a dependence of a light flux relaxation on a redshift value, and this dependence fits supernova observational data very well at least for $z < 0.5$. Further it is possible to find a pressure of single gravitons of the background which acts on any pair of bodies due to screening the graviton background with the bodies [8]. It turns out that the pressure is huge (a corresponding force is ~ 1000 times stronger than the Newtonian attraction) but it is compensated with a pressure of gravitons which are re-scattered by the bodies. The Newtonian attraction arises if a part of gravitons of the background forms pairs which are destructed by interaction with bodies. It is interesting that both Hubble's and Newton's constants may be computed in this approach with the ones being connected between themselves. It allows us to get a theoretical estimate of the Hubble constant. An unexpected feature of this mechanism of gravity is a necessity of "an atomic structure" of matter - the mechanism doesn't work without the one.

Collisions with gravitons should also call forth a deceleration of massive bodies of order $\sim Hc$ - namely the same as of NASA's probes. But at present stage it turns out unclear why such the deceleration has *not* been observed for planets. The situation reminds by something of the one that took place in physics before the creation of quantum mechanics when a motion of electrons should, as it seemed by canons of classical physics, lead to their fall to a nucleus.

Because the very unexpected hypothesis founded into the basis of this approach is a super-strong character of gravitational interaction on a quantum level, I would like to explain my motivation which conduced namely to such the choice. Learning symmetries of the quantum two-component composite system [9], I have found that its discrete symmetries in an 8-space may be interpreted by an observer from a 4-dimensional world as the exact global symmetries of the standard model of particle physics if internal coordinates of the system (the composite fermion) are rigidly fixed. This conclusion was hard for me and took much enough time. But to ensure almost full fixation of components of the system, an interaction connecting them should be very strong. Because of it, when a choice arise - an amount of gravitons or an intensity of the interaction, - I have remembered this overpassed earlier barrier and chose namely the super-strong interaction. Without this property, the graviton background would not be in the thermodynamical equilibrium with the cosmic microwave background that could entail big difficulties in the model.

So, in this approach we deal with the following small quantum effects of low-energy gravity: redshifts, its analog - a deceleration of massive bodies, and an additional relaxation of any light flux. The Newtonian attraction turns out to be the main statistical effect, with bodies themselves being not sources of gravitons - only correlational properties of *in* and *out* fluxes of gravitons in their neighbourhood are changed due to an interaction with bodies. There does still not exist a full and closed theory in this approach, but even the initial researches in this direction show that in this case quantum gravity cannot be described separately of other interactions, and also manifest the boundaries of applicability of a geometrical language in gravity.

3.2 Passing photons through the graviton background [4, 7]

Let us introduce the hypothesis, which is considered in this approach as independent from the standard cosmological model: there exists the isotropic graviton background. Photon scattering is possible on gravitons $\gamma + h \rightarrow \gamma + h$, where γ is a photon and h is a graviton, if one of the gravitons is virtual. The energy-momentum conservation law prohibits energy transfer to free gravitons. Due to forehead collisions with gravitons, an energy of any photon should decrease when it passes through the sea of gravitons.

From another side, none-forehead collisions of photons with gravitons of the background will lead to an additional relaxation of a photon flux, caused by transmission of a momentum transversal component to some photons. It will lead to an additional dimming of any remote objects, and may be connected with supernova dimming.

We deal here with the uniform non-expanding universe with the Euclidean space, and there are not any cosmological kinematic effects in this model.

3.2.1 Forehead collisions with gravitons: an alternative explanations of the redshift nature

We shall take into account that a gravitational "charge" of a photon must be proportional to E (it gives the factor E^2 in a cross-section) and a normalization of a photon wave function gives the factor E^{-1} in the cross-section. Also we assume here that a photon average energy loss $\bar{\epsilon}$ in one act of interaction is relatively small to a photon energy E. Then average energy losses of a photon with an energy E on a way dr will be equal to [4, 7]:

$$dE = -aEdr, \tag{3.2.1}$$

where a is a constant. If a *whole* redshift magnitude is caused by this effect, we must identify $a = H/c$, where c is the light velocity, to have the Hubble law for small distances [10].

A photon energy E should depend on a distance from a source r as

$$E(r) = E_0 \exp(-ar), \tag{3.2.2}$$

where E_0 is an initial value of energy.

The expression (2) is just only so far as the condition $\bar{\epsilon} << E(r)$ takes place. Photons with a very small energy may lose or acquire an energy changing their direction of propagation after scattering. Early or late such photons should turn out in the thermodynamic equilibrium with the graviton background, flowing into their own background. Decay of virtual gravitons should give photon pairs for this background, too. Perhaps, the last one is the cosmic microwave background [11, 12].

It follows from the expression (2) that an exact dependence $r(z)$ is the following one:

$$r(z) = ln(1+z)/a, \tag{3.2.3}$$

if an interaction with the graviton background is the only cause of redshifts. It is very important, that this redshift does not depend on a light frequency. For small z, the dependence $r(z)$ will be linear.

The expressions (1) - (3) are the same that appear in other tired-light models (compare with [13]). In this approach, the ones follow from a possible existence of the isotropic graviton background, from quantum electrodynamics, and from the fact that a gravitational "charge" of a photon must be proportional to E.

3.2.2 Non-forehead collisions with gravitons: an additional dimming of any light flux

Photon flux's average energy losses on a way dr due to non-forehead collisions with gravitons should be proportional to $badr$, where b is a new constant of the order 1. These losses are connected with a rejection of a part of photons from a source-observer direction. Such the relaxation together with the redshift will give a connection between visible object's diameter and its luminosity (i.e. the ratio of an object visible angular diameter to a square root of visible luminosity), distinguishing from the one of the standard cosmological model.

Let us consider that in a case of a non-forehead collision of a graviton with a photon, the latter leaves a photon flux detected by a remote observer (an assumption of a narrow beam of rays). The details of calculation of the theoretical value of relaxation factor b which was used in author's paper [4] were given later in the preprint [14]. So as both particles have velocities c, a cross-section of interaction, which is "visible" under an angle θ (see Fig. 1), will be equal to $\sigma_0|\cos\theta|$ if σ_0 is a cross-section by forehead collisions. The function $|\cos\theta|$ allows to take into account both front and back hemispheres for riding gravitons. Additionally, a graviton flux, which falls on a picked out area (cross-section), depends on the angle θ. We have for the ratio of fluxes:

$$\Phi(\theta)/\Phi_0 = S_s/\sigma_0,$$

where $\Phi(\theta)$ and Φ_0 are the fluxes which fall on σ_0 under the angle θ and normally, S_s is a square of side surface of a truncated cone with a base σ_0 (see Fig. 1).

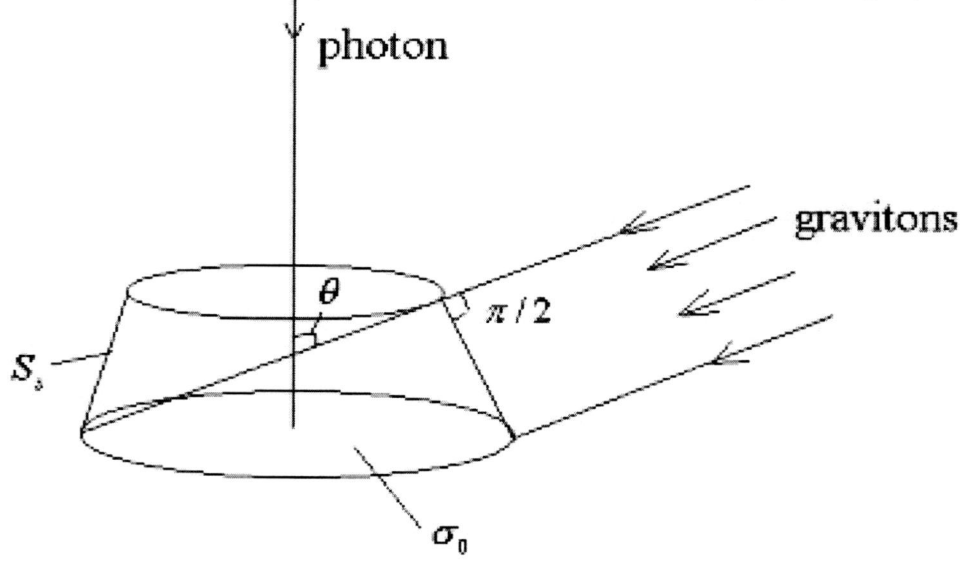

Figure 3.2.1: By non-forehead collisions of gravitons with a photon, it is necessary to calculate a cone's side surface square, S_s.

Finally, we get for the factor b :

$$b = 2\int_0^{\pi/2} \cos\theta \cdot (S_s/\sigma_0)\frac{d\theta}{\pi/2}. \tag{3.2.4}$$

By $0 < \theta < \pi/4$, a formed cone contains self-intersections, and it is $S_s = 2\sigma_0 \cdot \cos\theta$. By $\pi/4 \leq \theta \leq \pi/2$, we have $S_s = 4\sigma_0 \cdot \sin^2\theta\cos\theta$.

After computation of simple integrals, we get:

$$b = \frac{4}{\pi}(\int_0^{\pi/4} 2\cos^2\theta d\theta + \int_{\pi/4}^{\pi/2} \sin^2 2\theta d\theta) = \frac{3}{2} + \frac{2}{\pi} \simeq 2.137. \tag{3.2.5}$$

In the considered simplest case of the uniform non-expanding universe with the Euclidean space, we shall have the quantity

$$(1+z)^{(1+b)/2} \equiv (1+z)^{1.57}$$

in a visible object diameter-luminosity connection if a whole redshift magnitude would caused by such an interaction with the background (instead of $(1+z)^2$ for the expanding uniform universe). For near sources, the estimate of the factor b will be some increased one.

The luminosity distance (see [13]) is a convenient quantity for astrophysical observations. Both redshifts and the additional relaxation of any photonic flux due to non-forehead collisions of gravitons with photons lead in our model to the following luminosity distance D_L :

$$D_L = a^{-1}\ln(1+z) \cdot (1+z)^{(1+b)/2} \equiv a^{-1}f_1(z), \tag{3.2.6}$$

where $f_1(z) \equiv \ln(1+z) \cdot (1+z)^{(1+b)/2}$.

3.2.3 Comparison of the theoretical predictions with supernova data

To compare a form of this predicted dependence $D_L(z)$ by unknown, but constant H, with the latest observational supernova data by Riess et al. [15], one can introduce distance moduli $\mu_0 = 5\log D_L + 25 = 5\log f_1 + c_1$, where c_1 is an unknown constant (it is a single free parameter to fit the data); f_1 is the luminosity distance in units of c/H. In Figure 2, the Hubble diagram $\mu_0(z)$ is shown with $c_1 = 43$ to fit observations for low redshifts; observational data (82 points) are taken from Table 5 of [15]. The predictions fit observations very well for roughly $z < 0.5$. It excludes a need of any dark energy to explain supernovae dimming.

Discrepancies between predicted and observed values of $\mu_0(z)$ are obvious for higher z: we see that observations show brighter SNe that the theory allows, and a

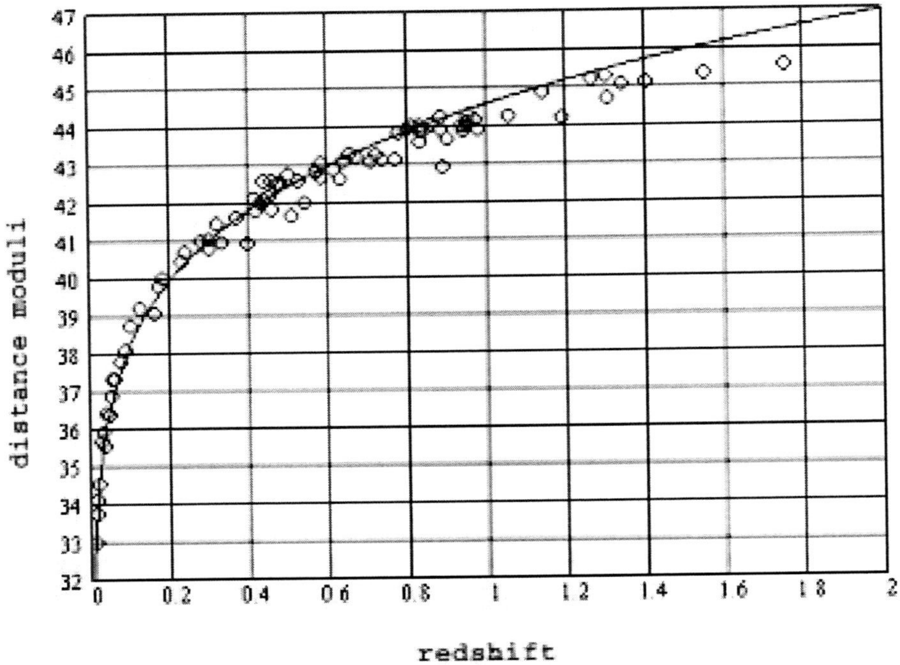

Figure 3.2.2: Comparison of the theoretical values of distance moduli $\mu_0(z)$ in this model (solid line) with observations (points) from [15] by Riess et al.

difference increases with z. It is better seen on Figure 3 with a linear scale for f_1; observations are transformed as $\mu_0 \rightarrow 10^{(\mu_0 - c_1)/5}$ with the same $c_1 = 43$.[1]

It would be explained in the model as a result of specific deformation of SN spectra due to a discrete character of photon energy losses. Today, a theory of this effect does not exist, and I explain its origin only qualitatively [16]. For very small redshifts z, only a small part of photons transmits its energy to the background (see below Fig. 8 in Section 6). Therefore any red-shifted narrow spectral strip will be a superposition of two strips. One of them has a form which is identical with an initial one, its space is proportional to $1 - n(r)$ where $n(r)$ is an average number of interactions of a single photon with the background, and its center's shift is negligible (for a narrow strip). Another part is expand, its space is proportional to $n(r)$, and its center's shift is equal to $\bar{\epsilon}_g/h$ where $\bar{\epsilon}_g$ is an average energy loss in one act of interaction. An amplitude of the red-shifted step should linear raise with a redshift. For big z, spectra of remote objects of the universe would be deformed. A deformation would appear because of multifold interactions of a initially-red-shifted part of photons with the graviton background. It means that the observed

[1]A spread of observations raises with z; it might be partially caused by quickly raising contribution of a dispersion of measured flux: it should be proportional to $f_1^6(z)$.

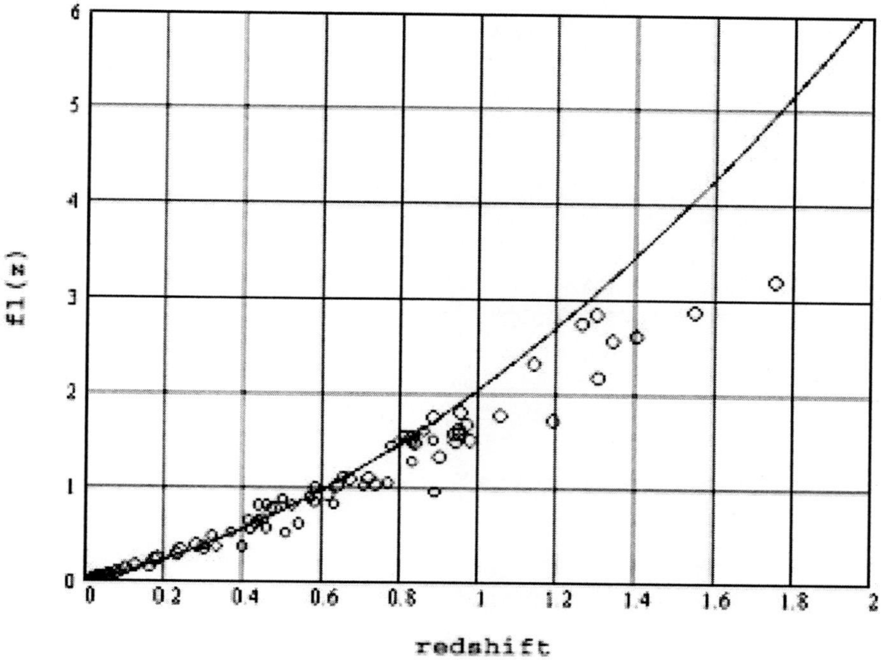

Figure 3.2.3: Predicted values of $f_1(z)$ (solid line) and observations (points) from [15] transformed to a linear scale

flux within a given passband would depend on a form of spectrum: the flux may be larger than an expected one without this effect if an initial flux within a next-blue neighbour band is big enough - due to a superposition of red-shifted parts of spectrum. Some other evidences of this effect would be an apparent variance of the fine structure constant [17] or of the CMB temperature [18] with epochs. In both cases, a ratio of red-shifted spectral line's intensities may be sensitive to the effect.

This comparison with observations is very important; to see some additional details, we can compute and graph the ratio $f_{1obs}(z)/f_1(z)$, where $f_{1obs}(z)$ is an observed analog of $f_1(z)$ (see Fig. 4) [19]. An expected value of the ratio should be equal to 1 for any z.

3.2.4 Computation of the Hubble constant

Let us consider that a full redshift magnitude is caused by an interaction with single gravitons. If $\sigma(E, \epsilon)$ is a cross-section of interaction by forehead collisions of a photon with an energy E with a graviton, having an energy ϵ, we consider really (see (1)), that

$$\frac{d\sigma(E, \epsilon)}{E d\Omega} = const(E),$$

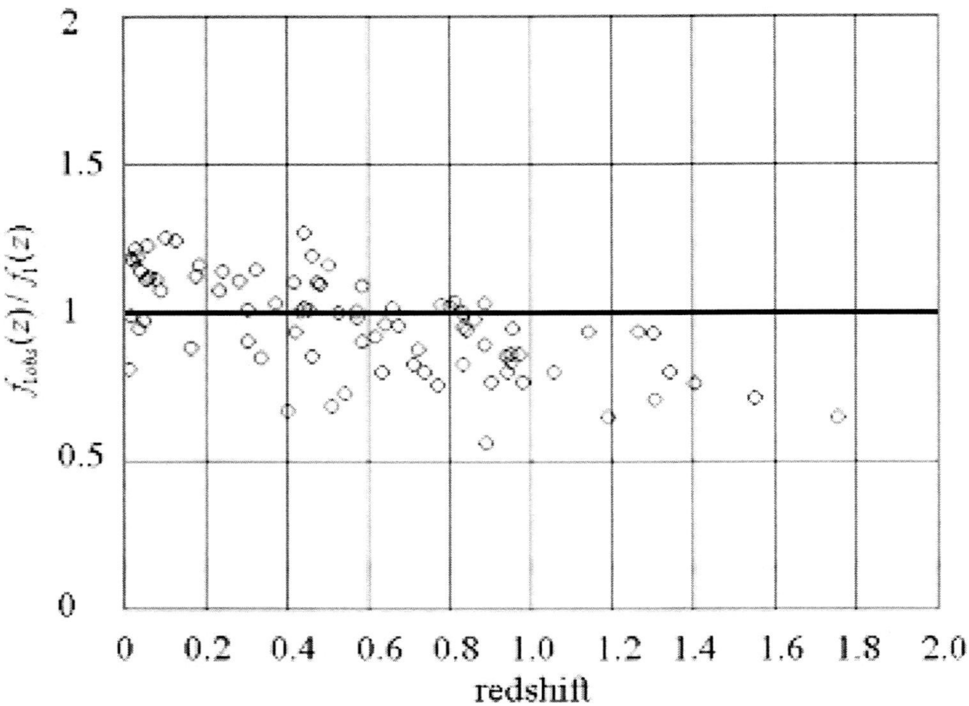

Figure 3.2.4: The ratio of observed to theoretical functions $f_{1obs}(z)/f_1(z)$ (dots); observational data are taken from Table 5 of [15]. If this model is true, the ratio should be equal to 1 for any z (solid line).

where $d\Omega$ is a space angle element, and the function $const(x)$ has a constant value for any x. If $f(\omega, T)d\Omega/2\pi$ is a spectral density of graviton flux in the limits of $d\Omega$ in some direction (ω is a graviton frequency, $\epsilon = \hbar\omega$), i.e. an intensity of a graviton flux is equal to the integral $(d\Omega/2\pi)\int_0^\infty f(\omega, T)d\omega$, T is an equivalent temperature of the graviton background, we can write for the Hubble constant $H = ac$, introduced in the expression (1):

$$H = \frac{1}{2\pi}\int_0^\infty \frac{\sigma(E,\epsilon)}{E} f(\omega, T)d\omega.$$

If $f(\omega, T)$ can be described by the Planck formula for equilibrium radiation, then

$$\int_0^\infty f(\omega, T)d\omega = \sigma T^4,$$

where σ is the Stephan- Boltzmann constant. As carriers of a gravitational "charge" (without consideration of spin properties), gravitons should be described in the same manner as photons (compare with [20]), i.e. one can write for them:

$$\frac{d\sigma(E,\epsilon)}{\epsilon d\Omega} = const(\epsilon).$$

Now let us introduce a new dimensional constant D, so that for forehead collisions:

$$\sigma(E,\epsilon) = D \cdot E \cdot \epsilon. \tag{3.2.7}$$

Then

$$H = \frac{1}{2\pi} D \cdot \bar{\epsilon} \cdot (\sigma T^4), \tag{3.2.8}$$

where $\bar{\epsilon}$ is an average graviton energy. Assuming $T \sim 3K$, $\bar{\epsilon} \sim 10^{-4}\ eV$, and $H = 1.6 \cdot 10^{-18}\ s^{-1}$, we get the following rough estimate for D :

$$D \sim 10^{-27}\ m^2/eV^2,$$

(see below Section 4.3 for more exact estimate of D and for a theoretical estimate of H) that gives us the phenomenological estimate of cross-section by the same and equal E and $\bar{\epsilon}$:

$$\sigma(E,\bar{\epsilon}) \sim 10^{-35}\ m^2.$$

3.2.5 Some new constants from dimensional reasonings

We can introduce the following new constants (see [4]): G_0, l_0, E_0, which are analogues, on this new scale, of classical constants: the Newton constant G, the Planck length l_{Pl}, and the Planck energy E_{Pl} correspondingly. Let us accept from dimensional reasonings that

$$D \equiv (l_0/E_0)^2 = (G_0/c^4)^2,$$

where $l_0 = \sqrt{G_0 \hbar / c^3}$, $E_0 = \sqrt{\hbar c^5 / G_0}$. Then we have for these new constants:

$$G_0 \sim 1.6 \cdot 10^{39} m^3/kg \cdot s^2,\ l_0 \sim 2.4 \cdot 10^{-12} m,\ E_0 \sim 1.6\ keV.$$

If one would replace G with G_0, then an electrostatic force, acting between two protons, will be only $\sim 2 \cdot 10^{13}$ times smaller than a gravitational one by the same distance (a theoretical finding of the Newton constant G is given below in Section 4.3).

Using E_0 instead of E_{Pl}, we can evaluate the new non-dimensional "constant" (a bilinear function of E and ϵ), k, which would characterize one act of interaction: $k \equiv E \cdot \epsilon / E_0^2$. We must remember here, that a universality of gravitational interaction allows to expect that this floating coupling "constant" k should characterize interactions of any particles with an energy E, including gravitons, with single gravitons. For $E \sim 1 eV$ and $\epsilon \sim 10^{-4} eV$, we have $k \sim 4 \cdot 10^{-9}$. But for $E \sim 25\ MeV$ and $\epsilon \sim 10^{-3}\ eV$, we shall have $k \sim 10^{-2}$, i.e. k will be comparable with QED's constant α. Already by $E \sim \epsilon \sim 5\ keV$, such an interaction would have the same intensity as the strong interaction ($k \sim 10$).

3.3 Deceleration of massive bodies: an analog of red-shifts

As it was reported by Anderson's team [12] , NASA deep-space probes (Pioneer 10/11, Galileo, and Ulysses) experience a small additional constant acceleration, directed towards the Sun (the Pioneer anomaly). Today, a possible origin of the effect is unknown. It must be noted here that the reported direction of additional acceleration may be a result of the simplest conjecture, which was accepted by the authors to provide a good fit for all probes. One should compare different conjectures to choose the one giving the best fit.

We consider here a deceleration of massive bodies, which would give a similar deformation of cosmic probes' trajectories [4]. The one would be a result of interaction of a massive body with the graviton background, but such an additional acceleration will be directed against a body velocity.

It follows from a universality of gravitational interaction, that not only photons, but all other objects, moving relative to the background, should lose their energy, too, due to such a quantum interaction with gravitons. If $a = H/c$, it turns out that massive bodies must feel a constant deceleration of the same order of magnitude as a small additional acceleration of cosmic probes.

Let us now denote as E a full energy of a moving body which has a velocity v relative to the background. Then energy losses of the body by an interaction with the graviton background (due to forehead collisions with gravitons) on the way dr must be expressed by the same formula (1):

$$dE = -aEdr,$$

where $a = H/c$. If $dr = vdt$, where t is a time, and $E = mc^2/\sqrt{1 - v^2/c^2}$, then we get for the body acceleration $w \equiv dv/dt$ by a non-zero velocity:

$$w = -ac^2(1 - v^2/c^2). \tag{3.3.9}$$

We assume here, that non-forehead collisions with gravitons give only stochastic deviations of a massive body's velocity direction, which are negligible. For small velocities:

$$w \simeq -Hc. \tag{3.3.10}$$

If the Hubble constant H is equal to $2.14 \cdot 10^{-18} s^{-1}$ (it is the theoretical estimate of H in this approach, see below Section 4.3), a modulus of the acceleration will be equal to

$$|w| \simeq Hc = 6.419 \cdot 10^{-10}\ m/s^2, \tag{3.3.11}$$

that has the same order of magnitude as a value of the observed additional acceleration $(8.74 \pm 1.33) \cdot 10^{-10} m/s^2$ for NASA probes.

We must emphasize here that the acceleration w is directed against a body velocity only in a special frame of reference (in which the graviton background is isotropic). In other frames, we may find its direction, using transformation formulae for an acceleration (see [24]). We can assume that the graviton background and the microwave one are isotropic in one frame (the Earth velocity relative to the microwave background was determined in [25]).

To verify my conjecture about the origin of probes' additional acceleration, one could re-analyze radio Doppler data for the probes. One should find a velocity of the special frame of reference and a constant probe acceleration w in this frame which must be negative, as it is described above. These two parameters must provide the best fit for all probes, if the conjecture is true. In such a case, one can get an independent estimate of the Hubble constant, based on the measured value of probe's additional acceleration: $H = \mid w \mid /c$. I would like to note that a deep-space mission to test the discovered anomaly is planned now at NASA by the authors of this very important discovery [26].

Under influence of such a small additional acceleration w, a probe must move on a deformed trajectory. Its view will be determined by small seeming deviations from exact conservation laws for energy and angular momentum of a not-fully reserved body system which one has in a case of neglecting with the graviton background. For example, Ulysses should go some nearer to the Sun when the one rounds it. It may be interpreted as an additional acceleration, directed towards the Sun, if we shall think that one deals with a reserved body system.

It is very important to understand, why such an acceleration has not been observed for planets. This acceleration will have different directions by motion of a body on a closed orbit, and one must take into account a solar system motion, too. As a result, an orbit should be deformed. The observed value of anomalous acceleration of Pioneer 10 should represent the vector difference of the two accelerations [8]: an acceleration of Pioneer 10 relative to the graviton background, and an acceleration of the Earth relative to the background. Possibly, the last is displayed as an annual periodic term in the residuals of Pioneer 10 [27]. If the solar system moves with a noticeable velocity relative to the background, the Earth's anomalous acceleration projection on the direction of this velocity will be smaller than for the Sun - because of the Earth's orbital motion. It means that in a frame of reference, connected with the Sun, the Earth should move with an anomalous acceleration having non-zero projections as well on the orbital velocity direction as on the direction of solar system motion relative to the background. Under some conditions, the Earth's anomalous acceleration in this frame of reference may be periodic. The axis of Earth's orbit should feel an annual precession by it. This question needs a further consideration.

3.4 Gravity as the screening effect

It was shown by the author [8, 21, 22] that screening the background of super-strong interacting gravitons creates for any pair of bodies both attraction and repulsion forces due to pressure of gravitons. For single gravitons, these forces are approximately balanced, but each of them is much bigger than a force of Newtonian attraction. If single gravitons are pairing, an attraction force due to pressure of such graviton pairs is twice exceeding a corresponding repulsion force if graviton pairs are destructed by collisions with a body. In such the model, the Newton constant is connected with the Hubble constant that gives a possibility to obtain a theoretical estimate of the last. We deal here with a flat non-expanding universe fulfilled with super-strong interacting gravitons; it changes the meaning of the Hubble constant which describes magnitudes of three small effects of quantum gravity but not any expansion or an age of the universe.

3.4.1 Pressure force of single gravitons

If gravitons of the background run against a pair of bodies with masses m_1 and m_2 (and energies E_1 and E_2) from infinity, then a part of gravitons is screened. Let $\sigma(E_1, \epsilon)$ is a cross-section of interaction of body 1 with a graviton with an energy $\epsilon = \hbar\omega$, where ω is a graviton frequency, $\sigma(E_2, \epsilon)$ is the same cross-section for body 2. In absence of body 2, a whole modulus of a gravitonic pressure force acting on body 1 would be equal to:

$$4\sigma(E_1, <\epsilon>) \cdot \frac{1}{3} \cdot \frac{4f(\omega, T)}{c}, \tag{3.4.12}$$

where $f(\omega, T)$ is a graviton spectrum with a temperature T (assuming to be Planckian), the factor 4 in front of $\sigma(E_1, <\epsilon>)$ is introduced to allow all possible directions of graviton running, $<\epsilon>$ is another average energy of running gravitons with a frequency ω taking into account a probability of that in a realization of flat wave a number of gravitons may be equal to zero, and that not all of gravitons ride at a body.

Body 2, placed on a distance r from body 1, will screen a portion of running against body 1 gravitons which is equal for big distances between the bodies (i.e. by $\sigma(E_2, <\epsilon>) \ll 4\pi r^2$) to:

$$\frac{\sigma(E_2, <\epsilon>)}{4\pi r^2.} \tag{3.4.13}$$

Taking into account all frequencies ω, the following attractive force will act between bodies 1 and 2 :

$$F_1 = \int_0^\infty \frac{\sigma(E_2, <\epsilon>)}{4\pi r^2} \cdot 4\sigma(E_1, <\epsilon>) \cdot \frac{1}{3} \cdot \frac{4f(\omega, T)}{c} d\omega. \tag{3.4.14}$$

Let $f(\omega, T)$ is described with the Planck formula:

$$f(\omega, T) = \frac{\omega^2}{4\pi^2 c^2} \frac{\hbar\omega}{\exp(\hbar\omega/kT) - 1}. \tag{3.4.15}$$

Let $x \equiv \hbar\omega/kT$, and $\bar{n} \equiv 1/(\exp(x) - 1)$ is an average number of gravitons in a flat wave with a frequency ω (on one mode of two distinguishing with a projection of particle spin). Let $P(n, x)$ is a probability of that in a realization of flat wave a number of gravitons is equal to n, for example $P(0, x) = \exp(-\bar{n})$.

A quantity $< \epsilon >$ must contain the factor $(1 - P(0, x))$, i.e. it should be:

$$< \epsilon > \sim \hbar\omega(1 - P(0, x)), \tag{3.4.16}$$

that let us to reject flat wave realizations with zero number of gravitons.

But attempting to define other factors in $< \epsilon >$, we find the difficult place in our reasoning. On this stage, it is necessary to introduce some new assumption to find the factors. Perhaps, this assumption will be well-founded in a future theory - or would be rejected. If a flat wave realization, running against a finite size body from infinity, contains one graviton, then one cannot consider that it must stringent ride at a body to interact with some probability with the one. It would break the uncertainty principle by W. Heisenberg. We should admit that we know a graviton trajectory. The same is pertaining to gravitons scattered by one of bodies by big distances between bodies. What is a probability that a single graviton will ride namely at the body? If one denotes this probability as P_1, then for a wave with n gravitons their chances to ride at the body must be equal to $n \cdot P_1$. Taking into account the probabilities of values of n for the Poisson flux of events, an additional factor in $< \epsilon >$ should be equal to $\bar{n} \cdot P_1$. I have admitted in [8] that

$$P_1 = P(1, x), \tag{3.4.17}$$

where $P(1, x) = \bar{n}\exp(-\bar{n})$; (below it is admitted for pairing gravitons: $P_1 = P(1, 2x)$ - see Section 4.3).

In such the case, we have for $< \epsilon >$ the following expression:

$$< \epsilon > = \hbar\omega(1 - P(0, x))\bar{n}^2 \exp(-\bar{n}). \tag{3.4.18}$$

Then we get for an attractive force F_1 :

$$F_1 = \frac{4}{3}\frac{D^2 E_1 E_2}{\pi r^2 c}\int_0^\infty \frac{\hbar^3\omega^5}{4\pi^2 c^2}(1 - P(0, x))^2 \bar{n}^5 \exp(-2\bar{n}) d\omega = \tag{3.4.19}$$

$$\frac{1}{3} \cdot \frac{D^2 c (kT)^6 m_1 m_2}{\pi^3 \hbar^3 r^2} \cdot I_1,$$

where

$$I_1 \equiv \int_0^\infty x^5(1 - \exp(-(\exp(x) - 1)^{-1}))^2(\exp(x) - 1)^{-5}\exp(-2(\exp(x) - 1)^{-1})dx = \tag{3.4.20}$$

$$5.636 \cdot 10^{-3}.$$

This and all other integrals were found with the MathCad software.

If $F_1 \equiv G_1 \cdot m_1 m_2 / r^2$, then the constant G_1 is equal to:

$$G_1 \equiv \frac{1}{3} \cdot \frac{D^2 c (kT)^6}{\pi^3 \hbar^3} \cdot I_1. \qquad (3.4.21)$$

By $T = 2.7\ K$:

$$G_1 = 1215.4 \cdot G, \qquad (3.4.22)$$

that is three order greater than the Newton constant, G.

But if single gravitons are elastically scattered with body 1, then our reasoning may be reversed: the same portion (13) of scattered gravitons will create a repulsive force $F_1^{'}$ acting on body 2 and equal to

$$F_1^{'} = F_1, \qquad (3.4.23)$$

if one neglects with small allowances which are proportional to D^3/r^4.

So, for bodies which elastically scatter gravitons, screening a flux of single gravitons does not ensure Newtonian attraction. But for gravitonic black holes which absorb any particles and do not re-emit them (by the meaning of a concept, the ones are usual black holes; I introduce a redundant adjective only from a caution), we will have $F_1^{'} = 0$. It means that such the object would attract other bodies with a force which is proportional to G_1 but not to G, i.e. Einstein's equivalence principle would be violated for them. This conclusion, as we shall see below, stays in force for the case of graviton pairing, too. The conclusion cannot be changed with taking into account of Hawking's quantum effect of evaporation of black holes [23].

3.4.2 Graviton pairing

To ensure an attractive force which is not equal to a repulsive one, particle correlations should differ for *in* and *out* flux. For example, single gravitons of running flux may associate in pairs [8]. If such pairs are destructed by collision with a body, then quantities $< \epsilon >$ will be distinguished for running and scattered particles. Graviton pairing may be caused with graviton's own gravitational attraction or gravitonic spin-spin interaction. Left an analysis of the nature of graviton pairing for the future; let us see that gives such the pairing.

To find an average number of pairs $\bar{n}_2$ in a wave with a frequency ω for the state of thermodynamic equilibrium, one may replace $\hbar \rightarrow 2\hbar$ by deducing the Planck formula. Then an average number of pairs will be equal to:

$$\bar{n}_2 = \frac{1}{\exp(2x) - 1}, \qquad (3.4.24)$$

and an energy of one pair will be equal to $2\hbar\omega$. It is important that graviton pairing does not change a number of stationary waves, so as pairs nucleate from existing gravitons. The question arises: how many different modes, i.e. spin projections, may graviton pairs have? We consider that the background of initial gravitons consists of two modes. For massless transverse bosons, it takes place as by spin 1 as by spin 2. If graviton pairs have maximum spin 2, then single gravitons should have spin 1. But from such particles one may constitute four combinations: $\uparrow\uparrow$, $\downarrow\downarrow$ (with total spin 2), and $\uparrow\downarrow$, $\downarrow\uparrow$ (with total spin 0). All these four combinations will be equiprobable if spin projections $\uparrow$ and $\downarrow$ are equiprobable in a flat wave (without taking into account a probable spin-spin interaction).

But it is happened that, if expression (24) is true, it follows from the energy conservation law that composite gravitons should be distributed only in two modes. So as

$$\lim_{x\to 0} \frac{\bar{n}_2}{\bar{n}} = 1/2, \tag{3.4.25}$$

then by $x \to 0$ we have $2\bar{n}_2 = \bar{n}$, i.e. all of gravitons are pairing by low frequencies. An average energy on every mode of pairing gravitons is equal to $2\hbar\omega\bar{n}_2$, the one on every mode of single gravitons - to $\hbar\omega\bar{n}$. These energies are equal by $x \to 0$, because of that, the numbers of modes are equal, too, if the background is in the thermodynamic equilibrium with surrounding bodies.

The above reasoning does not allow to choose a spin value 2 or 0 for composite gravitons. A choice of namely spin 2 would ensure the following proposition: all of gravitons in one realization of flat wave have the same spin projections. From another side, a spin-spin interaction would cause it.

The spectrum of composite gravitons is also the Planckian one, but with a smaller temperature; it has the view:

$$f_2(2\omega, T)d\omega = \frac{\omega^2}{4\pi^2 c^2} \cdot \frac{2\hbar\omega}{\exp(2x) - 1} d\omega \equiv \frac{(2\omega)^2}{32\pi^2 c^2} \cdot \frac{2\hbar\omega}{\exp(2x) - 1} d(2\omega). \tag{3.4.26}$$

It means that an absolute luminosity for the sub-system of composite gravitons is equal to:

$$\int_0^\infty f_2(2\omega, T)d(2\omega) = \frac{1}{8}\sigma T^4, \tag{3.4.27}$$

where σ is the Stephan-Boltzmann constant; i.e. an equivalent temperature of this sub-system is

$$T_2 \equiv (1/8)^{1/4} T = \frac{2^{1/4}}{2} T = 0.5946T. \tag{3.4.28}$$

The portion of pairing gravitons, $2\bar{n}_2/\bar{n}$, a spectrum of single gravitons, $f(x)$, and a spectrum of subsystem of pairing gravitons, $f_2(2x)$, are shown on Fig. 5 as functions of the dimensionless parameter $x \equiv \hbar\omega/kT$.

It is important that the graviton pairing effect does not change computed values of the Hubble constant and of anomalous deceleration of massive bodies: twice decreasing of a sub-system particle number due to the pairing effect is compensated with twice increasing the cross-section of interaction of a photon or any body with such the composite gravitons. Non-pairing gravitons with spin 1 give also its

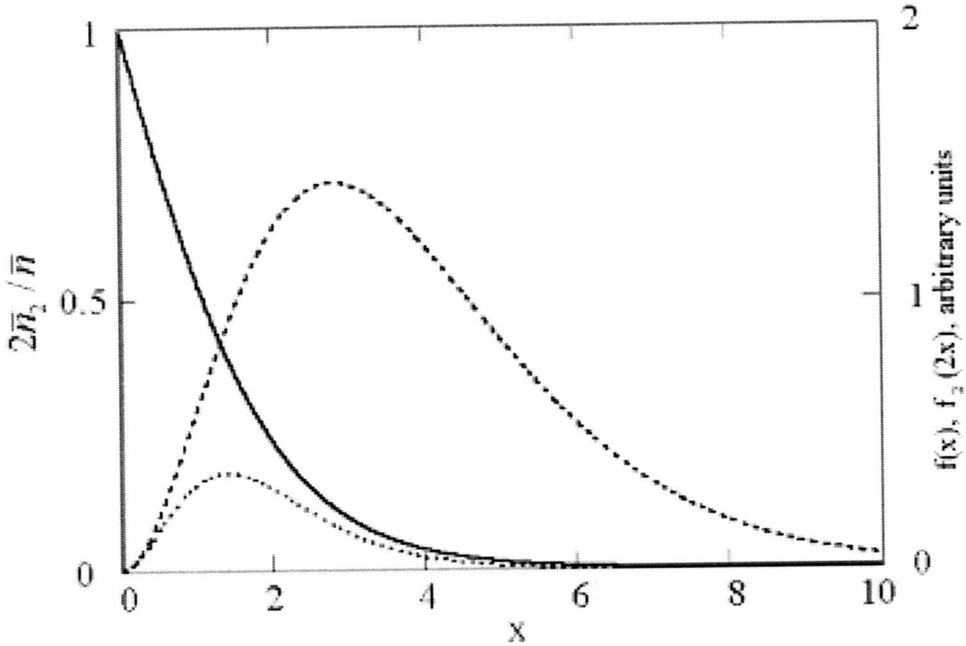

Figure 3.4.5: The portion of pairing gravitons, $2\bar{n}_2/\bar{n}$, (solid line), a spectrum of single gravitons, $f(x)$, (dashed line), and a spectrum of graviton pairs, $f_2(2x)$, (dotted line) versus the dimensionless parameter x.

contribution in values of redshifts, an additional relaxation of light intensity due to non-forehead collisions with gravitons, and anomalous deceleration of massive bodies moving relative to the background.

3.4.3 Computation of the Newton constant, and a connection between the two fundamental constants, G and H

If running graviton pairs ensure for two bodies an attractive force F_2, then a repulsive force due to re-emission of gravitons of a pair alone will be equal to $F_2^{'} = F_2/2$. It follows from that the cross-section for *single additional scattered* gravitons of destructed pairs will be twice smaller than for pairs themselves (the leading factor $2\hbar\omega$ for pairs should be replaced with $\hbar\omega$ for single gravitons). For pairs, we introduce here the cross-section $\sigma(E_2, <\epsilon_2>)$, where $<\epsilon_2>$ is an average pair energy with taking into account a probability of that in a realization of flat wave a number of graviton pairs may be equal to zero, and that not all of graviton pairs ride at a body ($<\epsilon_2>$ is an analog of $<\epsilon>$). This equality is true in neglecting with small allowances which are proportional to D^3/r^4 (see Section 4.4). Replacing $\bar{n} \to \bar{n}_2$, $\hbar\omega \to 2\hbar\omega$, and $P(n, x) \to P(n, 2x)$, where $P(0, 2x) = \exp(-\bar{n}_2)$, we get for graviton pairs:

$$< \epsilon_2 >\sim 2\hbar\omega(1 - P(0, 2x))\bar{n}_2^2 \exp(-\bar{n}_2). \tag{3.4.29}$$

This expression does not take into account only that beside pairs there may be single gravitons in a realization of flat wave. To reject cases when, instead of a pair, a single graviton runs against a body (a contribution of such gravitons in attraction and repulsion is the same), we add the factor $P(0, x)$ into $< \epsilon_2 >$:

$$< \epsilon_2 >= 2\hbar\omega(1 - P(0, 2x))\bar{n}_2^2 \exp(-\bar{n}_2) \cdot P(0, x). \tag{3.4.30}$$

Then a force of attraction of two bodies due to pressure of graviton pairs, F_2, - in the full analogy with (19) - will be equal to [2]:

$$F_2 = \int_0^\infty \frac{\sigma(E_2, < \epsilon_2 >)}{4\pi r^2} \cdot 4\sigma(E_1, < \epsilon_2 >) \cdot \frac{1}{3} \cdot \frac{4f_2(2\omega, T)}{c} d\omega = \tag{3.4.31}$$

$$\frac{8}{3} \cdot \frac{D^2 c (kT)^6 m_1 m_2}{\pi^3 \hbar^3 r^2} \cdot I_2,$$

where

$$I_2 \equiv \int_0^\infty \frac{x^5 (1 - \exp(-(\exp(2x) - 1)^{-1}))^2 (\exp(2x) - 1)^{-5}}{\exp(2(\exp(2x) - 1)^{-1}) \exp(2(\exp(x) - 1)^{-1})} dx = \tag{3.4.32}$$

$$2.3184 \cdot 10^{-6}.$$

The difference F between attractive and repulsive forces will be equal to:

$$F \equiv F_2 - F_2^{'} = \frac{1}{2} F_2 \equiv G_2 \frac{m_1 m_2}{r^2}, \tag{3.4.33}$$

where the constant G_2 is equal to:

$$G_2 \equiv \frac{4}{3} \cdot \frac{D^2 c (kT)^6}{\pi^3 \hbar^3} \cdot I_2. \tag{3.4.34}$$

Both G_1 and G_2 are proportional to T^6 (and $H \sim T^5$, so as $\bar{\epsilon} \sim T$).

If one assumes that $G_2 = G$, then it follows from (34) that by $T = 2.7K$ the constant D should have the value:

$$D = 0.795 \cdot 10^{-27} m^2/eV^2. \tag{3.4.35}$$

An average graviton energy of the background is equal to:

$$\bar{\epsilon} \equiv \int_0^\infty \hbar\omega \cdot \frac{f(\omega, T)}{\sigma T^4} d\omega = \frac{15}{\pi^4} I_4 kT, \tag{3.4.36}$$

---

[2]In initial version of this paper, factor 2 was lost in the right part of Eq. (31), and the theoretical values of D and H were overestimated of $\sqrt{2}$ times

where

$$I_4 \equiv \int_0^\infty \frac{x^4 dx}{\exp(x) - 1} = 24.866$$

(it is $\bar{\epsilon} = 8.98 \cdot 10^{-4} eV$ by $T = 2.7K$).

We can use (8) and (34) to establish a connection between the two fundamental constants, G and H, under the condition that $G_2 = G$. We have for D :

$$D = \frac{2\pi H}{\bar{\epsilon}\sigma T^4} = \frac{2\pi^5 H}{15k\sigma T^5 I_4}; \qquad (3.4.37)$$

then

$$G = G_2 = \frac{4}{3} \cdot \frac{D^2 c(kT)^6}{\pi^3 \hbar^3} \cdot I_2 = \frac{64\pi^5}{45} \cdot \frac{H^2 c^3 I_2}{\sigma T^4 I_4^2}. \qquad (3.4.38)$$

So as the value of G is known much better than the value of H, let us express H via G :

$$H = (G \frac{45}{64\pi^5} \frac{\sigma T^4 I_4^2}{c^3 I_2})^{1/2} = 2.14 \cdot 10^{-18} \ s^{-1}, \qquad (3.4.39)$$

or in the units which are more familiar for many of us: $H = 66.875 \ km \cdot s^{-1} \cdot Mpc^{-1}$.

This value of H is in the good accordance with the majority of present astrophysical estimations [13, 28, 29] (for example, the estimate (72 ± 8) km/s/Mpc has been got from SN1a cosmological distance determinations in [29]), but it is lesser than some of them [30] and than it follows from the observed value of anomalous acceleration of Pioneer 10 [12].

3.4.4 Restrictions on a geometrical language in gravity

The described quantum mechanism of classical gravity gives Newton's law with the constant G_2 value (34) and the connection (38) for the constants G_2 and H. We have obtained the rational value of H (39) by $G_2 = G$, if the condition of big distances is fulfilled:

$$\sigma(E_2, <\epsilon>) \ll 4\pi r^2. \qquad (3.4.40)$$

Because it is known from experience that for big bodies of the solar system, Newton's law is a very good approximation, one would expect that the condition (40) is fulfilled, for example, for the pair Sun-Earth. But assuming $r = 1 \ AU$ and $E_2 = m_\odot c^2$, we obtain assuming for rough estimation $<\epsilon> \to \bar{\epsilon}$:

$$\frac{\sigma(E_2, <\epsilon>)}{4\pi r^2} \sim 4 \cdot 10^{12}.$$

It means that in the case of interaction of gravitons or graviton pairs with the Sun in the aggregate, the considered quantum mechanism of classical gravity could not lead to Newton's law as a good approximation. This "contradiction" with experience is eliminated if one assumes that gravitons interact with "small particles" of matter

- for example, with atoms. If the Sun contains of N atoms, then $\sigma(E_2, <\epsilon>) = N\sigma(E_a, <\epsilon>)$, where E_a is an average energy of one atom. For rough estimation we assume here that $E_a = E_p$, where E_p is a proton rest energy; then it is $N \sim 10^{57}$, i.e. $\sigma(E_a, <\epsilon>)/4\pi r^2 \sim 10^{-45} \ll 1$.

This necessity of "atomic structure" of matter for working the described quantum mechanism is natural relative to usual bodies. But would one expect that black holes have a similar structure? If any radiation cannot be emitted with a black hole, a black hole should interact with gravitons as an aggregated object, i.e. the condition (40) for a black hole of sun mass has not been fulfilled even at distances $\sim 10^6$ AU.

For bodies without an atomic structure, the allowances, which are proportional to D^3/r^4 and are caused by decreasing a gravitonic flux due to the screening effect, will have a factor $m_1^2 m_2$ or $m_1 m_2^2$. These allowances break the equivalence principle for such the bodies.

For bodies with an atomic structure, a force of interaction is added up from small forces of interaction of their "atoms":

$$F \sim N_1 N_2 m_a^2/r^2 = m_1 m_2/r^2,$$

where N_1 and N_2 are numbers of atoms for bodies 1 and 2. The allowances to full forces due to the screening effect will be proportional to the quantity: $N_1 N_2 m_a^3/r^4$, which can be expressed via the full masses of bodies as $m_1^2 m_2/r^4 N_1$ or $m_1 m_2^2/r^4 N_2$. By big numbers N_1 and N_2 the allowances will be small. The allowance to the force F, acting on body 2, will be equal to:

$$\Delta F = \frac{1}{2N_2}\int_0^\infty \frac{\sigma^2(E_2, <\epsilon_2>)}{(4\pi r^2)^2} \cdot 4\sigma(E_1, <\epsilon_2>) \cdot \frac{1}{3} \cdot \frac{4f_2(2\omega, T)}{c} d\omega = \tag{3.4.41}$$

$$\frac{2}{3N_2} \cdot \frac{D^3 c^3 (kT)^7 m_1 m_2^2}{\pi^4 \hbar^3 r^4} \cdot I_3,$$

(for body 1 we shall have the similar expression if replace $N_2 \to N_1$, $m_1 m_2^2 \to m_1^2 m_2$), where

$$I_3 \equiv \int_0^\infty \frac{x^6 (1 - \exp(-(\exp(2x) - 1)^{-1}))^3 (\exp(2x) - 1)^{-7}}{\exp(3(\exp(x) - 1)^{-1})} dx = 1.0988 \cdot 10^{-7}.$$

Let us find the ratio:

$$\frac{\Delta F}{F} = \frac{D E_2 kT}{N_2 2\pi r^2} \cdot \frac{I_3}{I_2}. \tag{3.4.42}$$

Using this formula, we can find by $E_2 = E_\odot$, $r = 1$ AU :

$$\frac{\Delta F}{F} \sim 10^{-46}. \tag{3.4.43}$$

An analogical allowance to the force F_1 has by the same conditions the order $\sim 10^{-48} F_1$, or $\sim 10^{-45} F$. One can replace E_p with a rest energy of very big atom -

the geometrical approach will left a very good language to describe the solar system. We see that for bodies with an atomic structure the considered mechanism leads to very small deviations from Einstein's equivalence principle, if the condition (40) is fulfilled for microparticles, which prompt interact with gravitons.

For small distances we shall have:

$$\sigma(E_2, <\epsilon>) \sim 4\pi r^2. \tag{3.4.44}$$

It takes place by $E_a = E_p$, $<\epsilon>\sim 10^{-3}\ eV$ for $r \sim 10^{-11}\ m$. This quantity is many orders larger than the Planck length. The equivalence principle should be broken at such distances.

Under the condition (44), big digressions from Newton's law will be caused with two factors: 1) a screening portion of a running flux of gravitons is not small and it should be taken into account by computation of the repulsive force; 2) a value of this portion cannot be defined by the expression (13).

Instead of (13), one might describe this portion at small distances with an expression of the kind:

$$\frac{1}{2}(1 + \sigma(E_a, <\epsilon>)/\pi r^2 - (1 + \sigma(E_a, <\epsilon>)/\pi r^2)^{1/2}) \tag{3.4.45}$$

(the formula for a spheric segment area is used here [31]). Formally, by $\sigma(E_a, <\epsilon>)/\pi r^2 \to \infty$ we shall have for the portion (45):

$$\sim \frac{1}{2}(\sigma(E_a, <\epsilon>)/\pi r^2 - (\sigma(E_a, <\epsilon>)/\pi)^{1/2}/r),$$

where the second term shows that the interaction should be weaker at small distances. We might expect that a screening portion may tend to a fixing value at super-short distances, and it will be something similar to asymptotic freedom of strong interactions. But, of course, at such distances the interaction will be super-strong and our naive approach would be not valid.

3.5 Some cosmological consequences of the model

If the described model of redshifts is true, what is a picture of the universe? It is interesting that in a frame of this model, every observer has two own spheres of observability in the universe (two different cosmological horizons exist for any observer) [32, 33]. One of them is defined by maximum existing temperatures of remote sources - by big enough distances, all of them will be masked with the CMB radiation. Another, and much smaller, sphere depends on their maximum luminosity - the luminosity distance increases with a redshift much quickly than the geometrical one. The ratio of the luminosity distance to the geometrical one is the quickly increasing function of z :

$$D_L(z)/r(z) = (1 + z)^{(1+b)/2}, \tag{3.5.46}$$

which does not depend on the Hubble constant. An outer part of the universe will drown in a darkness.

By the found theoretical value of the Hubble constant: $H = 2.14 \cdot 10^{-18}\ s^{-1}$ (then a natural light unit of distances is equal to $1/H \simeq 14.85$ light GYR), plots of two theoretical functions of z in this model - the geometrical distance $r(z)$ and the luminosity distance $D_L(z)$ - are shown on Fig. 6 [32, 33]. As one can see, for objects with $z \sim 10$, which are observable now, we should anticipate geometrical distances of the order ~ 35 light GYR and luminosity distances of the order ~ 1555 light GYR in a frame of this model. An estimate of distances to objects with given z is changed, too: for example, the quasar with $z = 5.8$ [35] should be in a distance approximately of 2.8 times bigger than the one expected in the model based on the Doppler effect.

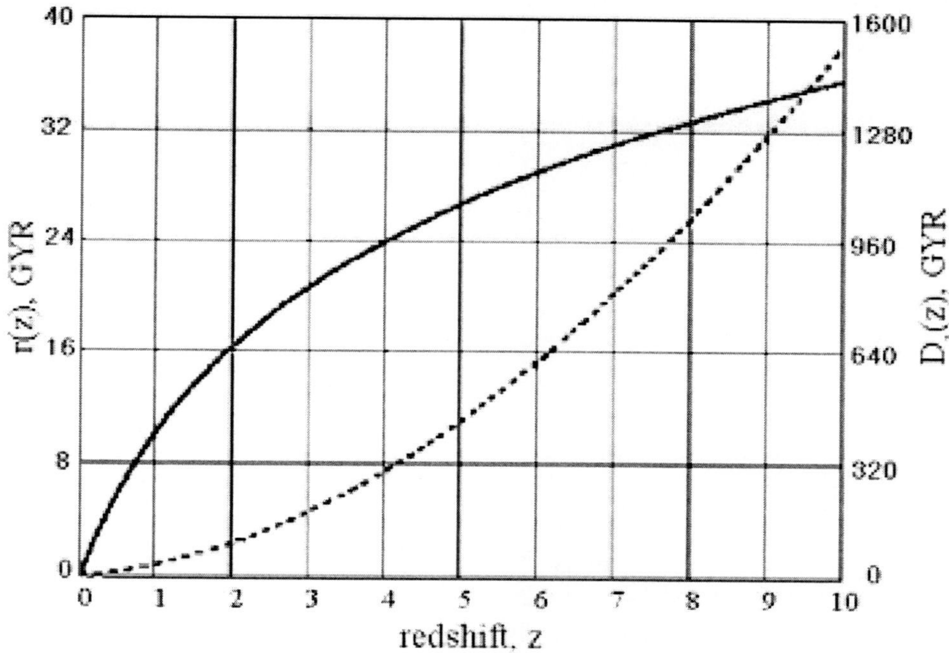

Figure 3.5.6: The geometrical distance, $r(z)$, (solid line) and the luminosity distance, $D_L(z)$, (dashed line) - both in light GYRs - in this model as functions of a redshift, z. The following theoretical value for H is accepted: $H = 2.14 \cdot 10^{-18} s^{-1}$.

We can assume that the graviton background and the cosmic microwave one are in a state of thermodynamical equilibrium, and have the same temperatures. CMB itself may arise as a result of cooling any light radiation up to reaching this equilibrium. Then it needs $z \sim 1000$ to get through the very edge of our cosmic "ecumene" (see Fig. 7).

Some other possible cosmological consequences of an existence of the graviton

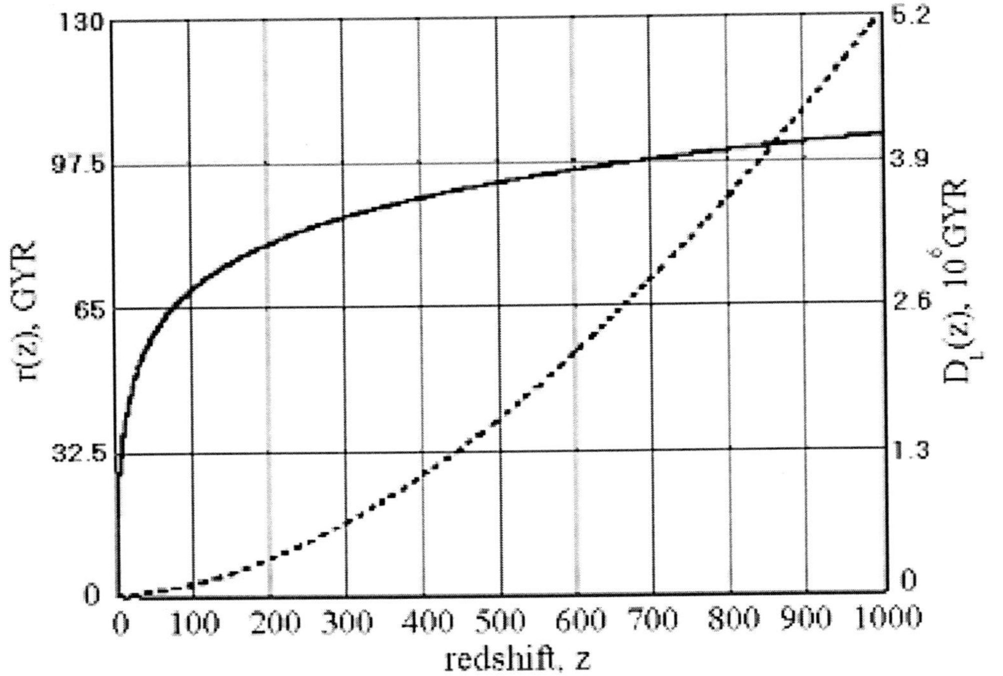

Figure 3.5.7: The same functions as on Fig. 6 (all notations are reserved), but for the huge range of z.

background were described in [34, 8]. Observations of last years give us strong evidences for supermassive and compact objects (named now supermassive black holes) in active and normal galactic nuclei [36, 37, 38, 39, 40]. Massive nuclear "black holes" of $10^6 - 10^9$ solar masses may be responsible for the energy production in quasars and active galaxies [36]. In a frame of this model, an existence of black holes contradicts to the equivalence principle. It means that these objects should have another nature; one must remember that we know only that these objects are supermassive and compact.

There should be two opposite processes of heating and cooling the graviton background [34] which may have a big impact on cosmology. Unlike models of expanding universe, in any tired light model one has a problem of utilization of energy, lost by radiation of remote objects. In the considered model, a virtual graviton forms under collision of a photon with a graviton of the graviton background. It should be massive if an initial graviton transfers its total momentum to a photon; it follows from the energy conservation law that its energy $\epsilon^{'}$ must be equal to 2ϵ if ϵ is an initial graviton energy. In force of the uncertainty relation, one has for a virtual graviton lifetime $\tau : \tau \leq \hbar/\epsilon^{'}$, i.e. for $\epsilon^{'} \sim 10^{-4}\ eV$ it is $\tau \leq 10^{-11}\ s$. In force of conservation laws for energy, momentum and angular momentum, a virtual graviton may decay into no less than three real gravitons. In a case of decay into three gravitons, its energies should be equal to ϵ, $\epsilon^{''}$, $\epsilon^{'''}$, with $\epsilon^{''} + \epsilon^{'''} = \epsilon$. So, after this decay, two new gravitons with $\epsilon^{''}$, $\epsilon^{'''} < \epsilon$ inflow into the graviton background. It is a source of adjunction of the graviton background.

From another side, an interaction of gravitons of the background between themselves should lead to the formation of virtual massive gravitons, too, with energies less than ϵ_{min} where ϵ_{min} is a minimal energy of one graviton of an initial interacting pair. If gravitons with energies $\epsilon^{''}$, $\epsilon^{'''}$ wear out a file of collisions with gravitons of the background, its lifetime increases. In every such a collision-decay cycle, an average energy of "redundant" gravitons will double decrease, and its lifetime will double increase. Only for $\sim$ 93 cycles, a lifetime will increase from 10^{-11} s to 10 Gyr. Such virtual massive gravitons, with a lifetime increasing from one collision to another, would duly serve dark matter particles. Having a zero (or near to zero) initial velocity relative to the graviton background, the ones will not interact with matter in any manner excepting usual gravitation. An ultra-cold gas of such gravitons will condense under influence of gravitational attraction into "black holes" or other massive objects. Additionally to it, even in absence of initial heterogeneity, the one will easy arise in such the gas that would lead to arising of super compact massive objects, which will be able to turn out "germs" of "black holes". It is a method "to cool" the graviton background.

So, the graviton background may turn up "a perpetual engine" of the universe, pumping energy from any radiation to massive objects. An equilibrium state of the background will be ensured by such a temperature T, for which an energy profit of the background due to an influx of energy from radiation will be equal to a loss of its energy due to a catch of virtual massive gravitons with "black holes" or other massive objects. In such the picture, the chances are that "black holes" would turn out "germs" of galaxies. After accumulation of a big enough energy by a "black hole" (to be more exact, by a super-compact massive object) by means of a catch of virtual massive gravitons, the one would be absolved from an energy excess in via ejection of matter, from which stars of galaxy should form. It awaits to understand else in such the approach how usual matter particles form from virtual massive gravitons.

There is a very interesting but non-researched possibility: due to relative decreasing of an intensity of graviton pair flux in an internal area of galaxies (pairs are destructed under collisions with matter particles), the effective Newton constant may turn out to be running on galactic scales. It might lead to something like to the modified Newtonian dynamics (MOND) by Mordehai Milgrom (about MOND, for example, see [41]). But to evaluate this effect, one should take into account a relaxation process for pairs, about which we know nothing today. It is obvious only that gravity should be stronger on a galactic periphery. The renormalization group approach to gravity leads to modifications of the theory of general relativity on galactic scales [42, 43], and a growth of Newton's constant at large distances takes place, too. Kepler's third law receives quantum corrections that may explain the flat rotation curves of the galaxies.

3.6 How to verify the main conjecture of this approach in a laser experiment on the Earth

I would like to show here (see [44]) a full realizability at present time of verifying my basic conjecture about the quantum gravitational nature of redshifts in a ground-based laser experiment. Of course, many details of this precision experiment will be in full authority of experimentalists.

It was not clear in 1995 how big is a temperature of the graviton background, and my proposal [16] to verify the conjecture about the described local quantum character of redshifts turned out to be very rigid: a laser with instability of $\sim 10^{-17}$ hasn't appeared after 9 years. But if $T = 2.7K$, the satellite of main laser line of frequency ν after passing the delay line will be red-shifted at $\sim 10^{-3}$ eV/h and its position will be fixed (see Fig. 8). It will be caused by the fact that on a very small

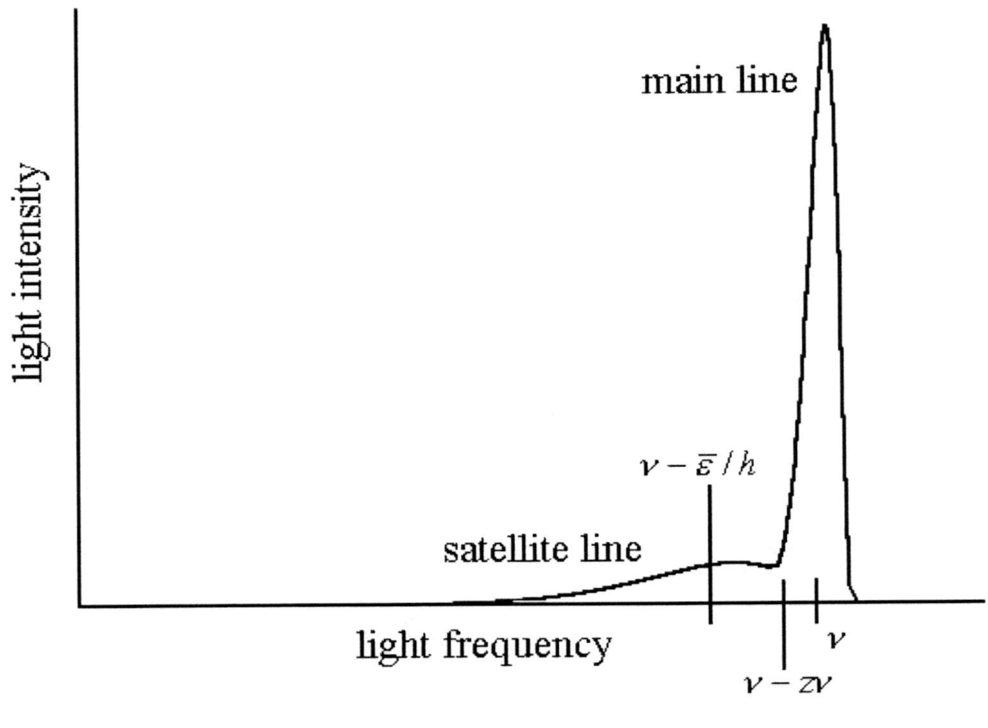

Figure 3.6.8: The main line and the expected red-shifted satellite line of a stable laser radiation spectrum after a delay line. Satellite's position should be fixed near $\nu - \bar{\epsilon}/h$, and its intensity should linear rise with a path of photons in a delay line, l. A center-of-mass of both lines is expected to be approximately near $\nu - z\nu$.

way in the delay line only a small part of photons may collide with gravitons of the background. The rest of them will have unchanged energies. The center-of-mass of laser radiation spectrum should be shifted proportionally to a photon path. Then due to the quantum nature of shifting process, the ratio of satellite's intensity to main line's intensity should have the order:

$$\sim \frac{h\nu}{\bar{\epsilon}}\frac{H}{c}l,$$

where l is a path of laser photons in a vacuum tube of delay line. It gives us a possibility to plan a laser-based experiment to verify the basic conjecture of this approach with much softer demands to the equipment. An instability of a laser of a power P must be only $\ll 10^{-3}$ if a photon energy is of $\sim 1\ eV$. It will be necessary to compare intensities of the red-shifted satellite at the very beginning of the path l and after it. Given a very low signal-to-noise ratio, one could use a single photon counter to measure the intensities. When q is a quantum output of a cathode of the used photomultiplier (a number of photoelectrons is q times smaller than a number of photons falling to the cathode), N_n is a frequency of its noise pulses, and n is a desired ratio of a signal to noise's standard deviation, then an evaluated time duration t of data acquisition would have the order:

$$t = \frac{\bar{\epsilon}^2 c^2}{H^2}\frac{n^2 N_n}{q^2 P^2 l^2}. \tag{3.6.47}$$

Assuming $n = 10,\ N_n = 10^3\ s^{-1},\ q = 0.3,\ P = 100\ mW,\ l = 100\ m$, we would have the estimate: $t = 200,000$ years, that is unacceptable. But given $P = 300\ W$, we get: $t \sim 8$ days, that is acceptable for the experiment of such the potential importance. Of course, one will rather choose a bigger value of l by a small laser power forcing a laser beam to whipsaw many times between mirrors in a delay line - it is a challenge for experimentalists.

3.7 Gravity in a frame of non-linear and non-local QED? - the question only to the Nature

From thermodynamic reasons, it is assumed here that the graviton background has the same temperature as the microwave background. Also it follows from the condition of detail equilibrium, that both backgrounds should have the Planckian spectra. Composite gravitons will have spin 2, if single gravitons have the same spin as photons. The question arise, of course: how are gravitons and photons connected? Has the conjecture by Adler et al. [45, 46] (that a graviton with spin 2 is composed with two photons) chances to be true? Intuitive demur calls forth a huge self-action, photons should be endued with which if one unifies the main conjecture of this approach with the one by Adler et al. - but one may get a unified theory on this way.

To verify this combined conjecture in experiment, one would search for transitions in interstellar gas molecules caused by the microwave background, with an angular momentum change corresponding to absorption of spin 2 particles (photon pairs). A frequency of such the transitions should correspond to an equivalent temperature of the sub-system of these composite particles $T_2 = 0.5946\ T$, if T is a temperature of the microwave background.

From another side, one might check this conjecture in a laser experiment, too. Taking two lasers with photon energies $h\nu_1$ and $h\nu_2$, one may force laser beams to collide on a way L (see Fig. 9). If photons are self-interacting particles, we might wait that photons with energies $h\nu_1 - h\nu_2$, if $h\nu_1 > h\nu_2$, would arise after collisions of initial photons. If we *assume (only here)* that single gravitons are identical to

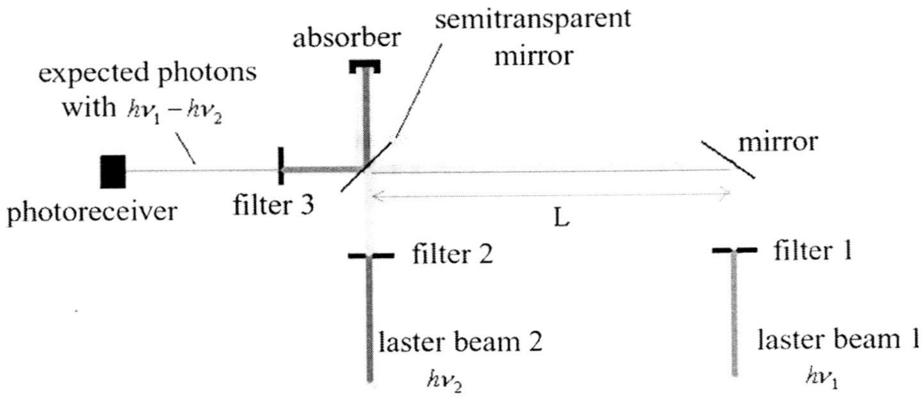

Figure 3.7.9: The scheme of laser beam passes. Two laser beams 1 and 2 collide into the area with a length L. An expected beam of photons with energies $h\nu_1 - h\nu_2$ falls to a photoreceiver.

photons, it will be necessary to take into account the following circumstances to calculate an analog of the Hubble constant for this experiment: an average graviton energy should be replaced with $h\nu_2$, the factor $1/2\pi$ in (8) should be replaced with $1/\varphi$, where φ is a divergence of laser beam 2, and one must use a quantity P/S instead of σT^4 in (8), where P is a laser 2 power and S is a cross-section of its beam. Together all it means that we should replace the Hubble constant with its analog for a laser beam collision, H_{laser}:

$$H \rightarrow H_{laser} = \frac{1}{\varphi} \cdot D \cdot h\nu_2 \cdot \frac{P}{S}. \qquad (3.7.48)$$

Taken $\varphi = 10^{-4}$, $h\nu_2 \sim 1\ eV$, $P \sim 10\ mW$, and $P/S \sim 10^3\ W/m^2$, that is characterizing a He-Ne laser, we get the estimate: $H_{laser} \sim 0.06\ s^{-1}$. Then photons with energies $h\nu_1 - h\nu_2$ would fall to a photoreceiver with a frequency which should linearly rise with L (proportionally to $\frac{H_{laser}}{c} \cdot L$), and it would be of $10^7\ s^{-1}$ if both lasers have equal powers $\sim 10\ mW$, and $L \sim 1\ m$. It is a big enough frequency to give us a possibility to detect easy a flux of these expected photons in IR band.

I think there is not any sense to try to analyze theoretically consequences of this conjecture - it will be easier to verify it experimentally. The Nature may answer the question if we ask correctly. All that was said in the above sections doesn't depend on the answer, but it would be very important for our understanding of known interactions. If this tentative non-linear vacuum effect exists, it would lead us far beyond standard quantum electrodynamics to take into account new non-linearities (which are not connected with electron-positron pair creation) and an essential impact of such a non-locally born object as the graviton background.

3.8 Conclusion

It follows from the above consideration that the geometrical description of gravity should be a good idealization for any pair of bodies at a big distance by the condition of an "atomic structure" of matter. This condition cannot be accepted only for black holes which must interact with gravitons as aggregated objects. In addition, the equivalence principle is roughly broken for black holes, if the described quantum mechanism of classical gravity is realized in the nature. Because attracting bodies are not initial sources of gravitons, a future theory must be non-local in this sense to describe gravitons running from infinity. Non-local models were considered by G.V. Efimov in his book [47]. The Le Sage's idea to describe gravity as caused by running *ab extra* particles was criticized by the great physicist Richard Feynman in his public lectures at Cornell University [48], but the Pioneer 10 anomaly [12], perhaps, is a good contra argument pro this idea.

The described quantum mechanism of classical gravity is obviously asymmetric relative to the time inversion. By the time inversion, single gravitons would run against bodies to form pairs after collisions with bodies. It would lead to replacing a body attraction with a repulsion. But such the change will do impossible the graviton pairing. Cosmological models with the inversion of the time arrow were considered by Sakharov [49]. Penrose has noted that a hidden physical law may determine the time arrow direction [50]; it will be very interesting if namely realization in the nature of Newton's law determines this direction.

A future theory dealing with gravitons as usual particles should have a number of features which are not characterizing any existing model to image the considered here features of the possible quantum mechanism of gravity. If this mechanism is realized in the nature, both the general relativity and quantum mechanics should be modified. Any divergencies, perhaps, would be not possible in such the model because of natural smooth cut-offs of the graviton spectrum from both sides. Gravity at short distances, which are much bigger than the Planck length, needs to be described only in some unified manner.

References

[1] Carlip, S. *Rept. Prog. Phys.* 2001, **64,** 885; [gr-qc/0108040].

[2] Anderson, J.D. et al. *Phys. Rev. Lett.* 1998, **81,** 2858; *Phys. Rev.* 2002, **D65,** 082004; [gr-qc/0104064 v4].

[3] Riess, A.G. et al. *AJ* 1998, **116,** 1009.

[4] Perlmutter, S. et al. *ApJ* 1999, **517,** 565.

[5] Nesvizhevsky, V.V. et al. *Nature* 2002, **415,** 297.

[6] Ivanov, M.A. *General Relativity and Gravitation* 2001, **33,** 479; Erratum: 2003, **35,** 939; [astro-ph/0005084 v2].

[7] Ivanov, M.A. [gr-qc/0009043]; *Proc. of the Int. Symp. "Frontiers of Fundamental Physics 4"* (9-13 Dec 2000, Hyderabad, India), Sidharth, B.G., Altaisky, M.V., Eds.; Kluwer Academic/Plenum Publishers: August 2001; *Proc. of the 4th Edoardo Amaldi Conference on Gravitational Waves* (Perth, Western Australia, 8-13 July 2001) *Class. Quantum Grav.* 2002, **19,** 1351.

[8] Ivanov, M.A. Screening the graviton background, graviton pairing, and Newtonian gravity [gr-qc/0207006 v2].

[9] Ivanov, M.A. *Acta Phys. Polonica* 1990, **B21,** 25 and 813; *Nuovo Cimento* 1992, **105A,** 77; [hep-th/0207210].

[10] Hubble, E.P., and Humason, M. *ApJ,* 1931, **74,** 43.

[11] Penzias, A.A., Wilson, R.W. *ApJ,* 1965, **142,** 419.

[12] Smoot, G.F. *ApJ (Letters),* 1992, **396,** L1.

[13] LaViolette, P.A. *ApJ* 1986, **301,** 544.

[14] Ivanov, M.A. Another origin of cosmological redshifts [astro-ph-0405083].

[15] Riess, A.G. et al. Type Ia Supernova Discoveries at $z > 1$ From the Hubble Space Telescope ... [astro-ph/0402512] (to appear in *ApJ,* 2004).

[16] Ivanov, M.A. Contribution to the Quantum Electronics and Laser Science Conference (QELS'95), May 21-26, 1995, Baltimore, USA; paper number: QThG1.

[17] Murphy, M. T. et al. Time evolution of the fine structure constant [astro-ph/0209488].

[18] Srianand, R. et al. *Nature* 2000, **408,** 931; [astro-ph/0012222].

[19] Ivanov, M.A. Contribution to The sixth international symposium "Frontiers of Fundamental and Computational Physics" (FFP6), 26-29 September 2004, Udine, Italy; [astro-ph/0409631].

[20] Choi, S.Y. et al. *Phys. Rev.* 1993, **D48,** 2953; **D48,** R5465; 1995, **D51,** 2751.

[21] Ivanov, M.A. Contribution to The Tenth Marcel Grossmann Meeting (MG10), 20-26 July 2003, Rio de Janeiro, Brazil; [gr-qc/0307093].

[22] Ivanov, M.A. Contribution to the Conference "Thinking, Observing and Mining the Universe" (Thinking2003), 22-27 Sep 2003, Sorrento, Italy; [astro-ph/0309566].

[23] Hawking, S.W. *Commun. Math. Phys.* 1975, **43,** 199.

[24] Tolman, R.C. *Relativity, Thermodynamics and Cosmology.* Clarendon Press: Oxford, 1969.

[25] Conklin, E.K. *Nature* 1969, **222,** 971.

[26] Anderson, J.D., Nieto, M.M., Turyshev, S.G. *Int. J. Mod. Phys.* 2002, **D11,** 1545; [gr-qc/0205059 v3].

[27] Turyshev, S.G. et al. XXXIV-th Rencontres de Moriond Meeting on Gravitational Waves and Experimental Gravity. Les Arcs, Savoi, France (January 23-30, 1999); [gr-qc/9903024 v2].

[28] Freedman, W. L. et al. *ApJ* 2001, **553,** 47.

[29] Filippenko, A.V. [astro-ph/0410609]; to be published in *White Dwarfs: Probes of Galactic Structure and Cosmology,* Sion, E. M., Shipman, H. L., Vennes, S., Eds.; Kluwer: Dordrecht.

[30] Willick, J.A., Puneet Batra. *ApJ* 2001, **548,** 564.

[31] Korn, G.A., Korn, T.M. *Mathematical Handbook,* Second Edition, McGraw-Hill Book Co.: New York, 1968, Table 1.10-2.

[32] Ivanov, M.A. In *Searches for a mechanism of gravity,* Ivanov, M.A., Savrov, L.A., Eds.; Nikolaev, Yu.A. Publisher: Nizhny Novgorod, 2004, pp 266-273 (in Russian).

[33] Ivanov, M.A. A quantum gravitational model of redshifts [astro-ph/0409111].

[34] Ivanov, M.A. Model of graviton-dusty universe [gr-qc/0107047].

[35] Fan, X. et al. *AJ* 2000, **120,** 1167.

[36] van der Marel, R. P. In *New Light on Galaxy Evolution,* Proc. IAU Symp. **171,** Heidelberg, June 1995, R. Bender and R.L. Davies, eds.; Kluwer Academic Publishers.

[37] Kleijn, G. A. V. et al. *AJ* 2000, **120,** 1221.

[38] Peterson, B. M., Wandel, A. *ApJ* 2000, **540,** L13.

[39] Di Matteo, T., Carilli, C.L., Fabian, A.C. *ApJ* 2001, **547,** 731.

[40] Kenji Bekki. *ApJ* 2000, **540,** L79.

[41] Milgrom, M. In the *Proceedings of the II Int. Workshop on the Identification of Dark Matter*, Buxton, England, 1998; World Scientific: Singapore, 1999; [astro-ph/9810302].

[42] Shapiro, I.L., Sola, J., Stefancic, H. Running G and Λ at low energies from physics at M_X: possible cosmological and astrophysical implications [hep-ph/0410095].

[43] Reuter, M., Weyer, H. Running Newton Constant, Improved Gravitational Actions, and Galaxy Rotation Curves [hep-th/0410117].

[44] Ivanov, M.A. In *the Proceedings of The 14th Workshop on General Relativity and Gravitation* (JGRG14), Kyoto, Japan, Nov 29 - Dec 3 2004; Hikida, W., Sasaki, M., Tanaka,T., Nakamura, T., Eds. Kyoto University: 2004; [gr-qc/0410076].

[45] Adler, S.L. et al. *Phys. Rev.* 1976, **D14,** 359.

[46] Adler, S.L. *Phys. Rev.* 1976, **D14,** 379.

[47] Efimov, G.V. *Non-local interactions of quantized fields.* Nauka: Moscow, 1968.

[48] Feynman, R. *The character of physical law.* Cox and Wyman Ltd.: London, 1965.

[49] Sakharov, A.D. *JETP* 1980, **79,** 689.

[50] Penrose. R. In the Einstein Survey *General Relativity,* Hawking, S.W., Israel, W., Eds.; Cambridge University Press: 1979.

In: Focus on Quantum Gravity Research
Editor: David C. Moore, pp. 121-169

ISBN 1-59454-660-6

Chapter 4

Quantum Gauge Theory of Gravity

Ning Wu
Institute of High Energy Physics, P.O.Box 918-1, Beijing 100039, P.R.China
E-mail address: wuning@mail.ihep.ac.cn

The quantum gravity is formulated based on the principle of local gauge invariance. The model discussed in this paper has local gravitational gauge symmetry and gravitational field is represented by gauge field. Then gravitational gauge interactions of pure gravitational gauge field is studied. It is found that the field equation of gravitational gauge field is just the Einstein's field equation. Path integral quantization of the theory is discussed in the paper. The quantum gauge gravity discussed in this paper is a perturbatively renormalizable quantum gravity, which is one of its most important advantage. In leading order approximation, the gravitational gauge theory gives out classical Newton's theory of gravity. In classical level, it gives out Einstein's general relativity. Another important advantage of the quantum gauge theory of gravity is that it can explain both classical tests of gravity and quantum effects of gravitational interactions, such as gravitational phase effects found in COW experiments and gravitational shielding effects found in Podkletnov experiments.

PACS Numbers: 11.15.-q, 04.60.-m, 04.20.Cv, 11.10.Gh
Keywords: gauge field, quantum gravity, general relativity, renormalization

4.1 Introduction

Gravity is an ancient topic in science, it is one kind of interaction which is known earliest by mankind. In ancient times, human has known the existence of weight. Now, we known that it is the gravity between an object and earth. Gravity exists everywhere in nature, it is met by us most commonly in our everyday life. It is the most familiar interactions to us. But it is the most mysterious and most profound interactions of nature.

Gravity is one of the key topics of physics. In the seventeenth century, Johannes Kepler found the famous Kepler's law. Later, Issac Newton explain these laws as the result that the gravity between the sun and planet obeys the inverse square law[1]. This study results in the establishment of the great Newton's theory of universal gravitation and Newtonian mechanics, which is one of the greatest milestone in the history of science. Newton's classical theory of gravity can explain the orbital motion of planet in solar system, the motion of the moon around the earth, and tide phenomenon in the earth. Based on Newton's theory of gravity, John Crouch Adams and Urbain Jean Joseph Le Verrier predict the existence and position of Neptune. The discovery of Neptune shortly thereafter was perhaps the most splendid verification of Newton's theory of gravity.

Newtonian mechanics and Newton's theory of universal gravitation met with a brilliant series of successes. It is known that Newton's theory is based on the concept of absolute space-time. Though the concept of absolute space-time is consistent with our experience, it is frequently attacked by many philosophers and physicists. Such attacks finally lead to a great revolution of physics: the foundation of the theory of relativity by Albert Einstein[2, 3]. A new marvelous concept of space-time, which is completely different from our ordinary experience, is established in the theory of relativity. In the general theory of relativity, gravity is treated in a novel manner: it is considered as the geometry of space-time. According to the general theory of relativity, we are living in a curved space-time. The existence of matter determines the curvature of space-time, and the effects of the curved space-time is just the gravity we usually feel. In this point of view, gravity and the electromagnetic force have some essential differences: the essence of gravity is the geometry of space-time, and the gravitational force we feel is only an effect of curved space-time; while the essence of electromagnetic force is a kind of physical interaction force which is transmitted by photon. In other words, the essence of gravity is geometry, while the essence of electromagnetic force is physics. We know that the general theory of relativity meet with a brilliant series of successes: it successfully explain the excess precession of Mercury, deflection of light by the sun and radar echo delay. Now, the application of the theory of relativity is becoming wider and wider, it is the basis of modern astrophysics and modern cosmology.

In general relativity, a new concept of space-time is established, and gravity is considered to be geometry of space-time, which is the essential distinction between Newton's classical theory of gravity and Einstein's general relativity. It is known that, in Newton's classical theory of gravity, gravity is considered to be a kind of physical interactions, space-time, which is the background that all matters exist and evolve, is always flat, and gravity can not change the structure of space-time. But in general relativity, gravity is no longer considered to be a kind of physical interactions. It is geometry of space-time, or it is a part of the structure of space-time. Gravity and space-time are essentially the same thing. So now, a fundamental issue is raised: Is the essence of gravity geometry of space-time or a kind of physical interactions?

One of the biggest revolution in human kind in the last century is the foundation of quantum theory. The quantum hypothesis was first introduced into physics by Max Plank in 1900. In 1916, Albert Einstein points out that quantum effects must lead to modifications in the theory of general relativity[4]. Soon after the foundation of quantum mechanics, physicists try to found a theory that could describe the quantum behavior of the full gravitational field. In the 80 years attempts in the twentieth century, physicists have found two theories based on quantum mechanics that attempt to unify general relativity and quantum mechanics, one is canonical quantum gravity and another is superstring theory. For quantum field theory, there are different kinds of mathematical infinities that naturally occur in quantum descriptions of fields. These infinities should be removed by the technique of perturbative renormalization. However, the perturbative renormalization does not work for in canonical quantum gravity. In superstring theory, in order to make perturbative renormalization to work, physicists have to introduce six extra dimensions. But up to now, none of the extra dimensions have been observed. To found a consistent theory that can unify general relativity and quantum mechanics is a long dream for physicists.

The "relativity revolution" and the "quantum revolution" are among the greatest successes of twentieth century physics, yet two theories appears to be fundamentally incompatible. General relativity remains a purely classical theory which describes the geometry of space and time as smooth and continuous, on the contrary, quantum mechanics divides everything into discrete quanta. The underlying theoretical incompatibility between two theories arises from the way that they treat the geometry of space and time. This situation makes some physicists still wonder whether quantum theory is a truly fundamental theory of Nature, or just a convenient description of some aspects of the microscopic world. Some physicists even consider the twentieth century as the century of the incomplete revolution. To set up a consistent quantum theory of gravity is considered to be the last challenge of quantum theory[5, 6]. In other words, combining general relativity with quantum mechanics is considered to be the last hurdle to be overcome in the "quantum revolution".

In 1921, H.Weyl introduced the concept of gauge transformation into physics[7, 8], which is one of the most important concepts in modern physics, though his original theory is not successful. Later, V.Fock, H.Weyl and W.Pauli found that quantum electrodynamics is a gauge invariant theory[9, 10, 11]. In 1954, Yang and Mills proposed non-Abelian gauge field theory[12]. This theory was soon applied to elementary particle physics. Unified electroweak theory[13, 14, 15] and quantum chromodynamics are all based on gauge field theory. The predictions of unified electroweak theory have been confirmed in a large number of experiments, and the intermediate gauge bosons $W^{\pm}$ and Z^0 which are predicted by unified electroweak model are also found in experiments. The $U(1)$ part of the unified electroweak model, quantum electrodynamics, now becomes one of the most accurate and best-tested theories of modern physics. All these achievements of gauge field theories suggest that gauge field theory is a fundamental theory that describes fundamental interactions in Nature. Now, it is generally believed that four kinds of fundamental interactions in Nature are all gauge interactions and they can be described by gauge field theory. From theoretical point of view, the principle of local gauge invariance plays a fundamental role in particle's interaction theory.

There are two way to establish the theory of quantum gravity. One way is to quantize general relativity. In this case, the starting point of the theory is general relativity. Another way is to pursue the idea of the unification of fundamental interactions. The idea of the unification of fundamental interactions was first proposed by Albert Einstein, but the real unification of fundamental interactions is accomplished in the frame work of quantum field theory. Quantum field theory is one of the most splendid achievement in the twentieth century physics. It is a powerful theory to describe fundamental interactions of elementary particles. Now, it is the framework for the realization of idea of the unification of fundamental interactions. We know that, there are four kinds of fundamental interactions: strong interactions, electromagnetic interactions, weak interactions and gravitational interactions. According general relativity, gravity is geometry of space-time, which is essentially different from other three kinds of fundamental interaction. In this point of view, it is difficult to unify gravity with other fundamental interactions, for they have different nature. But according to the idea of the grand unification, all fundamental interactions should have the same origin and the same essence. We know that strong interactions, electromagnetic interactions and weak interactions are all gauge interactions, gauge is their common essence, and the sound theory to describe them is gauge field theory. The great successes of gauge field theory enlighten us that the essence of gravity should also be gauge. Therefore, to establish quantum gauge theory of gravity is a direct inspiration from the idea of grand unification.

The essential distinction between quantum gauge theory of gravity and general relativity is that they have different transcendental principle. The transcendental principles of general relativity are equivalence principle and the the principle of general covariance, while the transcendental principle of quantum gauge theory of gravity is gauge principle, which states that the source of interactions is just the conserved charge of the corresponding global symmetry, gauge interactions need to be introduced when localize the global symmetry, and laws of gauge interactions are determined by local gauge symmetry. Different transcendental principles lead to that general relativity and quantum gauge theory of gravity have completely different mathematical expressions. Another essential distinction between quantum gauge theory of gravity and general relativity is that they have different concepts on space-time and gravity. In general relativity, gravity is considered to be geometry of space-time, and gravitational force is just an effect of curved space-time. But in quantum gauge theory of gravity, gravitational force is considered to be a kind of physical interactions, and the concept of curved space-time is only a convenient and equivalent description of the classical effects of gravitational interactions.

Because of physics-geometry duality, we need first select a picture of gravity when we perform any theoretical calculations in quantum gauge theory of gravity. There are mainly two pictures of gravity: one is physics picture, and another is geometry picture. The basic picture of gauge theory of gravity is the physics picture, and quantum gauge theory of gravity is established in the physics picture of gravity. As we have states above, in the physics picture of gravity, space-time is always flat, and gravitational force is transmitted by gravitational gauge field which exists in the flat space-time. However, we can also set up the geometry picture of gauge theory of gravity. That is, by performing a simple transformation, we can transform gauge theory of gravity from physics picture to geometry picture. When we go into the geometry picture of gravity, we need to define a equivalent "space-time metric", which is the equivalent of the metric tensor in the general relativity. This equivalent "space-time metric tensor" is defined in terms of gravitational gauge field. So, now things are very clear: gravity is put into the structure of space-time, "space-time" becomes "curved", and in this equivalent "curved space-time", gravity outwardly disappears. We find that, based on this equivalent "space-time metric tensor", we can formulate gauge theory of gravity again, which is just the gauge theory of gravity in geometry picture. In geometry picture, all mathematical formula are expressed in term of the equivalent "space-time metric tensor", and gravitational gauge field does not outwardly appear in any mathematical expressions. It means that all effects of gravitational interactions are contained in the structure of the "curved space-time". An important result of this transformation is that the field equation of gravitational gauge field in physics picture is changed into the Einstein's field equation in geometry picture. In fact, after transforming into geometry picture, gauge theory of gravity can give out all basic physical equations in general relativity. In other words, after transforming into geometry picture, gauge theory of gravity can "return" to general relativity. Here, the word "return" only means that it can mathematically return to general relativity and reproduce all physical results in general relativity. In this meaning, gauge theory of gravity can be considered to be a development of general relativity and a consistent unification of general relativity and quantum theory. General relativity can only describe classical motion of an object in gravity, but gauge theory of gravity can describe not only the classical motion of an object in gravity, but also evolution of a quantum gravitational field or a quantum states in gravity. For classical problems, gauge theory of gravity can "return" to general relativity after a transformation into geometry picture. But, gauge theory of gravity can not physically return to general relativity, for they have different transcendental principles and different concepts on gravity. For quantum problems, especially when there is quantum gravitational gauge field, the geometry picture is not valid and the equivalent "curved space-time" can not be correctly erected logically.

As we have stated above, the transcendental principle of gauge theory of gravity is gauge principle, and and transcendental principles of general relativity is equivalence principle and the principle of general covariance. Because gauge theory of gravity can "return" to general relativity after transforming into geometry picture, we have two different way to establish general relativity. One way is the Einstein's way, that is to start from the equivalence principle and the principle of general covariance. Another way is to start from gauge principle. But, gauge principle is not equivalent to the equivalence principle and the principle of general covariance. So, what should be the transcendental principle of the theory of gravity? For classical problems, equivalence principle and gauge principle can give out the same physical results. So, in order to answer the question that which one should be the transcendental principle of the theory of gravity, we need to study quantum problems. When we study quantum problems, we find that the spirit of equivalence principle is violated, while the spirit of gauge principle is conserved. The original expressions of the equivalence principle is that the gravitational mass of matter is equivalent to its inertial mass. Therefore, the mass term in Newton's equation of motion can be cancelled, and the orbit of a point particle in gravity does not depend on its mass. So, even if the initial position and initial velocity are the same, no matter how weight they are, these particles will have completely the same orbit. (Now, this expression is call weak equivalence principle.) It is this property that lead to the idea of the geometrization of gravity. But, this property is violated in quantum theory. The mass parameter in the Schrodinger equation with gravity potential can not be cancelled in any case.

Therefore, quantum states in gravitational field and its evolution are doomed to be dependent on its mass. In other words, the evolution of a quantum state in gravitational field depends on the mass of the particle, which violates weak equivalence principle. Or say that it violates the original spirit of the equivalence principle.

With regard to the above contradiction between quantum theory and weak equivalence principle, there are two distinct point of view striking opposite to each other. One states that weak equivalence principle holds in all cases, while Schrodinger equation is not applicable to the gravity dominated system. Another states that Schrodinger equation is applicable to the gravity dominated system also, but weak equivalence principle is violated in a quantum system. Experiments find that the latter viewpoint is correct. In 1970s, R. Covella, A.W.Overhauser and S.A.Werner performed the famous gravitational phase experiment, which is now called as COW experiment. Experimental result is highly consistent with the prediction given by the corresponding Schrodinger equation. COW experiments clearly manifest that Schrodinger equation holds for the gravity dominated system, and weak equivalence principle does not hold for a quantum system. COW experiment is an important experiment which help us to understand the essence of gravity and to determine the transcendental principles of a fundamental theory of gravity.

Gauge treatment of gravity was suggested immediately after the gauge theory birth itself[16, 17, 18, 19, 20, 21]. In the traditional gauge treatment of gravity, Lorentz group is localized, and the gravitational field is not represented by gauge potential. It is represented by metric field. There are also some other attempts to use Yang-Mills theory to reformulate quantum gravity[22, 23, 24, 25, 26]. In these new approaches, the importance of gauge fields is emphasized. Some physicists also try to use gauge potential to represent gravitational field, some suggest that we should pay more attention on translation group.

In 2001, quantum gauge theory of gravity is proposed By N.Wu [27, 28, 29, 30, 31, 32]. In quantum gauge theory of gravity, the gravitational interactions are treated as a kind of fundamental interactions in flat Minkowski space-time, not space-time geometry. So, in quantum gauge theory of gravity, the basic physical space-time is always flat. In other words, it is formulated in the physics picture of gravity. In this approach, the unification of fundamental interactions can be formulated in a simple and beautiful way[33, 34, 35]. If we use the mass generation mechanism which is proposed in literature [36], we can propose a new theory on gravity which contains massive graviton and the introduction of massive graviton does not affect the strict local gravitational gauge symmetry of the Lagrangian and does not affect the traditional long-range gravitational force[37]. The existence of massive graviton will help us to understand the possible origin of dark energy and dark matter in the Universe. Because in classical level, quantum gauge gravity just gives out Einstein's general relativity, it gives out the same theoretical predictions for classical tests of gravity as those of general relativity. Quantum gauge theory of gravity can also be used to explain quantum effects of gravitational interactions, such as gravitational phase effects in COW experiments[38, 39, 40], and the gravitational shielding effects in Podkletnov experiments[41, 42, 43, 44].

4.2 Gauge Principle and Gravitational Gauge Group

As we have stated above, the foundations of quantum gauge theory of gravity is gauge principle. Gauge principle can be formulated as follows: Any kind of fundamental interactions has a gauge symmetry corresponding to it; the gauge symmetry completely determines the forms of interactions. In principle, the gauge principle has the following four different contents:

1. **Conservation Law:** the global gauge symmetry gives out conserved current and conserved charge;

2. **Interactions:** the requirement of the local gauge symmetry requires introduction of gauge field or a set of gauge fields; the interactions between gauge fields and matter fields are completely determined by the requirement of local gauge symmetry; these gauge fields transmit the corresponding interactions;

3. **Source:** qualitative speaking, the conserved charge given by global gauge symmetry is the source of gauge field; for non-Abel gauge field, gauge field is also the source of itself;

4. **Quantum Transformation:** the conserved charges given by global gauge symmetry become generators of quantum gauge transformation after quantization, and for this kind of interactions, the quantum transformation can not have generators other than quantum conserved charges given by global gauge symmetry.

It is known that conservation law is the objective origin of gauge symmetry, so gauge symmetry is the exterior exhibition of the interior conservation law. The conservation law is the law that exists in fundamental interactions, so fundamental interactions are the logic precondition and foundation of the conservation law. Gauge principle tells us how to study conservation law and fundamental interactions through symmetry. Gauge principle is one of the most important transcendental fundamental principles for all kinds of fundamental interactions in Nature; it reveals the common nature of all kinds of fundamental interactions in Nature. It is also the transcendental foundation of the quantum gravity which is formulated in this paper. It will help us to select the gauge symmetry for quantum gravitational theory and help us to determine the Lagrangian of the system. In a meaning, we can say that without gauge principle, we can not set up the renormalizable quantum gauge theory of gravity.

The essence of gauge is to reveal the essential and positive connection among conservation, symmetry and interactions. Grasping this connection, we will grasp the crux of the interaction theory, which is to determine the law of interactions through conservation and symmetry. Therefore, the key to the work of the establishment of quantum gauge theory of gravity is to determine the symmetry and the conserved charge of gravitational interactions, which is the starting point of the whole theory. It is know that, in the traditional gauge theory of gravity, Lorentz group is localized. We will not follow this way. Now, we use gauge principle to determine which group is the exact group for quantum gauge theory of gravity.

Some of the most important properties of gravity can be obtained by studying classical Newton's gravity. In the classical Newton's theory of gravity, the gravitational force between two point objects is given by:

$$f = G\frac{m_1 m_2}{r^2} \tag{4.2.1}$$

with m_1 and m_2 masses of two objects, r the distance between two objects. So, gravity is proportional to the masses of both objects, in other words, mass is the source of gravitational field. In general relativity, the Einstein's field equation is the equation which gives out the relation between energy-momentum tensor and space-time curvature, which is essentially the relation between energy-momentum tensor and gravitational field. In the Einstein's field equation, energy-momentum is treated as the source of gravity. This points of view is inherited in the quantum gauge theory of gravity. In other words, the starting point of the quantum gauge theory of gravity is that the energy-momentum is the source of gravitational field. According to rule 3 and rule 1 of gauge principle, we know that, energy-momentum should be the conserved charges of the corresponding global symmetry, which is just the symmetry for gravity. It is known that energy-momentum is the conserved charge of global space-time translation, the corresponding conserved current is energy-momentum tensor. Therefore, the global space-time translation is the global gravitational gauge transformation.

Suppose that there is an arbitrary function $\phi(x)$ of space-time coordinates x^μ. The global space-time translation is:

$$x^\mu \to x'^\mu = x^\mu + \epsilon^\mu. \tag{4.2.2}$$

The corresponding transformation for function $\phi(x)$ is

$$\phi(x) \to \phi'(x') = \phi(x) = \phi(x' - \epsilon). \tag{4.2.3}$$

According to Taylor series expansion, we have:

$$\phi(x-\epsilon) = \left(1+\sum_{n=1}^{\infty}\frac{(-1)^n}{n!}\epsilon^{\mu_1}\cdots\epsilon^{\mu^n}\partial_{\mu_1}\cdots\partial_{\mu_n}\right)\phi(x), \tag{4.2.4}$$

where

$$\partial_{\mu_i} = \frac{\partial}{\partial x^{\mu_i}}. \tag{4.2.5}$$

Let's define a special exponential operation here. Define

$$E^{a^\mu\cdot b_\mu} \triangleq 1+\sum_{n=1}^{\infty}\frac{1}{n!}a^{\mu_1}\cdots a^{\mu_n}\cdot b_{\mu_1}\cdots b_{\mu_n}. \tag{4.2.6}$$

This definition is quite different from that of ordinary exponential function. In general cases, operators a^μ and b_μ do not commutate each other, so

$$E^{a^\mu\cdot b_\mu} \neq E^{b_\mu\cdot a^\mu}, \tag{4.2.7}$$

$$E^{a^\mu\cdot b_\mu} \neq e^{a^\mu\cdot b_\mu}, \tag{4.2.8}$$

where $e^{a^\mu\cdot b_\mu}$ is the ordinary exponential function whose definition is

$$e^{a^\mu\cdot b_\mu} \equiv 1+\sum_{n=1}^{\infty}\frac{1}{n!}(a^{\mu_1}\cdot b_{\mu_1})\cdots(a_{\mu_n}\cdot b_{\mu_n}). \tag{4.2.9}$$

If operators a^μ and b_μ commutate each other, we will have

$$E^{a^\mu\cdot b_\mu} = E^{b_\mu\cdot a^\mu}, \tag{4.2.10}$$

$$E^{a^\mu\cdot b_\mu} = e^{a^\mu\cdot b_\mu}. \tag{4.2.11}$$

The translation operator $\hat{U}_\epsilon$ can be defined by

$$\hat{U}_\epsilon \equiv 1+\sum_{n=1}^{\infty}\frac{(-1)^n}{n!}\epsilon^{\mu_1}\cdots\epsilon^{\mu_n}\partial_{\mu_1}\cdots\partial_{\mu_n}. \tag{4.2.12}$$

Then we have

$$\phi(x-\epsilon) = (\hat{U}_\epsilon\phi(x)). \tag{4.2.13}$$

In order to have a good form which is similar to ordinary gauge transformation operators, the form of $\hat{U}_\epsilon$ can also be written as

$$\hat{U}_\epsilon = E^{-i\epsilon^\mu\cdot\hat{P}_\mu}, \tag{4.2.14}$$

where

$$\hat{P}_\mu = -i\frac{\partial}{\partial x^\mu}. \tag{4.2.15}$$

$\hat{P}_\mu$ is just the energy-momentum operator in space-time coordinate space. In the definition of $\hat{U}_\epsilon$ given by eq.(4.2.14), ϵ^μ can be independent of space-time coordinate or a function of space-time coordinate, in a ward, it can be an arbitrary function of space time coordinate x^μ.

Some operation properties of the translation operator $\hat{U}_\epsilon$ are summarized below.

1. Operator $\hat{U}_\epsilon$ translate the space-time point of a field from x to $x-\epsilon$,

$$\phi(x-\epsilon) = (\hat{U}_\epsilon\phi(x)), \tag{4.2.16}$$

where ϵ^μ can be any function of space-time coordinate. This relation can also be regarded as the definition of the translation operator $\hat{U}_\epsilon$.

2. Suppose that $\phi_1(x)$ and $\phi_2(x)$ are two arbitrary functions of space-time coordinate, then we have

$$\left(\hat{U}_\epsilon(\phi_1(x)\cdot\phi_2(x))\right) = \left(\hat{U}_\epsilon\phi_1(x)\right)\cdot\left(\hat{U}_\epsilon\phi_2(x)\right) \tag{4.2.17}$$

3. Suppose that $\hat{U}_{\epsilon_1}$ and $\hat{U}_{\epsilon_2}$ are two arbitrary translation operators, define

$$\hat{U}_{\epsilon 3} = \hat{U}_{\epsilon_2}\cdot\hat{U}_{\epsilon_1}, \tag{4.2.18}$$

then,

$$\epsilon_3^\mu(x) = \epsilon_2^\mu(x) + \epsilon_1^\mu(x-\epsilon_2(x)). \tag{4.2.19}$$

This property means that the product of two translation operator satisfy closure property, which is one of the conditions that any group must satisfy.

4. Suppose that $\hat{U}_\epsilon$ is a non-singular translation operator, then

$$\hat{U}_\epsilon^{-1} = E^{i\epsilon^\mu(f(x))\cdot\hat{P}_\mu}, \tag{4.2.20}$$

where $f(x)$ is defined by the following relations:

$$f(x-\epsilon(x)) = x. \tag{4.2.21}$$

$\hat{U}_\epsilon^{-1}$ is the inverse operator of $\hat{U}_\epsilon$, so

$$\hat{U}_\epsilon^{-1}\hat{U}_\epsilon = \hat{U}_\epsilon\hat{U}_\epsilon^{-1} = \mathbf{1}, \tag{4.2.22}$$

where $\mathbf{1}$ is the unit element of the gravitational gauge group.

5. The product operation of translation also satisfies associative law. Suppose that $\hat{U}_{\epsilon 1}$, $\hat{U}_{\epsilon 2}$ and $\hat{U}_{\epsilon 3}$ are three arbitrary translation operators, then

$$\hat{U}_{\epsilon 3} \cdot (\hat{U}_{\epsilon_2} \cdot \hat{U}_{\epsilon_1}) = (\hat{U}_{\epsilon 3} \cdot \hat{U}_{\epsilon_2}) \cdot \hat{U}_{\epsilon_1}. \tag{4.2.23}$$

6. Suppose that $\hat{U}_\epsilon$ is an arbitrary translation operator and $\phi(x)$ is an arbitrary function of space-time coordinate, then

$$\hat{U}_\epsilon \phi(x) \hat{U}_\epsilon^{-1} = (\hat{U}_\epsilon \phi(x)) = \phi(x - \epsilon(x)). \tag{4.2.24}$$

This relation is quite useful in the discussions of gravitational gauge symmetry.

7. Suppose that $\hat{U}_\epsilon$ is an arbitrary translation operator. Define

$$\Lambda^\alpha{}_\beta = \frac{\partial x^\alpha}{\partial (x - \epsilon(x))^\beta}, \tag{4.2.25}$$

$$\Lambda_\alpha{}^\beta = \frac{\partial (x - \epsilon(x))^\beta}{\partial x^\alpha}. \tag{4.2.26}$$

They satisfy

$$\Lambda_\alpha{}^\mu \Lambda^\alpha{}_\nu = \delta^\mu_\nu, \tag{4.2.27}$$

$$\Lambda_\mu{}^\alpha \Lambda^\nu{}_\alpha = \delta^\nu_\mu. \tag{4.2.28}$$

Then we have following relations:

$$\hat{U}_\epsilon \hat{P}_\alpha \hat{U}_\epsilon^{-1} = \Lambda^\beta{}_\alpha \hat{P}_\beta, \tag{4.2.29}$$

$$\hat{U}_\epsilon \mathrm{d}x^\alpha \hat{U}_\epsilon^{-1} = \Lambda_\beta{}^\alpha \mathrm{d}x^\beta, \tag{4.2.30}$$

which give out the the transformation laws of $\hat{P}_\alpha$ and $\mathrm{d}x^\alpha$ under local gravitational gauge transformations.

Gravitational gauge group (GGG) is a transformation group which consists of all non-singular translation operators $\hat{U}_\epsilon$. We can easily see that gravitational gauge group is indeed a group, for

1. the product of two arbitrary non-singular translation operators is also a non-singular translation operator, which is also an element of the gravitational gauge group. So, the product of the group satisfies closure property which is expressed in eq(4.2.18);

2. the product of the gravitational gauge group also satisfies the associative law which is expressed in eq(4.2.23);

3. the gravitational gauge group has its unit element $\mathbf{1}$, which satisfies

$$\mathbf{1} \cdot \hat{U}_\epsilon = \hat{U}_\epsilon \cdot \mathbf{1} = \hat{U}_\epsilon; \tag{4.2.31}$$

4. every non-singular element $\hat{U}_\epsilon$ has its inverse element which is given by eqs(4.2.20) and (4.2.22).

According to gauge principle, the gravitational gauge group is the symmetry of gravitational interactions. The global invariance of gravitational gauge transformation will give out conserved charges which is just the ordinary inertial energy-momentum; the requirement of local gravitational gauge invariance needs introducing gravitational gauge field, and gravitational interactions are completely determined by the local gravitational gauge invariance.

The generators of global gravitational gauge group is just the energy-momentum operators $\hat{P}_\alpha$. This is required by gauge principle. It can also be seen from the form of infinitesimal transformations. Suppose that ϵ is an infinitesimal quantity, then we have

$$\hat{U}_\epsilon \simeq 1 - i\epsilon^\alpha \hat{P}_\alpha. \tag{4.2.32}$$

Therefore,

$$i\frac{\partial \hat{U}_\epsilon}{\partial \epsilon^\alpha} |_{\epsilon=0} \tag{4.2.33}$$

gives out generators $\hat{P}_\alpha$ of gravitational gauge group. It is known that generators of gravitational gauge group commute each other

$$[\hat{P}_\alpha \ , \ \hat{P}_\beta] = 0. \tag{4.2.34}$$

However, the commutation property of generators does not mean that gravitational gauge group is an Abelian group, because two general elements of gravitational gauge group do not commute:

$$[\hat{U}_{\epsilon_1} \ , \ \hat{U}_{\epsilon_2}] \neq 0. \tag{4.2.35}$$

Gravitational gauge group is a kind of non-Abelian gauge group. The non-Able nature of gravitational gauge group will cause self-interactions of gravitational gauge field.

Suppose that $\phi(x)$ is an arbitrary scalar field. Its gravitational gauge transformation is

$$\phi(x) \to \phi'(x) = (\hat{U}_\epsilon \phi(x)). \tag{4.2.36}$$

Similar to ordinary $SU(N)$ non-Abelian gauge field theory, there are two kinds of scalars. For example, in chiral perturbative theory, the ordinary π mesons are scalar fields, but they are vector fields in isospin space. Similar case exists in gravitational gauge field theory. A Lorentz scalar can be a scalar or a vector or a tensor in the space of gravitational gauge group. For the sake of simplicity, we call this space gravitational gauge space or gravitational gauge Lie algebra. If $\phi(x)$ is a scalar in the gravitational gauge space, we just simply denote it as $\phi(x)$ in gauge group space. If it is a vector in the gravitational gauge space, it can be expanded in the gravitational gauge space in the following way:

$$\phi(x) = \phi^{\alpha}(x) \cdot \hat{P}_{\alpha}. \tag{4.2.37}$$

The transformation of component field is

$$\phi^{\alpha}(x) \to \phi'^{\alpha}(x) = \Lambda^{\alpha}{}_{\beta} \hat{U}_{\epsilon} \phi^{\beta}(x) \hat{U}_{\epsilon}^{-1}. \tag{4.2.38}$$

Please note that the α index is not a Lorentz index, it is just a group index. For gravitation gauge group, it is quite special that a group index looks like a Lorentz index. This will cause some fundamental changes on quantum gravity. Lorentz scalar $\phi(x)$ can also be a tensor in gauge group space. Suppose that it is a nth order tensor in gauge group space, then it can be expanded as

$$\phi(x) = \phi^{\alpha_1 \cdots \alpha_n}(x) \cdot \hat{P}_{\alpha_1} \cdots \hat{P}_{\alpha_n}. \tag{4.2.39}$$

The transformation of component field is

$$\phi^{\alpha_1 \cdots \alpha_n}(x) \to \phi'^{\alpha_1 \cdots \alpha_n}(x) = \Lambda^{\alpha_1}{}_{\beta_1} \cdots \Lambda^{\alpha_n}{}_{\beta_n} \hat{U}_{\epsilon} \phi^{\beta_1 \cdots \beta_n}(x) \hat{U}_{\epsilon}^{-1}. \tag{4.2.40}$$

If $\phi(x)$ is a spinor field, the above discussion is also valid. That is, a spinor can also be a scalar or a vector or a tensor in the space of gravitational gauge group. The gravitational gauge transformations of the component fields are also given by eqs.(4.2.36-4.2.40). There is no transformations in spinor space, which is different from the Lorentz transformation of a spinor.

Suppose that $A_{\mu}(x)$ is an arbitrary vector field. Here, the index μ is a Lorentz index. Its gravitational gauge transformation is:

$$A_{\mu}(x) \to A'_{\mu}(x) = (\hat{U}_{\epsilon} A_{\mu}(x)). \tag{4.2.41}$$

There is no rotation in the space of Lorentz index μ, while in the general coordinates transformations in general relativity, there is rotation in the space of Lorentz index μ. The reason is that gravitational gauge transformation is a kind of system transformation, while in general relativity, the general coordinates transformation is a kind of coordinates transformation. If $A_{\mu}(x)$ is a scalar in the gravitational gauge space, eq(4.2.41) is all for its gauge transformation. If $A_{\mu}(x)$ is a vector in the gravitational gauge space, it can be expanded as:

$$A_{\mu}(x) = A^{\alpha}_{\mu}(x) \cdot \hat{P}_{\alpha}. \tag{4.2.42}$$

The transformation of component field is

$$A_\mu^\alpha(x) \to A_\mu^{\prime\alpha}(x) = \Lambda^\alpha{}_\beta \hat{U}_\epsilon A_\mu{}^\beta(x) \hat{U}_\epsilon^{-1}. \tag{4.2.43}$$

If $A_\mu(x)$ is a nth order tensor in the gravitational gauge space, then

$$A_\mu(x) = A_\mu^{\alpha_1 \cdots \alpha_n}(x) \cdot \hat{P}_{\alpha_1} \cdots \hat{P}_{\alpha_n}. \tag{4.2.44}$$

The transformation of component fields is

$$A_\mu^{\alpha_1 \cdots \alpha_n}(x) \to A_\mu^{\prime\alpha_1 \cdots \alpha_n}(x) = \Lambda^{\alpha_1}{}_{\beta_1} \cdots \Lambda^{\alpha_n}{}_{\beta_n} \hat{U}_\epsilon A_\mu^{\beta_1 \cdots \beta_n}(x) \hat{U}_\epsilon^{-1}. \tag{4.2.45}$$

Therefore, under gravitational gauge transformations, the behavior of a group index is quite different from that of a Lorentz index. However, they have the same behavior in global Lorentz transformations.

$\eta^{\mu\nu}$ is a second order Lorentz tensor, but it is a scalar in gravitational gauge space. It is the metric of the coordinate space. A Lorentz index can be raised or descended by this metric tensor. In a Minkowski space-time, it is selected to be:

$$\left\{ \begin{array}{lcl} \eta^{0\,0} & = & -1, \\ \eta^{1\,1} & = & 1, \\ \eta^{2\,2} & = & 1, \\ \eta^{3\,3} & = & 1, \end{array} \right. \tag{4.2.46}$$

and other components of $\eta^{\mu\nu}$ vanish. $\eta^{\mu\nu}$ is the traditional Minkowski metric.

4.3 Pure Gravitational Gauge Fields

Now, let's begin to construct the Lagrangian of quantum gauge theory of gravity. For the sake of simplicity, let's suppose that $\phi(x)$ is a Lorentz scalar and gauge group scalar. According to above discussions, its gravitational gauge transformation is

$$\phi(x) \to \phi'(x) = (\hat{U}_\epsilon \phi(x)). \tag{4.3.1}$$

Because

$$(\partial_\mu \hat{U}_\epsilon) \neq 0, \tag{4.3.2}$$

partial derivative of $\phi(x)$ does not transform covariantly under gravitational gauge transformation

$$\partial_\mu \phi(x) \to \partial_\mu \phi'(x) \neq (\hat{U}_\epsilon \partial_\mu \phi(x)). \tag{4.3.3}$$

In order to construct an action which is invariant under local gravitational gauge transformation, gravitational gauge covariant derivative is needed. The gravitational gauge covariant derivative is defined by

$$D_\mu = \partial_\mu - igC_\mu(x), \tag{4.3.4}$$

where $C_\mu(x)$ is the gravitational gauge field and g is the coupling constant of gravitational gauge interactions. It is a Lorentz vector. Under gravitational gauge transformations, it transforms as

$$C_\mu(x) \to C'_\mu(x) = \hat{U}_\epsilon(x) C_\mu(x) \hat{U}_\epsilon^{-1}(x) + \frac{i}{g} \hat{U}_\epsilon(x)(\partial_\mu \hat{U}_\epsilon^{-1}(x)). \tag{4.3.5}$$

Using the original definition of $\hat{U}_\epsilon$, we can strictly proved that

$$[\partial_\mu \ , \ \hat{U}_\epsilon] = (\partial_\mu \hat{U}_\epsilon). \tag{4.3.6}$$

Therefor, we have

$$\hat{U}_\epsilon \partial_\mu \hat{U}_\epsilon^{-1} = \partial_\mu + \hat{U}_\epsilon(\partial_\mu \hat{U}_\epsilon^{-1}), \tag{4.3.7}$$

$$\hat{U}_\epsilon D_\mu \hat{U}_\epsilon^{-1} = \partial_\mu - igC'_\mu(x). \tag{4.3.8}$$

So, under local gravitational gauge transformations,

$$D_\mu \phi(x) \to D'_\mu \phi'(x) = (\hat{U}_\epsilon D_\mu \phi(x)), \tag{4.3.9}$$

$$D_\mu(x) \to D'_\mu(x) = \hat{U}_\epsilon D_\mu(x) \hat{U}_\epsilon^{-1}. \tag{4.3.10}$$

Gravitational gauge field $C_\mu(x)$ is vector field, it is a Lorentz vector. It is also a vector in gravitational gauge space, so it can be expanded as linear combinations of generators of gravitational gauge group

$$C_\mu(x) = C_\mu^\alpha(x) \cdot \hat{P}_\alpha. \tag{4.3.11}$$

C_μ^α are component fields of gravitational gauge field. It looks like a second rank tensor. The index α is not a Lorentz index, it is just a gauge group index. Gravitational gauge field C_μ^α has one Lorentz index, and one gauge group index. The gravitational gauge transformation of component field is

$$C_\mu^\alpha(x) \to C_\mu'^\alpha(x) = \Lambda^\alpha{}_\beta(\hat{U}_\epsilon C_\mu^\beta(x)) - \frac{1}{g}(\hat{U}_\epsilon \partial_\mu \epsilon^\alpha(y)), \tag{4.3.12}$$

where y is a function of space-time coordinates which satisfy

$$(\hat{U}_\epsilon y(x)) = x. \tag{4.3.13}$$

Define matrix G as

$$G = (G_\mu^\alpha) = (\delta_\mu^\alpha - gC_\mu^\alpha). \tag{4.3.14}$$

A simple form for matrix G is

$$G = I - gC, \tag{4.3.15}$$

where I is a unit matrix and $C = (C_\mu^\alpha)$. Therefore,

$$G^{-1} = \frac{1}{I - gC}. \tag{4.3.16}$$

G^{-1} is the inverse matrix of G, it satisfies

$$(G^{-1})_\beta^\mu G_\mu^\alpha = \delta_\beta^\alpha, \tag{4.3.17}$$

$$G_\mu^\alpha (G^{-1})_\alpha^\nu = \delta_\mu^\nu. \tag{4.3.18}$$

Define

$$g^{\alpha\beta} \triangleq \eta^{\mu\nu} G_\mu^\alpha G_\nu^\beta. \tag{4.3.19}$$

$$g_{\alpha\beta} \triangleq \eta_{\mu\nu} (G^{-1})_\alpha^\mu (G^{-1})_\beta^\nu. \tag{4.3.20}$$

It can be easily proved that

$$g_{\alpha\beta} g^{\beta\gamma} = \delta_\alpha^\gamma, \tag{4.3.21}$$

$$g^{\alpha\beta} g_{\beta\gamma} = \delta_\gamma^\alpha. \tag{4.3.22}$$

Under gravitational gauge transformations, they transform as

$$G_\mu^\alpha(x) \to G_\mu^{\prime\alpha}(x) = \Lambda^\alpha{}_{\alpha_1} (\hat{U}_\epsilon G_\mu^{\alpha_1}(x)), \tag{4.3.23}$$

$$G_\alpha^{-1\mu}(x) \to G_\alpha^{\prime -1\mu}(x) = \Lambda_\alpha{}^{\alpha_1} (\hat{U}_\epsilon G_{\alpha_1}^{-1\mu}(x)), \tag{4.3.24}$$

$$g_{\alpha\beta}(x) \to g'_{\alpha\beta}(x) = \Lambda_\alpha{}^{\alpha_1} \Lambda_\beta{}^{\beta_1} (\hat{U}_\epsilon g_{\alpha_1\beta_1}(x)), \tag{4.3.25}$$

$$g^{\alpha\beta}(x) \to g^{\prime\alpha\beta}(x) = \Lambda^\alpha{}_{\alpha_1} \Lambda^\beta{}_{\beta_1} (\hat{U}_\epsilon g^{\alpha_1\beta_1}(x)), \tag{4.3.26}$$

The strength of gravitational gauge field is defined by

$$F_{\mu\nu} = \frac{1}{-ig} [D_\mu \ , \ D_\nu], \tag{4.3.27}$$

or

$$F_{\mu\nu} = \partial_\mu C_\nu(x) - \partial_\nu C_\mu(x) - igC_\mu(x)C_\nu(x) + igC_\nu(x)C_\mu(x). \tag{4.3.28}$$

$F_{\mu\nu}$ is a second order Lorentz tensor. It is a vector in group space, so it can be expanded in group space,

$$F_{\mu\nu}(x) = F_{\mu\nu}^\alpha(x) \cdot \hat{P}_\alpha. \tag{4.3.29}$$

The explicit form of component field strength is

$$F_{\mu\nu}^\alpha = \partial_\mu C_\nu^\alpha - \partial_\nu C_\mu^\alpha - gC_\mu^\beta \partial_\beta C_\nu^\alpha + gC_\nu^\beta \partial_\beta C_\mu^\alpha \tag{4.3.30}$$

The strength of gravitational gauge field transforms covariantly under gravitational gauge transformation

$$F_{\mu\nu} \to F'_{\mu\nu} = \hat{U}_\epsilon F_{\mu\nu} \hat{U}_\epsilon^{-1}. \tag{4.3.31}$$

The gravitational gauge transformation of the component field strength is

$$F^\alpha_{\mu\nu} \to F'^\alpha_{\mu\nu} = \Lambda^\alpha{}_\beta (\hat{U}_\epsilon F^\beta_{\mu\nu}). \tag{4.3.32}$$

In physics picture of gravity, $g_{\alpha\beta}$ defined by (4.3.20) and $g^{\alpha\beta}$ defined by (4.3.19) are not space-time metric, for the space-time metric are always the Minkowski metric. They are only two composite field operators which are composed of gravitational gauge field. Using these two operators, we can calculate another important operator

$$\Gamma^\gamma_{\alpha\beta} = \frac{1}{2} g^{\gamma\delta} \left(\frac{\partial g_{\alpha\delta}}{\partial x^\beta} + \frac{\partial g_{\beta\delta}}{\partial x^\alpha} - \frac{\partial g_{\alpha\beta}}{\partial x^\delta} \right). \tag{4.3.33}$$

From above definition, we can see that $\Gamma^\gamma_{\alpha\beta}$ looks like the affine connection in general relativity. But now, it has no geometric meaning, for now we are in the physics picture of gravity. It is only a composite field operator. Using the following relation,

$$-gF^\gamma_{\mu\nu} = G^\alpha_\mu G^\beta_\nu [(G^{-1}\partial_\alpha G)^\gamma_\beta - (G^{-1}\partial_\beta G)^\gamma_\alpha], \tag{4.3.34}$$

where $F^\gamma_{\mu\nu}$ is the component field strength of gravitational gauge field, we get

$$\begin{aligned} \Gamma^\gamma_{\alpha\beta} &= -\tfrac{1}{2}[G^{-1\mu}_\beta \partial_\alpha G^\gamma_\mu + G^{-1\mu}_\alpha \partial_\beta G^\gamma_\mu] \\ &\quad + \tfrac{g}{2} g^{\gamma\delta} (g_{\beta\alpha_1} G^{-1\mu}_\alpha + g_{\alpha\alpha_1} G^{-1\mu}_\beta) G^{-1\nu}_\delta F^{\alpha_1}_{\mu\nu}. \end{aligned} \tag{4.3.35}$$

Operator $R^\delta_{\alpha\beta\gamma}$ is defined by

$$R^\delta_{\alpha\beta\gamma} \triangleq \partial_\gamma \Gamma^\delta_{\alpha\beta} - \partial_\beta \Gamma^\delta_{\alpha\gamma} + \Gamma^\eta_{\alpha\beta} \Gamma^\delta_{\gamma\eta} - \Gamma^\eta_{\alpha\gamma} \Gamma^\delta_{\beta\eta}, \tag{4.3.36}$$

the operator $R_{\alpha\gamma}$ is defined by

$$R_{\alpha\gamma} \triangleq R^\beta_{\alpha\beta\gamma}, \tag{4.3.37}$$

and the scalar operator R is defined by

$$R \triangleq g^{\alpha\gamma} R_{\alpha\gamma}. \tag{4.3.38}$$

Operator $R_{\alpha\gamma}$ can also be calculated from the following relation

$$\begin{aligned} R_{\alpha\gamma} &= \tfrac{1}{2} g^{\beta\delta} (\partial_\beta \partial_\delta g_{\alpha\gamma} - \partial_\alpha \partial_\delta g_{\beta\gamma} - \partial_\beta \partial_\gamma g_{\alpha\delta} + \partial_\alpha \partial_\gamma g_{\beta\delta}) \\ &\quad + g^{\beta\delta} g_{\alpha_1\beta_1} (\Gamma^{\alpha_1}_{\alpha\gamma} \Gamma^{\beta_1}_{\beta\delta} - \Gamma^{\alpha_1}_{\alpha\delta} \Gamma^{\beta_1}_{\beta\gamma}). \end{aligned} \tag{4.3.39}$$

After rather lengthy and extremely complicated calculations, we can proved that the operator $R_{\alpha\beta}$ can be explicitly expressed as

$$\begin{aligned} R_{\alpha\beta} \;=\; & -\tfrac{g^2}{4}\eta^{\mu\rho}\eta^{\nu\sigma}g_{\alpha\alpha_1}g_{\beta\beta_1}F^{\alpha_1}_{\mu\nu}F^{\beta_1}_{\rho\sigma} \\ & +\tfrac{g^2}{2}\eta^{\nu\sigma}g_{\alpha_1\beta_1}G^{-1\rho}_{\alpha}G^{-1\mu}_{\beta}F^{\alpha_1}_{\mu\nu}F^{\beta_1}_{\rho\sigma} \\ & +\tfrac{g^2}{2}G^{-1\mu}_{\beta}G^{-1\rho}_{\alpha}G^{-1\nu}_{\beta_1}G^{-1\sigma}_{\alpha_1}F^{\alpha_1}_{\mu\nu}F^{\beta_1}_{\rho\sigma} \\ & +\tfrac{g^2}{2}\eta^{\nu\sigma}(g_{\alpha_1\alpha}G^{-1\mu}_{\beta}+g_{\alpha_1\beta}G^{-1\mu}_{\alpha})G^{-1\rho}_{\beta_1}F^{\alpha_1}_{\mu\nu}F^{\beta_1}_{\rho\sigma} \\ & +\tfrac{g}{2}(G^{-1\mu}_{\beta}\partial_\alpha+G^{-1\mu}_{\alpha}\partial_\beta)(G^{-1\nu}_{\gamma}F^{\gamma}_{\mu\nu}) \\ & +\tfrac{g}{2}\eta_{\mu_1\nu_1}g^{\delta\beta_1}(G^{-1\mu_1}_{\alpha}G^{-1\nu}_{\beta}+G^{-1\mu_1}_{\beta}G^{-1\nu}_{\alpha})G^{-1\mu}_{\beta_1}\partial_\delta(G^{-1\nu_1}_{\gamma}F^{\gamma}_{\mu\nu}). \end{aligned} \tag{4.3.40}$$

Operator R can be calculated from the following relation

$$R = g^{\alpha\gamma}g^{\beta\delta}(\partial_\beta\partial_\delta g_{\alpha\gamma}-\partial_\beta\partial_\gamma g_{\alpha\delta})+g^{\alpha\gamma}g^{\beta\delta}g_{\alpha_1\beta_1}(\Gamma^{\alpha_1}_{\alpha\gamma}\Gamma^{\beta_1}_{\beta\delta}-\Gamma^{\alpha_1}_{\alpha\delta}\Gamma^{\beta_1}_{\beta\gamma}). \tag{4.3.41}$$

The explicit expression for the operator R is

$$R = R_0 + \frac{2g}{J(C)}\partial_\beta\left(J(C)g^{\alpha\beta}G^{-1\mu}_{\alpha}G^{-1\nu}_{\gamma}F^{\gamma}_{\mu\nu}\right), \tag{4.3.42}$$

where

$$\begin{aligned} R_0 \;=\; & \tfrac{g^2}{4}\eta^{\mu\rho}\eta^{\nu\sigma}g_{\alpha\beta}F^{\alpha}_{\mu\nu}F^{\beta}_{\rho\sigma} \\ & +\tfrac{g^2}{2}\eta^{\mu\rho}G^{-1\nu}_{\beta}G^{-1\sigma}_{\alpha}F^{\alpha}_{\mu\nu}F^{\beta}_{\rho\sigma} \\ & -g^2\eta^{\mu\rho}G^{-1\nu}_{\alpha}G^{-1\sigma}_{\beta}F^{\alpha}_{\mu\nu}F^{\beta}_{\rho\sigma}, \end{aligned} \tag{4.3.43}$$

and

$$J(C) = \sqrt{-\det(g_{\alpha\beta})}. \tag{4.3.44}$$

It can be strictly proved that, under gravitational gauge transformations, these operators transforms as

$$\Gamma^{\gamma}_{\alpha\beta}(x) \to \Gamma'^{\gamma}_{\alpha\beta}(x) = \Lambda_{\alpha}^{\ \alpha_1}\Lambda_{\beta}^{\ \beta_1}\Lambda^{\gamma}_{\ \gamma_1}(\hat{U}_\epsilon\Gamma^{\gamma_1}_{\alpha_1\beta_1}(x)) - \Lambda_{\alpha}^{\ \alpha_1}\Lambda_{\beta}^{\ \beta_1}(\partial_{\beta_1}\Lambda^{\gamma}_{\ \alpha_1}) \tag{4.3.45}$$

$$R_{\alpha\beta\gamma\delta}(x) \to R'_{\alpha\beta\gamma\delta}(x) = \Lambda_{\alpha}^{\ \alpha_1}\Lambda_{\beta}^{\ \beta_1}\Lambda_{\gamma}^{\ \gamma_1}\Lambda_{\delta}^{\ \delta_1}(\hat{U}_\epsilon R_{\alpha_1\beta_1\gamma_1\delta_1}(x)), \tag{4.3.46}$$

$$R_{\alpha\beta}(x) \to R'_{\alpha\beta}(x) = \Lambda_{\alpha}^{\ \alpha_1}\Lambda_{\beta}^{\ \beta_1}(\hat{U}_\epsilon R_{\alpha_1\beta_1}(x)), \tag{4.3.47}$$

$$R(x) \to R'(x) = (\hat{U}_\epsilon R(x)). \tag{4.3.48}$$

Now, we need to determine the Lagrangian of the model. Define

$$\begin{aligned} \mathcal{L}_0 &= -\tfrac{1}{16\pi G} R_0 = -\tfrac{1}{4g^2} R_0 \\ &= -\tfrac{1}{16}\eta^{\mu\rho}\eta^{\nu\sigma} g_{\alpha\beta} F^\alpha_{\mu\nu} F^\beta_{\rho\sigma} \\ & \quad -\tfrac{1}{8}\eta^{\mu\rho} G^{-1\nu}_\beta G^{-1\sigma}_\alpha F^\alpha_{\mu\nu} F^\beta_{\rho\sigma} \\ & \quad +\tfrac{1}{4}\eta^{\mu\rho} G^{-1\nu}_\alpha G^{-1\sigma}_\beta F^\alpha_{\mu\nu} F^\beta_{\rho\sigma}. \end{aligned} \tag{4.3.49}$$

If we define

$$C^{\mu\nu\rho\sigma}_{\alpha\beta} = \frac{1}{4}\eta^{\mu\rho}\eta^{\nu\sigma} g_{\alpha\beta} + \frac{1}{2}\eta^{\mu\rho} G^{-1\nu}_\beta G^{-1\sigma}_\alpha - \eta^{\mu\rho} G^{-1\nu}_\alpha G^{-1\sigma}_\beta, \tag{4.3.50}$$

then the above lagrangian $\mathcal{L}_0$ can be written into a simpler form:

$$\mathcal{L}_0 = -\frac{1}{4} C^{\mu\nu\rho\sigma}_{\alpha\beta} F^\alpha_{\mu\nu} F^\beta_{\rho\sigma}. \tag{4.3.51}$$

The first term in the Lagrangian $\mathcal{L}_0$ of (4.3.49) is just the Lagrangian in the previous work[30]. In fact, every term in the above Lagrangian has gravitational gauge symmetry. So, the most general Lagrangian for quantum gauge theory of gravity is

$$\begin{aligned} \mathcal{L}_0 &= c_1 \eta^{\mu\rho}\eta^{\nu\sigma} g_{\alpha\beta} F^\alpha_{\mu\nu} F^\beta_{\rho\sigma} \\ & \quad + c_2 \eta^{\mu\rho} G^{-1\nu}_\beta G^{-1\sigma}_\alpha F^\alpha_{\mu\nu} F^\beta_{\rho\sigma} \\ & \quad + c_3 \eta^{\mu\rho} G^{-1\nu}_\alpha G^{-1\sigma}_\beta F^\alpha_{\mu\nu} F^\beta_{\rho\sigma}. \end{aligned} \tag{4.3.52}$$

Because the action given by this Lagrangian has strict local gravitational gauge symmetry, the model based on this Lagrangian is perturbatively renormalizable. The strict formal proof on the renormalizability of the quantum gauge theory of gravity is suitable to the model based on this most general Lagrangian. For quantum gauge theory of gravity, the parameters c_1, c_2 and c_3 are

$$c_1 = -\frac{1}{16}, \quad c_2 = -\frac{1}{8}, \quad c_3 = \frac{1}{4}. \tag{4.3.53}$$

For the model discussed in the literature [30], the parameters c_1, c_2 and c_3 are

$$c_1 = -\frac{1}{4}, \quad c_2 = 0, \quad c_3 = 0. \tag{4.3.54}$$

Different selection of the parameters c_1, c_2 and c_3 gives out different model for quantum gravity, and therefore different dynamics of gravitational field. This situation is quite special for gravitational interactions. We know that, for ordinary $SU(N)$ gauge field theory, using gauge field strength, we can only construct one gauge invariant Lagrangian which is a quadratic form of field strength. In other words, the Lagrangian for pure gauge field in ordinary $SU(N)$ gauge field theory is unique. However, in quantum gauge theory of gravity, we can construct three different gauge invariant terms which are quadratic forms of field strength of gravitational gauge field. Indeed, in many aspects, quantum gauge theory of gravity is different from the traditional gauge field theory.

Using relations (4.3.24), (4.3.25) and (4.3.32), we can easily prove that the Lagrangian defined by eq.(4.3.49) transforms covariantly under gravitational gauge transformation

$$\mathcal{L}_0 \to \mathcal{L}'_0 = (\hat{U}_\epsilon \mathcal{L}_0). \tag{4.3.55}$$

In order to resume the gravitational gauge symmetry of the action, the factor $J(C)$ which is defined by eq.(4.3.44) is needed. The gravitational gauge transformations of $g_{\alpha\beta}$ is given by eq.(4.3.25). Then $J(C)$ transforms as

$$J(C) \to J'(C') = J \cdot (\hat{U}_\epsilon J(C)), \tag{4.3.56}$$

where J is the Jacobian of the transformation,

$$J = det\left(\frac{\partial(x-\epsilon)^\mu}{\partial x^\nu}\right). \tag{4.3.57}$$

The full Lagrangian of gravitational gauge field is selected as

$$\mathcal{L} = J(C)\mathcal{L}_0 = \sqrt{-\det g_{\alpha\beta}} \cdot \mathcal{L}_0, \tag{4.3.58}$$

and the action for gravitational gauge field is

$$S = \int d^4x \mathcal{L}. \tag{4.3.59}$$

It can be proved that this action has gravitational gauge symmetry. In other words, it is invariant under gravitational gauge transformation,

$$S \to S' = S. \tag{4.3.60}$$

In order to prove the gravitational gauge symmetry of the action, the following relation is used

$$\int d^4x J\left(\hat{U}_\epsilon f(x)\right) = \int d^4x f(x), \tag{4.3.61}$$

where $f(x)$ is an arbitrary function of space-time coordinate.

According to gauge principle, the global gauge symmetry will give out conserved charges. Now, let's discuss the conserved charges of global gravitational gauge transformation. Suppose that ϵ^{α} is an infinitesimal constant 4-vector. Then, in the first order approximation, we have

$$\hat{U}_{\epsilon} = 1 - \epsilon^{\alpha}\partial_{\alpha} + o(\epsilon^2). \tag{4.3.62}$$

The first order variation of the gravitational gauge field is

$$\delta C_{\mu}^{\alpha}(x) = -\epsilon^{\nu}\partial_{\nu}C_{\mu}^{\alpha}, \tag{4.3.63}$$

and the first order variation of action is:

$$\delta S = \int \mathrm{d}^4x \; \epsilon^{\alpha}\partial_{\mu}T_{i\alpha}^{\mu}, \tag{4.3.64}$$

where $T_{i\alpha}^{\mu}$ is the inertial energy-momentum tensor, whose definition is

$$T_{i\alpha}^{\mu} \equiv J(C)\left(-\frac{\partial\mathcal{L}_0}{\partial\partial_{\mu}C_{\nu}^{\beta}}\partial_{\alpha}C_{\nu}^{\beta} + \delta_{\alpha}^{\mu}\mathcal{L}_0\right). \tag{4.3.65}$$

Its explicit expression is

$$\begin{aligned}
T_{i\alpha}^{\mu} = J(C)[& \tfrac{1}{4}\eta^{\lambda\rho}\eta^{\nu\sigma}g_{\beta\gamma}G_{\lambda}^{\mu}F_{\rho\sigma}^{\gamma}(\partial_{\alpha}C_{\nu}^{\beta}) - \tfrac{1}{4}\eta^{\nu\rho}G_{\beta}^{-1\sigma}F_{\rho\sigma}^{\mu}(\partial_{\alpha}C_{\nu}^{\beta}) \\
& + \tfrac{1}{4}\eta^{\lambda\rho}G_{\gamma}^{-1\nu}G_{\beta}^{-1\sigma}G_{\lambda}^{\mu}F_{\rho\sigma}^{\gamma}(\partial_{\alpha}C_{\nu}^{\beta}) - \tfrac{1}{2}\eta^{\lambda\rho}G_{\beta}^{-1\nu}G_{\gamma}^{-1\sigma}G_{\lambda}^{\mu}F_{\rho\sigma}^{\gamma}(\partial_{\alpha}C_{\nu}^{\beta}) \\
& + \tfrac{1}{2}\eta^{\nu\rho}\delta_{\beta}^{\mu}G_{\gamma}^{-1\sigma}F_{\rho\sigma}^{\gamma}(\partial_{\alpha}C_{\nu}^{\beta}) - \tfrac{1}{16}\delta_{\alpha}^{\mu}\eta^{\lambda\rho}\eta^{\nu\sigma}g_{\beta\gamma}F_{\lambda\nu}^{\gamma}F_{\rho\sigma}^{\beta} \\
& - \tfrac{1}{8}\delta_{\alpha}^{\mu}\eta^{\lambda\rho}G_{\beta}^{-1\nu}G_{\gamma}^{-1\sigma}F_{\lambda\nu}^{\gamma}F_{\rho\sigma}^{\beta} + \tfrac{1}{4}\delta_{\alpha}^{\mu}\eta^{\lambda\rho}G_{\gamma}^{-1\nu}G_{\beta}^{-1\sigma}F_{\lambda\nu}^{\gamma}F_{\rho\sigma}^{\beta}].
\end{aligned} \tag{4.3.66}$$

The global gravitational gauge symmetry of the system gives out the conservation law of inertial energy-momentum tensor

$$\partial_{\mu}T_{i\alpha}^{\mu} = 0. \tag{4.3.67}$$

Except for the factor $J(C)$, the form of the inertial energy-momentum tensor is almost completely the same as that in the traditional quantum field theory. It means that gravitational interactions will change energy-momentum of matter fields, which is what we expected in Einstein's general relativity.

The Euler-Lagrange equation of gravitational gauge field is

$$\partial_{\mu}\frac{\partial\mathcal{L}}{\partial\partial_{\mu}C_{\nu}^{\alpha}} = \frac{\partial\mathcal{L}}{\partial C_{\nu}^{\alpha}}. \tag{4.3.68}$$

This form is completely the same as what we have ever seen in quantum field theory. But if we insert eq.(4.3.58) into it, we will get

$$\partial_\mu \frac{\partial \mathcal{L}_0}{\partial \partial_\mu C_\nu^\alpha} = \frac{\partial \mathcal{L}_0}{\partial C_\nu^\alpha} + g G_\alpha^{-1\nu} \mathcal{L}_0 - g G_\alpha^{-1\nu} (\partial_\mu C_\nu^\alpha) \frac{\partial \mathcal{L}_0}{\partial \partial_\mu C_\nu^\alpha}. \tag{4.3.69}$$

Eq.(4.3.30) can be changed into

$$F_{\mu\nu}^\alpha = (D_\mu C_\nu^\alpha) - (D_\nu C_\mu^\alpha), \tag{4.3.70}$$

so the Lagrangian $\mathcal{L}_0$ depends on gravitational gauge fields through its covariant derivative, $g_{\alpha\beta}$ and $G_\alpha^{-1\mu}$. Therefore, we have

$$\begin{aligned} \frac{\partial \mathcal{L}_0}{\partial C_\nu^\alpha} = & \tfrac{g}{4}\eta^{\nu\rho}\eta^{\lambda\sigma} g_{\beta\gamma} F_{\rho\sigma}^\beta (\partial_\alpha C_\lambda^\gamma) + \tfrac{g}{4}\eta^{\nu\rho} G_\beta^{-1\lambda} G_\gamma^{-1\sigma} F_{\rho\sigma}^\beta (\partial_\alpha C_\lambda^\gamma) \\ & -\tfrac{g}{2}\eta^{\nu\rho} G_\gamma^{-1\lambda} G_\beta^{-1\sigma} F_{\rho\sigma}^\beta (\partial_\alpha C_\lambda^\gamma) - \tfrac{g}{4}\eta^{\lambda\rho} G_\beta^{-1\nu} G_\gamma^{-1\sigma} F_{\rho\sigma}^\beta (\partial_\alpha C_\lambda^\gamma) \\ & +\tfrac{g}{2}\eta^{\lambda\rho} G_\gamma^{-1\nu} G_\beta^{-1\sigma} F_{\rho\sigma}^\beta (\partial_\alpha C_\lambda^\gamma) + \tfrac{g}{8}\eta^{\lambda\rho}\eta^{\kappa\sigma} g_{\alpha\gamma} G_\beta^{-1\nu} F_{\rho\sigma}^\beta F_{\kappa\lambda}^\gamma \\ & -\tfrac{g}{4}\eta^{\kappa\rho} G_\beta^{-1\nu} G_\alpha^{-1\lambda} G_\gamma^{-1\sigma} F_{\rho\sigma}^\beta F_{\kappa\lambda}^\gamma + \tfrac{g}{2}\eta^{\kappa\rho} G_\gamma^{-1\nu} G_\alpha^{-1\lambda} G_\beta^{-1\sigma} F_{\rho\sigma}^\beta F_{\kappa\lambda}^\gamma, \end{aligned} \tag{4.3.71}$$

and

$$\begin{aligned} \frac{\partial \mathcal{L}_0}{\partial \partial_\mu C_\nu^\alpha} = & -\tfrac{1}{4}\eta^{\lambda\rho}\eta^{\nu\sigma} g_{\alpha\beta} G_\lambda^\mu F_{\rho\sigma}^\beta + \tfrac{1}{4}\eta^{\nu\rho} G_\alpha^{-1\sigma} F_{\rho\sigma}^\mu - \tfrac{1}{4}\eta^{\lambda\rho} G_\beta^{-1\nu} G_\alpha^{-1\sigma} G_\lambda^\mu F_{\rho\sigma}^\beta \\ & +\tfrac{1}{2}\eta^{\lambda\rho} G_\alpha^{-1\nu} G_\beta^{-1\sigma} G_\lambda^\mu F_{\rho\sigma}^\beta - \tfrac{1}{2}\eta^{\nu\rho}\delta_\alpha^\mu G_\beta^{-1\sigma} F_{\rho\sigma}^\beta . \end{aligned} \tag{4.3.72}$$

The above field equation of gravitational gauge fields are changed into

$$\begin{aligned} \partial_\mu (& \tfrac{1}{4}\eta^{\mu\rho}\eta^{\nu\sigma} g_{\alpha\beta} F_{\rho\sigma}^\beta - \tfrac{1}{4}\eta^{\nu\rho} F_{\rho\alpha}^\mu + \tfrac{1}{4}\eta^{\mu\rho} F_{\rho\alpha}^\nu \\ & -\tfrac{1}{2}\eta^{\mu\rho}\delta_\alpha^\nu F_{\rho\beta}^\beta + \tfrac{1}{2}\eta^{\nu\rho}\delta_\alpha^\mu F_{\rho\beta}^\beta) = -g T_{g\alpha}^\nu, \end{aligned} \tag{4.3.73}$$

where

$$\begin{aligned} T^{\nu}_{g\alpha} = & \tfrac{1}{4}\eta^{\nu\rho}\eta^{\lambda\sigma}g_{\beta\gamma}F^{\beta}_{\rho\sigma}(\partial_{\alpha}C^{\gamma}_{\lambda}) + \tfrac{1}{4}\eta^{\nu\rho}G^{-1\lambda}_{\beta}G^{-1\sigma}_{\gamma}F^{\beta}_{\rho\sigma}(\partial_{\alpha}C^{\gamma}_{\lambda}) \\ & -\tfrac{1}{2}\eta^{\nu\rho}G^{-1\lambda}_{\gamma}G^{-1\sigma}_{\beta}F^{\beta}_{\rho\sigma}(\partial_{\alpha}C^{\gamma}_{\lambda}) - \tfrac{1}{4}\eta^{\lambda\rho}G^{-1\nu}_{\beta}G^{-1\sigma}_{\gamma}F^{\beta}_{\rho\sigma}(\partial_{\alpha}C^{\gamma}_{\lambda}) \\ & +\tfrac{1}{2}\eta^{\lambda\rho}G^{-1\nu}_{\gamma}G^{-1\sigma}_{\beta}F^{\beta}_{\rho\sigma}(\partial_{\alpha}C^{\gamma}_{\lambda}) + \tfrac{1}{4}\eta^{\lambda\rho}\eta^{\nu\sigma}g_{\alpha\beta}G^{-1\kappa}_{\gamma}F^{\beta}_{\rho\sigma}(D_{\lambda}C^{\gamma}_{\kappa}) \\ & -\tfrac{1}{4}\eta^{\nu\rho}G^{-1\sigma}_{\alpha}G^{-1\kappa}_{\gamma}F^{\mu}_{\rho\sigma}(\partial_{\mu}C^{\gamma}_{\kappa}) + \tfrac{1}{2}\eta^{\nu\rho}G^{-1\sigma}_{\beta}G^{-1\kappa}_{\gamma}F^{\beta}_{\rho\sigma}(\partial_{\alpha}C^{\gamma}_{\kappa}) \\ & +\tfrac{1}{4}\eta^{\lambda\rho}G^{-1\nu}_{\beta}G^{-1\sigma}_{\alpha}G^{-1\kappa}_{\gamma}F^{\beta}_{\rho\sigma}(D_{\lambda}C^{\gamma}_{\kappa}) - \tfrac{1}{2}\eta^{\lambda\rho}G^{-1\nu}_{\alpha}G^{-1\sigma}_{\beta}G^{-1\kappa}_{\gamma}F^{\beta}_{\rho\sigma}(D_{\lambda}C^{\gamma}_{\kappa}) \\ & -\tfrac{1}{4}\eta^{\lambda\rho}\eta^{\nu\sigma}\partial_{\mu}(g_{\alpha\beta}C^{\mu}_{\lambda}F^{\beta}_{\rho\sigma}) - \tfrac{1}{4}\eta^{\nu\rho}\partial_{\beta}(C^{\sigma}_{\lambda}G^{-1\lambda}_{\alpha}F^{\beta}_{\rho\sigma}) + \tfrac{1}{2}\eta^{\nu\rho}\partial_{\alpha}(C^{\sigma}_{\lambda}G^{-1\lambda}_{\beta}F^{\beta}_{\rho\sigma}) \\ & +\tfrac{1}{4g}\eta^{\lambda\rho}\partial_{\mu}[(G^{-1\nu}_{\beta}G^{-1\sigma}_{\alpha}G^{\mu}_{\lambda} - \delta^{\nu}_{\beta}\delta^{\sigma}_{\alpha}\delta^{\mu}_{\lambda})F^{\beta}_{\rho\sigma}] \\ & -\tfrac{1}{2g}\eta^{\lambda\rho}\partial_{\mu}[(G^{-1\nu}_{\alpha}G^{-1\sigma}_{\beta}G^{\mu}_{\lambda} - \delta^{\nu}_{\alpha}\delta^{\sigma}_{\beta}\delta^{\mu}_{\lambda})F^{\beta}_{\rho\sigma}] \\ & -\tfrac{1}{4}\eta^{\kappa\rho}G^{-1\nu}_{\beta}G^{-1\lambda}_{\alpha}G^{-1\sigma}_{\gamma}F^{\beta}_{\rho\sigma}F^{\gamma}_{\kappa\lambda} + \tfrac{1}{2}\eta^{\kappa\rho}G^{-1\nu}_{\gamma}G^{-1\lambda}_{\alpha}G^{-1\sigma}_{\beta}F^{\beta}_{\rho\sigma}F^{\gamma}_{\kappa\lambda} \\ & -\tfrac{1}{8}\eta^{\mu\rho}\eta^{\lambda\sigma}g_{\alpha\gamma}G^{-1\nu}_{\beta}F^{\beta}_{\rho\sigma}F^{\gamma}_{\mu\lambda} - \tfrac{1}{16}\eta^{\mu\rho}\eta^{\lambda\sigma}g_{\beta\gamma}G^{-1\nu}_{\alpha}F^{\beta}_{\rho\sigma}F^{\gamma}_{\mu\lambda} \\ & -\tfrac{1}{8}\eta^{\mu\rho}G^{-1\nu}_{\alpha}G^{-1\lambda}_{\beta}G^{-1\sigma}_{\gamma}F^{\beta}_{\rho\sigma}F^{\gamma}_{\mu\lambda} + \tfrac{1}{4}\eta^{\mu\rho}G^{-1\nu}_{\alpha}G^{-1\lambda}_{\gamma}G^{-1\sigma}_{\beta}F^{\beta}_{\rho\sigma}F^{\gamma}_{\mu\lambda}. \end{aligned} \tag{4.3.74}$$

$T^{\nu}_{g\alpha}$ is also a conserved current, that is

$$\partial_{\nu}T^{\nu}_{g\alpha} = 0, \tag{4.3.75}$$

because of the following identity

$$\begin{aligned} \partial_{\nu}\partial_{\mu}(& \tfrac{1}{4}\eta^{\mu\rho}\eta^{\nu\sigma}g_{\alpha\beta}F^{\beta}_{\rho\sigma} - \tfrac{1}{4}\eta^{\nu\rho}F^{\mu}_{\rho\alpha} + \tfrac{1}{4}\eta^{\mu\rho}F^{\nu}_{\rho\alpha} \\ & -\tfrac{1}{2}\eta^{\mu\rho}\delta^{\nu}_{\alpha}F^{\beta}_{\rho\beta} + \tfrac{1}{2}\eta^{\nu\rho}\delta^{\mu}_{\alpha}F^{\beta}_{\rho\beta}) = 0. \end{aligned} \tag{4.3.76}$$

$T^{\nu}_{g\alpha}$ is called gravitational energy-momentum tensor, which is the source of gravitational gauge field. Now we get two different energy-momentum tensors, one is the inertial energy-momentum tensor $T^{\nu}_{i\alpha}$ and another is the gravitational energy-momentum tensor $T^{\nu}_{g\alpha}$. They are similar, but they are different. The inertial energy-momentum tensor $T^{\nu}_{i\alpha}$ is given by conservation law which is associate with global gravitational gauge symmetry, it gives out an energy-momentum 4-vector:

$$P_{i\alpha} = \int \mathrm{d}^3\, \vec{x}\; T^{0}_{i\alpha}. \tag{4.3.77}$$

It is a conserved charges,

$$\frac{\mathrm{d}}{\mathrm{d}t}P_{i\alpha} = 0. \tag{4.3.78}$$

The time component of $P_{i\alpha}$, that is P_{i0}, gives out the Hamiltonian H of the system,

$$H = -P_{i\,0} = \int \mathrm{d}^3\, \vec{x}\; J(C)\left(\pi^{\mu}_{\alpha}\, \dot{C}^{\alpha}_{\mu} - \mathcal{L}_0\right). \tag{4.3.79}$$

According to our conventional belief, H should be the inertial energy of the system, therefore $P_{i\alpha}$ is the inertial energy-momentum of the system. The gravitational energy-momentum is given by the field equation of gravitational gauge field, it is also a conserved current. The space integration of the time component of it gives out a conserved energy-momentum 4-vector,

$$P_{g\alpha} = \int \mathrm{d}^3 \vec{x}\, T^0_{g\alpha}. \tag{4.3.80}$$

It is also a conserved charge,

$$\frac{\mathrm{d}}{\mathrm{d}t} P_{g\alpha} = 0. \tag{4.3.81}$$

Its time component just gives out the gravitational energy of the system, which is the source of gravitational gauge field. This can be easily seen. For a static system, if we Set ν and α in eq.(4.3.73) to 0, select harmonic gauge and make the leading term approximation, we can get(details on this deducing can be found in the chapter on classical limit of quantum gauge theory of gravity)

$$\nabla^2 C^0_0 = -g T^0_{g0}. \tag{4.3.82}$$

Define

$$E^i = -\partial_i C^0_0. \tag{4.3.83}$$

E^i is just the field strength of gravitational gauge field for a static system. The space integration of eq.(4.3.82) gives out

$$\oint \mathrm{d}\vec{\sigma} \cdot \vec{E} = g \int \mathrm{d}^3 \vec{x}\, T^0_{g0}. \tag{4.3.84}$$

According to Newton's classical theory of gravity, $\int \mathrm{d}^3 \vec{x}\, T^0_{g0}$ in the right hand term is just the gravitational mass of the system. Denote the gravitational mass of the system as M_g, that is

$$M_g = -\int \mathrm{d}^3 \vec{x}\, T^0_{g0}. \tag{4.3.85}$$

Then eq(4.3.84) is changed into

$$\oint \mathrm{d}\vec{\sigma} \cdot \vec{E} = -g M_g. \tag{4.3.86}$$

This is just the classical Newton's law of universal gravitation. It can be strictly proved that gravitational mass is different from inertial mass. They are not equivalent. But their difference is at least first order infinitesimal quantity if the gravitational field gC^α_μ, for this difference is proportional to gC^α_μ. So, this difference is too small to be detected in experiments. But in the environment of strong gravitational field, the difference will become relatively larger and will be easier to be detected. Much more highly precise measurement of this difference is strongly needed to test this prediction and to test the validity of the equivalence principle.

In the chapter of classical limit of quantum gauge theory of gravity, we will return to discuss this problem again.

If we use periodic boundary conditions, all surface terms in the action must vanish, that is

$$\int \mathrm{d}^4x \; J(C) \cdot \frac{1}{J(C)} \partial_\beta \left[2J(C) g^{\alpha\beta} G^{-1\mu}_{\alpha} (\partial_\gamma G^\gamma_\mu) - 2J(C) g^{\beta\delta} G^{-1\mu}_{\alpha} (\partial_\delta G^\alpha_\mu) \right] = 0. \tag{4.3.87}$$

Combine the above equation with the equation (4.3.42), we have

$$\int \mathrm{d}^4x \; J(C) \cdot R_0 = \int \mathrm{d}^4x \; J(C) \cdot R. \tag{4.3.88}$$

So, the action of the model can be written into

$$S = -\frac{1}{16\pi G} \int \mathrm{d}^4x \; \sqrt{-\mathrm{det} g_{\alpha\beta}} \cdot R. \tag{4.3.89}$$

Denote the action of matter field as S_M. If we consider the gravitational interactions of matter fields, the total action of the system should be

$$S = -\frac{1}{16\pi G} \int \mathrm{d}^4x \; \sqrt{-\mathrm{det} g_{\alpha\beta}} \cdot R + S_M. \tag{4.3.90}$$

Make a variation of gravitational gauge field C^α_μ, the variation of the operator $g_{\alpha\beta}$ is

$$\delta g_{\alpha\beta} = g(g_{\gamma\beta} G^{-1\mu}_{\alpha} + g_{\gamma\alpha} G^{-1\mu}_{\beta}) \delta C^\gamma_\mu, \tag{4.3.91}$$

and the change in the action is

$$\delta S = \frac{g}{16\pi G} \times \tag{4.3.92}$$
$$\int \mathrm{d}^4x \; \sqrt{-\mathrm{det} g_{\alpha_1\beta_1}} \left[R^{\alpha\beta} - \frac{1}{2} g^{\alpha\beta} R + 8\pi G T^{\alpha\beta} \right] (g_{\gamma\beta} G^{-1\mu}_{\alpha} + g_{\gamma\alpha} G^{-1\mu}_{\beta}) \delta C^\gamma_\mu,$$

where $T^{\alpha\beta}$ is the symmetric energy-momentum tensor which is defined by

$$T^{\alpha\beta} = \frac{2}{\sqrt{-\mathrm{det} g_{\alpha_1\beta_1}}} \frac{\delta S_M}{\delta g_{\alpha\beta}}. \tag{4.3.93}$$

The total action S is stationary with respect to arbitrary variations in C^γ_μ if and only if

$$\left[R^{\alpha_1\beta_1} - \frac{1}{2} g^{\alpha_1\beta_1} R + 8\pi G T^{\alpha_1\beta_1} \right] (g_{\gamma\beta_1} G^{-1\mu}_{\alpha_1} + g_{\gamma\alpha_1} G^{-1\mu}_{\beta_1}) = 0. \tag{4.3.94}$$

Multiply both side of the above equation with G^{α}_{μ}, we get

$$\left[R^{\alpha\beta_1} - \frac{1}{2}g^{\alpha\beta_1}R + 8\pi G T^{\alpha\beta_1}\right] g_{\gamma\beta_1} = 0. \tag{4.3.95}$$

Then multiply both side of the above equation with $g^{\gamma\beta}$, we get

$$R^{\alpha\beta} - \frac{1}{2}g^{\alpha\beta}R + 8\pi G T^{\alpha\beta} = 0, \tag{4.3.96}$$

which is just the Einstein's field equation. Now, from the same action, the least action principle gives out two equations (4.3.73) and (4.3.96). They are different in forms, but they are essentially the same, for one action can only give out one field equation. So, in quantum gauge theory of gravity, the field equation of gravitational gauge field is the Einstein's field equation.

Now, let's discuss self-coupling of gravitational field. The Lagrangian of gravitational gauge field is given by eq(4.3.58). Because

$$J(C) = 1 + \sum_{m=1}^{\infty} \frac{1}{m!}\left(\sum_{n=1}^{\infty} \frac{g^n}{n}\mathrm{tr}(C^n)\right)^m \tag{4.3.97}$$

there are vertexes of n gravitational gauge fields in tree diagram where n can be arbitrary integer number that is greater than 3. This property is important for renormalization of the theory. Because the coupling constant of the gravitational gauge interactions has negative mass dimension, any kind of regular vertex exists divergence. In order to cancel these divergences, we need to introduce the corresponding counterterms. Because of the existence of the vertex of n gravitational gauge fields in tree diagram in the non-renormalized Lagrangian, we need not introduce any new counterterm which does not exist in the non-renormalized Lagrangian, what we need to do is to redefine gravitational coupling constant g and gravitational gauge field C^{α}_{μ} in renormalization. If there is no $J(C)$ term in the original Lagrangian, then we will have to introduce infinite counterterms in renormalization, and therefore the theory is non-renormalizable. Because of the existence of the factor $J(C)$, though quantum gauge theory of gravity looks like a non-renormalizable theory according to the power counting law, it is indeed renormalizable. In a word, the factor $J(C)$ is highly important for the quantum gauge theory of gravity.

4.4 Classical Limit of Quantum Gauge theory of gravity

In this chapter, we mainly discuss leading order approximation of quantum gauge theory of gravity, which will give out classical Newton's theory of gravity.

Suppose that the gravitational field is very weak, so that gC_μ^α is a first order infinitesimal quantity. In leading order approximation, both inertial energy-momentum tensor and gravitational energy-momentum tensor give the same results, which we denoted as T_α^μ. The space integration of the time component of the current T_α^μ gives out the energy-momentum of the system. First, we discuss classical limit of the field equation of gravitational gauge field. Then, we discuss classical limit of the Einstein's field equation. We will find that two field equations give out the same classical limit, which is what we expected, for these two field equations are essentially the same.

The field equation (4.3.73) of gravitational gauge field in the leading order is:

$$\begin{aligned} -\tfrac{1}{4}\eta^{\nu\sigma}\eta_{\alpha\beta}\partial^\mu\partial_\mu C_\sigma^\beta + \tfrac{1}{4}\eta_{\alpha\beta}\partial^\nu\partial^\mu C_\mu^\beta + \tfrac{1}{4}\eta^{\nu\rho}\partial_\mu\partial_\rho C_\alpha^\mu & \\ -\tfrac{1}{4}\partial^\mu\partial_\mu C_\alpha^\nu + \tfrac{1}{4}\partial^\mu\partial_\alpha C_\mu^\nu + \tfrac{1}{2}\delta_\alpha^\nu\partial^\mu\partial_\mu C_\beta^\beta & \\ -\tfrac{1}{2}\delta_\alpha^\nu\partial^\mu\partial_\beta C_\mu^\beta - \tfrac{1}{2}\partial^\nu\partial_\alpha C_\beta^\beta + \tfrac{1}{4}\eta^{\nu\rho}\partial_\alpha\partial_\mu C_\rho^\mu &= gT_\alpha^\nu, \end{aligned} \tag{4.4.1}$$

where $\partial^\mu = \eta^{\mu\nu}\partial_\nu$. In order to simplify the above field equation, harmonic gauge will be used. In general relativity, the harmonic coordinate conditions are represented by

$$\Gamma^\gamma \triangleq g^{\alpha\beta}\Gamma_{\alpha\beta}^\gamma = 0, \tag{4.4.2}$$

where $\Gamma_{\alpha\beta}^\gamma$ is given by (4.3.35). In leading order approximation, it gives out the following restriction

$$\partial_\mu C_\nu^\mu - \partial_\nu C_\mu^\mu = -\eta_{\lambda\nu}\partial^\mu C_\mu^\lambda. \tag{4.4.3}$$

After the harmonic coordinate conditions are considered, eq. (4.4.1) is changed into

$$-\frac{1}{4}\eta^{\nu\sigma}\eta_{\alpha\beta}\partial^\mu\partial_\mu C_\sigma^\beta - \frac{1}{4}\partial^\mu\partial_\mu C_\alpha^\nu + \frac{1}{4}\delta_\alpha^\nu\partial^\mu\partial_\mu C_\beta^\beta = gT_\alpha^\nu. \tag{4.4.4}$$

Multiply both side of the above equation with δ_ν^α, we can get

$$\partial^\mu\partial_\mu C_\beta^\beta = 2gT_\beta^\beta. \tag{4.4.5}$$

Then, eq. (4.4.4) is changed into

$$-\frac{1}{4}\eta^{\nu\sigma}\eta_{\alpha\beta}\partial^\mu\partial_\mu C_\sigma^\beta - \frac{1}{4}\partial^\mu\partial_\mu C_\alpha^\nu = g\left(T_\alpha^\nu - \frac{1}{2}\delta_\alpha^\nu T_\beta^\beta\right). \tag{4.4.6}$$

As a classical limit approximation, let's consider static gravitational interactions between two static objects. In this case, the leading order component of energy-momentum tensor is T_0^0, other components of energy-momentum tensor is a first order infinitesimal quantity. So, we only need to consider the field equation (4.4.4) of $\nu = \alpha = 0$, which now becomes

$$\partial_\mu \partial^\mu C_0^0 = -g T_0^0. \tag{4.4.7}$$

For static problems, all time derivatives vanish. Therefor, the above equation is changed into

$$\nabla^2 C_0^0 = -g T_0^0. \tag{4.4.8}$$

This is just the Newton's equation of gravitational field. Suppose that there is only one point object at the origin of the coordinate system. Because T_0^0 is the negative value of energy density, we can let

$$T_0^0 = -M\delta(\vec{x}). \tag{4.4.9}$$

Applying

$$\nabla^2 \frac{1}{r} = -4\pi\delta(\vec{x}), \tag{4.4.10}$$

with $r = \mid \vec{x} \mid$, we get

$$C_0^0 = -\frac{gM}{4\pi r}. \tag{4.4.11}$$

This is just the gravitational potential which is expected in Newton's theory of gravity.

Suppose that there is another point object at the position of point $\vec{x}$ with mass m. The gravitational potential energy between these two objects is that

$$V(r) = \int \mathrm{d}^3 \vec{y}\, \mathcal{H}_I = -g \int \mathrm{d}^3 \vec{y}\, T_{2\ 0}^{0}(\vec{x}) C_0^0, \tag{4.4.12}$$

with C_0^0 is the gravitational potential generated by the first point object, and $T_{2\ 0}^{0}$ is the $(0,0)$ component of the energy-momentum tensor of the second object,

$$T_{2\ 0}^{0}(\vec{y}) = -m\delta(\vec{y} - \vec{x}). \tag{4.4.13}$$

The final result for gravitational potential energy between two point objects is

$$V(r) = -\frac{g^2 Mm}{4\pi r}. \tag{4.4.14}$$

The gravitational potential energy between two point objects is always negative, which is what expected by Newton's theory of gravity and is the inevitable result of the attractive nature of gravitational interactions.

The gravitational force that the first point object acts on the second point object is

$$\vec{f} = -\nabla V(r) = -\frac{g^2 Mm}{4\pi r^2}\hat{r}, \tag{4.4.15}$$

where $\hat{r} = \vec{r}\ /r$. This is the famous formula of Newton's gravitational force. Combine the above relation with the tradition relation, we have

$$G = \frac{g^2}{4\pi}, \tag{4.4.16}$$

where G is the Newtonian gravitational constant. Therefore, in the classical limit, the gravitational gauge theory can return to Newton's theory of gravity. Besides, from eq.(4.4.15), we can clearly see that the gravitational interaction force between two point objects is attractive.

Another form of the field equation of gravitational gauge field is the Einstein's field equation (4.3.96)

$$R_{\alpha\beta} - \frac{1}{2} g_{\alpha\beta} R = -8\pi G T_{\alpha\beta}. \tag{4.4.17}$$

Its classical limit should gives out the same results, for these two field equations are essentially the same. Now, we discuss its classical limit. Multiply both side of the above equation with $g^{\alpha\beta}$, we get

$$R = 8\pi G T_{\gamma}^{\gamma}. \tag{4.4.18}$$

Therefore, eq.(4.4.17) can be changed into

$$R_{\alpha\beta} = -8\pi G S_{\alpha\beta}, \tag{4.4.19}$$

where

$$S_{\alpha\beta} = T_{\alpha\beta} - \frac{1}{2} g_{\alpha\beta} T_{\gamma}^{\gamma}. \tag{4.4.20}$$

$R_{\alpha\beta}$ is given by eq. (4.3.40). In first order approximation, it is

$$\begin{aligned} R_{\alpha\beta} \ = \ & g\partial_\alpha\partial_\beta C_\mu^\mu - \tfrac{g}{2}\partial_\alpha\partial_\lambda C_\beta^\lambda - \tfrac{g}{2}\partial_\beta\partial_\lambda C_\alpha^\lambda \\ & + \tfrac{g}{2}\eta^{\lambda\nu}\eta_{\beta\gamma}\partial_\nu\partial_\lambda C_\alpha^\gamma + \tfrac{g}{2}\eta^{\lambda\nu}\eta_{\alpha\gamma}\partial_\nu\partial_\lambda C_\beta^\gamma \\ & - \tfrac{g}{2}\eta^{\mu\nu}\eta_{\beta\gamma}\partial_\nu\partial_\alpha C_\mu^\gamma - \tfrac{g}{2}\eta^{\mu\nu}\eta_{\alpha\gamma}\partial_\nu\partial_\beta C_\mu^\gamma + o((gC)^2). \end{aligned} \tag{4.4.21}$$

Applying harmonic coordinate conditions (4.4.3), we can simplify the above equation into

$$R_{\alpha\beta} = \frac{g}{2}\eta^{\lambda\nu}\eta_{\beta\gamma}\partial_\nu\partial_\lambda C_\alpha^\gamma + \frac{g}{2}\eta^{\lambda\nu}\eta_{\alpha\gamma}\partial_\nu\partial_\lambda C_\beta^\gamma. \tag{4.4.22}$$

For classical limit, we only need to consider the field equation (4.4.19) with $\alpha = \beta = 0$, which is

$$-g\partial^{\nu}\partial_{\nu}C_0^0 = -8\pi G S_{00} \tag{4.4.23}$$

Using the following relation

$$g^2 = 4\pi G, \tag{4.4.24}$$

$$S_{00} = \frac{1}{2}T_{00} = -\frac{1}{2}T_0^0, \tag{4.4.25}$$

we can change eq. (4.4.26) into

$$\partial^{\nu}\partial_{\nu}C_0^0 = -gT_0^0, \tag{4.4.26}$$

which is the same as eq.(4.4.7). Therefore, the Einstein's field equation (4.3.96) gives out the same classical limit as that of field equation (4.3.73), which is what we expected.

4.5 Path Integral Quantization of Gravitational Gauge Fields

For the sake of simplicity, in this chapter, we only discuss quantization of pure gravitational gauge field. For pure gravitational gauge field, its Lagrangian function is

$$\mathcal{L} = -\frac{1}{4}J(C)C^{\mu\nu\rho\sigma}_{\alpha\beta}F^{\alpha}_{\mu\nu}F^{\beta}_{\rho\sigma}. \tag{4.5.1}$$

Its space-time integration gives out the action of the system

$$S = \int \mathrm{d}^4x\mathcal{L}. \tag{4.5.2}$$

This action has local gravitational gauge symmetry. Gravitational gauge field C^{α}_{μ} has $4 \times 4 = 16$ degrees of freedom. But, if gravitons are massless, the system has only $2 \times 4 = 8$ degrees of freedom. There are gauge degrees of freedom in the theory. Because only physical degrees of freedom can be quantized, in order to quantize the system, we have to introduce gauge conditions to eliminate un-physical degrees of freedom. For the sake of convenience, we take temporal gauge conditions

$$C^{\alpha}_0 = 0, \quad (\alpha = 0, 1, 2, 3). \tag{4.5.3}$$

In temporal gauge, the generating functional $W[J]$ is given by

$$W[J] = N\int [\mathcal{D}C]\left(\prod_{\alpha,x}\delta(C^{\alpha}_0(x))\right) exp\left\{i\int \mathrm{d}^4x(\mathcal{L} + J^{\mu}_{\alpha}C^{\alpha}_{\mu}),\right\} \tag{4.5.4}$$

where N is the normalization constant, J^{μ}_{α} is a fixed external source and $[\mathcal{D}C]$ is the integration measure,

$$[\mathcal{D}C] = \prod_{\mu=0}^{3}\prod_{\alpha=0}^{3}\prod_{j}\left(\varepsilon \mathrm{d}C^{\alpha}_{\mu}(\tau_j)/\sqrt{2\pi i\hbar}\right). \tag{4.5.5}$$

We use this generation functional as our starting point of the path integral quantization of gravitational gauge field.

Generally speaking, the action of the system has local gravitational gauge symmetry, but the gauge condition has no local gravitational gauge symmetry. If we make a local gravitational gauge transformations, the action of the system is kept unchanged while gauge condition will be changed. Therefore, through local gravitational gauge transformation, we can change one gauge condition into another gauge condition. The most general gauge condition is

$$f^{\alpha}(C(x)) - \varphi^{\alpha}(x) = 0, \tag{4.5.6}$$

where $\varphi^{\alpha}(x)$ is an arbitrary space-time function. The Fadeev-Popov determinant $\Delta_f(C)$ [49] is defined by

$$\Delta_f^{-1}(C) \equiv \int [\mathcal{D}g]\prod_{x,\alpha}\delta\left(f^{\alpha}({}^{g}C(x)) - \varphi^{\alpha}(x)\right), \tag{4.5.7}$$

where g is an element of gravitational gauge group, ${}^{g}C$ is the gravitational gauge field after gauge transformation g and $[\mathcal{D}g]$ is the integration measure on gravitational gauge group

$$[\mathcal{D}g] = \prod_{x}\mathrm{d}^4\epsilon(x), \tag{4.5.8}$$

where $\epsilon(x)$ is the transformation parameter of $\hat{U}_{\epsilon}$. Both $[\mathcal{D}g]$ and $[\mathcal{D}C]$ are not invariant under gravitational gauge transformation. Suppose that,

$$[\mathcal{D}(gg')] = J_1(g')[\mathcal{D}g], \tag{4.5.9}$$

$$[\mathcal{D}\,{}^{g}C] = J_2(g)[\mathcal{D}C]. \tag{4.5.10}$$

$J_1(g)$ and $J_2(g)$ satisfy the following relations

$$J_1(g)\cdot J_1(g^{-1}) = 1, \tag{4.5.11}$$

$$J_2(g)\cdot J_2(g^{-1}) = 1. \tag{4.5.12}$$

It can be proved that, under gravitational gauge transformations, the Fadeev-Popov determinant transforms as

$$\Delta_f^{-1}({}^{g'}C) = J_1^{-1}(g')\Delta_f^{-1}(C). \tag{4.5.13}$$

Insert eq.(4.5.7) into eq.(4.5.4), we get

$$
\begin{aligned}
W[J] &= N\int[\mathcal{D}g]\int[\mathcal{D}C] \; \left[\prod_{\alpha,y}\delta(C_0^\alpha(y))\right]\cdot\Delta_f(C) \\
&\cdot\left[\prod_{\beta,z}\delta(f^\beta({}^gC(z))-\varphi^\beta(z))\right]\cdot exp\left\{i\int \mathrm{d}^4x(\mathcal{L}+J_\alpha^\mu C_\mu^\alpha)\right\}.
\end{aligned}
\tag{4.5.14}
$$

Make a gravitational gauge transformation,

$$
C(x) \;\; \rightarrow \;\; {}^{g^{-1}}C(x), \tag{4.5.15}
$$

then,

$$
{}^gC(x) \;\; \rightarrow \;\; {}^{gg^{-1}}C(x). \tag{4.5.16}
$$

After this transformation, the generating functional is changed into

$$
\begin{aligned}
W[J] &= N\int[\mathcal{D}g]\int[\mathcal{D}C] \;\; J_1(g)J_2(g^{-1})\cdot\left[\prod_{\alpha,y}\delta({}^{g^{-1}}C_0^\alpha(y))\right]\cdot\Delta_f(C) \\
&\cdot\left[\prod_{\beta,z}\delta(f^\beta(C(z))-\varphi^\beta(z))\right]\cdot exp\left\{i\int \mathrm{d}^4x(\mathcal{L}+J_\alpha^\mu\cdot{}^{g^{-1}}C_\mu^\alpha)\right\}.
\end{aligned}
\tag{4.5.17}
$$

Suppose that the gauge transformation $g_0(C)$ transforms general gauge condition $f^\beta(C)-\varphi^\beta=0$ to temporal gauge condition $C_0^\alpha=0$, and suppose that this transformation $g_0(C)$ is unique. Then two δ-functions in eq.(4.5.17) require that the integration on gravitational gauge group must be in the neighborhood of $g_0^{-1}(C)$. Therefore eq.(4.5.17) is changed into

$$
\begin{aligned}
W[J] &= N\int[\mathcal{D}C] \;\; \Delta_f(C)\cdot\left[\prod_{\beta,z}\delta(f^\beta(C(z))-\varphi^\beta(z))\right] \\
&\cdot exp\left\{i\int \mathrm{d}^4x(\mathcal{L}+J_\alpha^\mu\cdot{}^{g_0}C_\mu^\alpha)\right\} \\
&\cdot J_1(g_0^{-1})J_2(g_0)\cdot\int[\mathcal{D}g]\left[\prod_{\alpha,y}\delta({}^{g^{-1}}C_0^\alpha(y))\right].
\end{aligned}
\tag{4.5.18}
$$

The last line in eq.(4.5.18) will cause no trouble in renormalization, and if we consider the contribution from ghost fields which will be introduced below, it will becomes a quantity which is independent of gravitational gauge field. So, we put it into normalization constant N and still denote the new normalization constant as N. We also change $J_\alpha^\mu\,{}^{g_0}C_\mu^\alpha$ into $J_\alpha^\mu C_\mu^\alpha$, this will cause no trouble in renormalization. Then we get

$$
\begin{aligned}
W[J] &= N\int[\mathcal{D}C] \;\; \Delta_f(C)\cdot[\prod_{\beta,z}\delta(f^\beta(C(z))-\varphi^\beta(z))] \\
&\cdot exp\{i\int \mathrm{d}^4x(\mathcal{L}+J_\alpha^\mu C_\mu^\alpha)\}.
\end{aligned}
\tag{4.5.19}
$$

In fact, we can use this formula as our start-point of path integral quantization of gravitational gauge field, so we need not worried about the influences of the third line in eq.(4.5.18).

Use another functional

$$exp\left\{-\frac{i}{2\alpha}\int \mathrm{d}^4x\eta_{\alpha\beta}\varphi^{\alpha}(x)\varphi^{\beta}(x)\right\}, \tag{4.5.20}$$

times both sides of eq.(4.5.19) and then make functional integration $\int[\mathcal{D}\varphi]$, we get

$$W[J]=N\int[\mathcal{D}C]\ \ \Delta_f(C)\cdot exp\left\{i\int \mathrm{d}^4x(\mathcal{L}-\frac{1}{2\alpha}\eta_{\alpha\beta}f^{\alpha}f^{\beta}+J^{\mu}_{\alpha}C^{\alpha}_{\mu})\right\}. \tag{4.5.21}$$

Now, let's discuss the contribution from $\Delta_f(C)$ which is related to the ghost fields. Suppose that $g=\hat{U}_{\epsilon}$ is an infinitesimal gravitational gauge transformation. Then eq.(4.3.12) gives out

$${}^{g}C^{\alpha}_{\mu}(x)=C^{\alpha}_{\mu}(x)-\frac{1}{g}\mathbf{D}^{\alpha}_{\mu\ \sigma}\epsilon^{\sigma}, \tag{4.5.22}$$

where

$$\mathbf{D}^{\alpha}_{\mu\ \sigma}=\delta^{\alpha}_{\sigma}\partial_{\mu}-g\delta^{\alpha}_{\sigma}C^{\beta}_{\mu}\partial_{\beta}+g\partial_{\sigma}C^{\alpha}_{\mu}. \tag{4.5.23}$$

In order to deduce eq.(4.5.22), the following relation is used

$$\Lambda^{\alpha}{}_{\beta}=\delta^{\alpha}_{\beta}+\partial_{\beta}\epsilon^{\alpha}+o(\epsilon^2). \tag{4.5.24}$$

$\mathbf{D}_{\mu}$ can be regarded as the covariant derivative in adjoint representation, for

$$\mathbf{D}_{\mu}\epsilon=[D_{\mu}\quad,\quad\epsilon], \tag{4.5.25}$$

$$(\mathbf{D}_{\mu}\epsilon)^{\alpha}=\mathbf{D}^{\alpha}_{\mu\ \sigma}\epsilon^{\sigma}. \tag{4.5.26}$$

Using all these relations, we have,

$$f^{\alpha}({}^{g}C(x))=f^{\alpha}(C)-\frac{1}{g}\int \mathrm{d}^4y\frac{\delta f^{\alpha}(C(x))}{\delta C^{\beta}_{\mu}(y)}\mathbf{D}^{\beta}_{\mu\ \sigma}(y)\epsilon^{\sigma}(y)+o(\epsilon^2). \tag{4.5.27}$$

Therefore, according to eq.(4.5.7) and eq.(4.5.6), we get

$$\Delta_f^{-1}(C)=\int[\mathcal{D}\epsilon]\prod_{x,\alpha}\delta\left(-\frac{1}{g}\int \mathrm{d}^4y\frac{\delta f^{\alpha}(C(x))}{\delta C^{\beta}_{\mu}(y)}\mathbf{D}^{\beta}_{\mu\ \sigma}(y)\epsilon^{\sigma}(y)\right). \tag{4.5.28}$$

Define

$$\begin{aligned}\mathbf{M}^{\alpha}{}_{\sigma}(x,y) &= -g\frac{\delta}{\delta\epsilon^{\sigma}(y)}f^{\alpha}({}^{g}C(x))\\ &= \int \mathrm{d}^4z\frac{\delta f^{\alpha}(C(x))}{\delta C^{\beta}_{\mu}(z)}\mathbf{D}^{\beta}_{\mu\ \sigma}(z)\delta(z-y).\end{aligned} \tag{4.5.29}$$

Then eq.(4.5.28) is changed into

$$\begin{aligned} \Delta_f^{-1}(C) &= \int[\mathcal{D}\epsilon]\prod_{x,\alpha}\delta\left(-\frac{1}{g}\int \mathrm{d}^4y \mathbf{M}^{\alpha}{}_{\sigma}(x,y)\epsilon^{\sigma}(y)\right) \\ &= const. \times (det\mathbf{M})^{-1}. \end{aligned} \tag{4.5.30}$$

Therefore,

$$\Delta_f(C) = const. \times det\mathbf{M}. \tag{4.5.31}$$

Put the above constant into normalization constant, then generating functional eq.(4.5.21) is changed into

$$W[J] = N\int[\mathcal{D}C] \;\; det\mathbf{M}\cdot exp\left\{i\int \mathrm{d}^4x(\mathcal{L} - \frac{1}{2\alpha}\eta_{\alpha\beta}f^{\alpha}f^{\beta} + J^{\mu}_{\alpha}C^{\alpha}_{\mu})\right\}. \tag{4.5.32}$$

In order to evaluate the contribution from $det\mathbf{M}$, we introduce ghost fields $\eta^{\alpha}(x)$ and $\bar{\eta}_{\alpha}(x)$. Using the following relation

$$\int[\mathcal{D}\eta][\mathcal{D}\bar{\eta}]exp\left\{i\int \mathrm{d}^4x\mathrm{d}^4y\;\bar{\eta}_{\alpha}(x)\mathbf{M}^{\alpha}{}_{\beta}(x,y)\eta^{\beta}(y)\right\} = const. \times det\mathbf{M} \tag{4.5.33}$$

and put the constant into the normalization constant, we can get

$$W[J] = N\int[\mathcal{D}C][\mathcal{D}\eta][\mathcal{D}\bar{\eta}]exp\left\{i\int \mathrm{d}^4x(\mathcal{L} - \frac{1}{2\alpha}\eta_{\alpha\beta}f^{\alpha}f^{\beta} + \bar{\eta}\mathbf{M}\eta + J^{\mu}_{\alpha}C^{\alpha}_{\mu})\right\}, \tag{4.5.34}$$

where $\int \mathrm{d}^4x\bar{\eta}\mathbf{M}\eta$ is a simplified notation, whose explicit expression is

$$\int \mathrm{d}^4x\bar{\eta}\mathbf{M}\eta = \int \mathrm{d}^4x\mathrm{d}^4y\;\bar{\eta}_{\alpha}(x)\mathbf{M}^{\alpha}{}_{\beta}(x,y)\eta^{\beta}(y). \tag{4.5.35}$$

The appearance of the non-trivial ghost fields is a inevitable result of the non-Able nature of the gravitational gauge group.

Set external source J^{μ}_{α} to zero, we get,

$$W[0] = N\int[\mathcal{D}C][\mathcal{D}\eta][\mathcal{D}\bar{\eta}]exp\left\{i\int \mathrm{d}^4x(\mathcal{L} - \frac{1}{2\alpha}\eta_{\alpha\beta}f^{\alpha}f^{\beta} + \bar{\eta}\mathbf{M}\eta)\right\}, \tag{4.5.36}$$

Now, let's take Lorentz covariant gauge condition,

$$f^{\alpha}(C) = \partial^{\mu}C^{\alpha}_{\mu}. \tag{4.5.37}$$

Then

$$\int \mathrm{d}^4x\bar{\eta}\mathbf{M}\eta = -\int \mathrm{d}^4x\,(\partial^{\mu}\bar{\eta}_{\alpha}(x))\,\mathbf{D}^{\alpha}_{\mu\;\beta}(x)\eta^{\beta}(x). \tag{4.5.38}$$

And eq.(4.5.36) is changed into

$$W[0] = N \int [\mathcal{D}C][\mathcal{D}\eta][\mathcal{D}\bar{\eta}]exp\left\{ i \int \mathrm{d}^4x(\mathcal{L} - \frac{1}{2\alpha}\eta_{\alpha\beta}f^\alpha f^\beta - (\partial^\mu\bar{\eta}_\alpha)\mathbf{D}_{\mu}^{\alpha}{}_{\sigma}\eta^\sigma) \right\}. \tag{4.5.39}$$

For quantum gauge theory of gravity, the external source of gravitational gauge field should be introduced in a special way. Define the generating functional with external sources as

$$\begin{aligned} W[J,\beta,\bar{\beta}] &= N \int[\mathcal{D}C][\mathcal{D}\eta][\mathcal{D}\bar{\eta}]exp\left\{ i \int \mathrm{d}^4x(\mathcal{L} - \tfrac{1}{2\alpha}\eta_{\alpha\beta}f^\alpha f^\beta \right. \\ & \left. -(\partial^\mu\bar{\eta}_\alpha)\mathbf{D}_{\mu}^{\alpha}{}_{\sigma}\eta^\sigma + C_\mu^\alpha \stackrel{\sim\mu\beta}{\delta}_{\alpha\nu}(x) J_{0\beta}^\nu + \bar{\eta}_\alpha\beta^\alpha + \bar{\beta}_\alpha\eta^\alpha)\right\} \\ &= N \int[\mathcal{D}C][\mathcal{D}\eta][\mathcal{D}\bar{\eta}]exp\left\{ i \int \mathrm{d}^4x(\mathcal{L} - \tfrac{1}{2\alpha}\eta_{\alpha\beta}f^\alpha f^\beta \right. \\ & \left. -(\partial^\mu\bar{\eta}_\alpha)\mathbf{D}_{\mu}^{\alpha}{}_{\sigma}\eta^\sigma + C_\mu^\alpha J_\alpha^\mu + \bar{\eta}_\alpha\beta^\alpha + \bar{\beta}_\alpha\eta^\alpha)\right\}, \end{aligned} \tag{4.5.40}$$

where $\stackrel{\sim\mu\gamma}{\delta}_{\alpha\rho}(x)$ is defined by

$$\stackrel{\sim\mu\gamma}{\delta}_{\alpha\rho}(x) \triangleq \frac{1}{2}\left(\stackrel{\sim\mu}{\delta}_{\rho}(x) \stackrel{\sim\gamma}{\delta}_{\alpha}(x) + \stackrel{\sim\mu\gamma}{\eta}(x) \stackrel{\sim}{\eta}_{\alpha\rho}(x) \right), \tag{4.5.41}$$

and

$$J_\alpha^\mu \triangleq \stackrel{\sim\mu\beta}{\delta}_{\alpha\nu}(x) J_{0\beta}^\nu. \tag{4.5.42}$$

In the above definition, $\stackrel{\sim\mu}{\delta}_{\rho}(x)$, $\stackrel{\sim\mu\gamma}{\eta}(x)$ and $\stackrel{\sim}{\eta}_{\mu\gamma}(x)$ are defined by

$$\stackrel{\sim\mu}{\delta}_{\rho}(x) = \delta_\rho^\mu - \frac{\partial^\mu\partial_\rho}{\Box + i\epsilon}, \tag{4.5.43}$$

$$\stackrel{\sim\mu\gamma}{\eta}(x) = \eta^{\mu\gamma} - \frac{\partial^\mu\partial^\gamma}{\Box + i\epsilon}, \tag{4.5.44}$$

$$\stackrel{\sim}{\eta}_{\mu\gamma}(x) = \eta_{\mu\gamma} - \frac{\partial_\mu\partial_\gamma}{\Box + i\epsilon}, \tag{4.5.45}$$

where

$$\Box \triangleq \partial^2 = \partial^\mu\partial_\mu = \eta^{\mu\nu}\partial_\mu\partial_\nu. \tag{4.5.46}$$

Using these relations, we can prove that

$$J_\alpha^\mu = \stackrel{\sim\mu\beta}{\delta}_{\alpha\nu} J_\beta^\nu. \tag{4.5.47}$$

The effective Lagrangian $\mathcal{L}_{eff}$ is defined by

$$\mathcal{L}_{eff} \equiv \mathcal{L} - \frac{1}{2\alpha}\eta_{\alpha\beta} f^{\alpha} f^{\beta} - (\partial^{\mu}\bar{\eta}_{\alpha})\mathbf{D}_{\mu\ \sigma}^{\alpha}\eta^{\sigma}. \tag{4.5.48}$$

$\mathcal{L}_{eff}$ can be separated into free Lagrangian $\mathcal{L}_F$ and interaction Lagrangian $\mathcal{L}_I$,

$$\mathcal{L}_{eff} = \mathcal{L}_F + \mathcal{L}_I, \tag{4.5.49}$$

where

$$\begin{aligned} \mathcal{L}_F &= -\tfrac{1}{16}\eta^{\mu\rho}\eta^{\nu\sigma}\eta_{\alpha\beta}F_{0\mu\nu}^{\alpha}F_{0\rho\sigma}^{\beta} - \tfrac{1}{8}\eta^{\mu\rho}F_{0\mu\beta}^{\alpha}F_{0\rho\alpha}^{\beta} + \tfrac{1}{4}\eta^{\mu\rho}F_{0\mu\alpha}^{\alpha}F_{0\rho\beta}^{\beta} \\ &\quad -\tfrac{1}{2\alpha}\eta_{\alpha\beta}(\partial^{\mu}C_{\mu}^{\alpha})(\partial^{\nu}C_{\nu}^{\beta}) - (\partial^{\mu}\bar{\eta}_{\alpha})(\partial_{\mu}\eta^{\alpha}), \end{aligned} \tag{4.5.50}$$

$$\begin{aligned} \mathcal{L}_I &= +g(\partial^{\mu}\bar{\eta}_{\alpha})C_{\mu}^{\beta}(\partial_{\beta}\eta^{\alpha}) - g(\partial^{\mu}\bar{\eta}_{\alpha})(\partial_{\sigma}C_{\mu}^{\alpha})\eta^{\sigma} \\ &\quad + self\ interaction\ terms\ of\ Gravitational\ gauge\ field. \end{aligned} \tag{4.5.51}$$

From the interaction Lagrangian, we can see that ghost fields do not couple to $J(C)$. This is the reflection of the fact that ghost fields are not physical fields, they are virtual fields. Besides, the gauge fixing term does not couple to $J(C)$ either. Using effective Lagrangian $\mathcal{L}_{eff}$, the generating functional $W[J,\beta,\bar{\beta}]$ can be simplified to

$$W[J,\beta,\bar{\beta}] = N\int[\mathcal{D}C][\mathcal{D}\eta][\mathcal{D}\bar{\eta}]exp\left\{i\int \mathrm{d}^4x(\mathcal{L}_{eff} + J_{\alpha}^{\mu}C_{\mu}^{\alpha} + \bar{\eta}_{\alpha}\beta^{\alpha} + \bar{\beta}_{\alpha}\eta^{\alpha})\right\}, \tag{4.5.52}$$

Using eq.(4.5.50), we can deduce propagator of gravitational gauge fields and ghost fields. First, after a partial integration, we change the form of eq. (4.5.50) into

$$\int \mathrm{d}^4x\mathcal{L}_F = \int \mathrm{d}^4x\left\{\frac{1}{2}C_{\mu}^{\alpha}\mathbb{M}_{\alpha\beta}^{\mu\nu}(x)C_{\nu}^{\beta} + \bar{\eta}_{\alpha}\partial^2\eta^{\alpha}\right\}, \tag{4.5.53}$$

where the operator $\mathbb{M}_{\alpha\beta}^{\mu\nu}(x)$ is defined by

$$\begin{aligned} \mathbb{M}_{\alpha\beta}^{\mu\nu}(x) &= \tfrac{1}{4}\eta^{\mu\nu}\eta_{\alpha\beta}\partial^{\rho}\partial_{\rho} - \tfrac{1}{4}\eta_{\alpha\beta}(1-\tfrac{4}{\alpha})\partial^{\mu}\partial^{\nu} - \tfrac{1}{4}\delta_{\beta}^{\mu}\partial^{\nu}\partial_{\alpha} \\ &\quad +\tfrac{1}{4}\delta_{\beta}^{\mu}\delta_{\alpha}^{\nu}\partial^{\rho}\partial_{\rho} - \tfrac{1}{4}\delta_{\alpha}^{\nu}\partial^{\mu}\partial_{\beta} + \tfrac{1}{2}\delta_{\beta}^{\nu}\partial^{\mu}\partial_{\alpha} \\ &\quad -\tfrac{1}{4}\eta^{\mu\nu}\partial_{\alpha}\partial_{\beta} - \tfrac{1}{2}\delta_{\alpha}^{\mu}\delta_{\beta}^{\nu}\partial^{\rho}\partial_{\rho} + \tfrac{1}{2}\delta_{\alpha}^{\mu}\partial^{\nu}\partial_{\beta}. \end{aligned} \tag{4.5.54}$$

Denote the propagator of gravitational gauge field as

$$-i\Delta_{F\mu\nu}^{\alpha\beta}(x), \tag{4.5.55}$$

and denote the propagator of ghost field as

$$-i\Delta^{\alpha}_{F\beta}(x). \tag{4.5.56}$$

They satisfy the following equation,

$$-\mathbb{M}^{\mu\nu}_{\alpha\beta}(x)\Delta^{\beta\gamma}_{F\nu\rho}(x) = \overset{\sim\mu\gamma}{\delta}_{\alpha\rho}(x)\delta(x), \tag{4.5.57}$$

$$-\partial^2\Delta^{\alpha}_{F\beta}(x) = \delta^{\alpha}_{\beta}\delta(x), \tag{4.5.58}$$

where $\overset{\sim\mu\gamma}{\delta}_{\alpha\rho}(x)$ is defined by (4.5.41).

Make Fourier transformations to momentum space

$$\Delta^{\alpha\beta}_{F\mu\nu}(x) = \int \frac{\mathrm{d}^4 k}{(2\pi)^4} \overset{\sim\alpha\beta}{\Delta}_{F\mu\nu}(k) \cdot e^{ikx}, \tag{4.5.59}$$

$$\Delta^{\alpha}_{F\beta}(x) = \int \frac{\mathrm{d}^4 k}{(2\pi)^4} \overset{\sim\alpha}{\Delta}_{F\beta}(k) \cdot e^{ikx}, \tag{4.5.60}$$

where $\overset{\sim\alpha\beta}{\Delta}_{F\mu\nu}(k)$ and $\overset{\sim\alpha}{\Delta}_{F\beta}(k)$ are corresponding propagators in momentum space. They satisfy the following equations,

$$-\mathbb{M}^{\mu\nu}_{\alpha\beta} \overset{\sim\beta\gamma}{\Delta}_{F\nu\rho}(k) = \overset{\sim\mu\gamma}{\delta}_{\alpha\rho}, \tag{4.5.61}$$

$$k^2 \overset{\sim\alpha}{\Delta}_{F\beta}(k) = \delta^{\alpha}_{\beta}, \tag{4.5.62}$$

where the operator $\mathbb{M}^{\mu\nu}_{\alpha\beta}$ is defined by

$$\begin{aligned}
\mathbb{M}^{\mu\nu}_{\alpha\beta} &\triangleq \mathbb{M}^{\mu\nu}_{\alpha\beta}(k) \\
&\triangleq \tfrac{1}{4}\eta^{\mu\nu}\eta_{\alpha\beta}k^2 - \tfrac{1}{4}\eta_{\alpha\beta}(1-\tfrac{4}{\alpha})k^{\mu}k^{\nu} - \tfrac{1}{4}\delta^{\mu}_{\beta}k^{\nu}k_{\alpha} \\
&\quad +\tfrac{1}{4}\delta^{\mu}_{\beta}\delta^{\nu}_{\alpha}k^2 - \tfrac{1}{4}\delta^{\nu}_{\alpha}k^{\mu}k_{\beta} + \tfrac{1}{2}\delta^{\nu}_{\beta}k^{\mu}k_{\alpha} \\
&\quad -\tfrac{1}{4}\eta^{\mu\nu}k_{\alpha}k_{\beta} - \tfrac{1}{2}\delta^{\mu}_{\alpha}\delta^{\nu}_{\beta}k^2 + \tfrac{1}{2}\delta^{\mu}_{\alpha}k^{\nu}k_{\beta},
\end{aligned} \tag{4.5.63}$$

and $\overset{\sim\mu\gamma}{\delta}_{\alpha\rho}$ is defined by

$$\overset{\sim\mu\gamma}{\delta}_{\alpha\rho} \triangleq \overset{\sim\mu\gamma}{\delta}_{\alpha\rho}(k) = \frac{1}{2}\left(\overset{\sim\mu}{\delta}_{\rho}\overset{\sim\gamma}{\delta}_{\alpha} + \overset{\sim\mu\gamma}{\eta}\overset{\sim}{\eta}_{\alpha\rho}\right). \tag{4.5.64}$$

The operator $\mathbb{M}^{\mu\nu}_{\alpha\beta}$ has the following symmetric property

$$\mathbb{M}^{\mu\nu}_{\alpha\beta} = \mathbb{M}^{\nu\mu}_{\beta\alpha}. \tag{4.5.65}$$

In the above relation, $\tilde{\delta}^{\mu}_{\rho}$, $\tilde{\eta}^{\mu\gamma}$ and $\tilde{\eta}_{\mu\gamma}$ are defined by

$$\tilde{\delta}^{\mu}_{\rho} = \delta^{\mu}_{\rho} - \frac{k^{\mu}k_{\rho}}{k^2 - i\epsilon}, \tag{4.5.66}$$

$$\tilde{\eta}^{\mu\gamma} = \eta^{\mu\gamma} - \frac{k^{\mu}k^{\gamma}}{k^2 - i\epsilon}, \tag{4.5.67}$$

$$\tilde{\eta}_{\mu\gamma} = \eta_{\mu\gamma} - \frac{k_{\mu}k_{\gamma}}{k^2 - i\epsilon}. \tag{4.5.68}$$

It can be proved that $\tilde{\delta}^{\mu\gamma}_{\alpha\rho}$ satisfies the following relation

$$\tilde{\delta}^{\mu\gamma}_{\alpha\rho} \cdot \tilde{\delta}^{\rho\beta}_{\gamma\nu} = \tilde{\delta}^{\mu\beta}_{\alpha\nu} \cdot \tag{4.5.69}$$

The solutions to the two propagator equations (4.5.61) and (4.5.62) give out the propagators in momentum space,

$$-i\,\tilde{\Delta}^{\alpha\beta}_{F\mu\nu}(k) = \frac{-i}{k^2 - i\epsilon}\left[\tilde{\eta}_{\mu\nu}\tilde{\eta}^{\alpha\beta} + \tilde{\delta}^{\beta}_{\mu}\tilde{\delta}^{\alpha}_{\nu} - \tilde{\delta}^{\alpha}_{\mu}\tilde{\eta}^{\beta}_{\nu}\right], \tag{4.5.70}$$

$$-i\,\tilde{\Delta}^{\alpha}_{F\beta}(k) = \frac{-i}{k^2 - i\epsilon}\delta^{\alpha}_{\beta}. \tag{4.5.71}$$

The forms of these propagators are quite beautiful and symmetric. $\tilde{\Delta}^{\alpha\beta}_{F\mu\nu}(k)$ satisfies the following relation

$$\tilde{\Delta}^{\alpha\beta}_{F\mu\nu}(k) \cdot \tilde{\delta}^{\nu\gamma}_{\beta\rho} = \tilde{\Delta}^{\alpha\gamma}_{F\mu\rho}(k). \tag{4.5.72}$$

Its Fourier transformation gives out the following relation

$$\tilde{\delta}^{\nu\gamma}_{\beta\rho}(x) \cdot \Delta^{\alpha\beta}_{F\mu\nu}(x) = \Delta^{\alpha\gamma}_{F\mu\rho}(x). \tag{4.5.73}$$

The above solution (4.5.70) is not the most general solution to equation (4.5.61). The most general solution to the equation (4.5.61) is

$$\begin{aligned} -i\,\tilde{\Delta}^{\alpha\beta}_{F\mu\nu}(k) \;=\; & \tfrac{-i}{k^2-i\epsilon}\Big[c_1\eta_{\mu\nu}\eta^{\alpha\beta} - c_1\eta^{\alpha\beta}\tfrac{k_{\mu}k_{\nu}}{k^2-i\epsilon} + c_3\eta_{\mu\nu}\tfrac{k^{\alpha}k^{\beta}}{k^2-i\epsilon} \\ & + c_2\delta^{\beta}_{\mu}\delta^{\alpha}_{\nu} - c_2\delta^{\alpha}_{\nu}\tfrac{k_{\mu}k^{\beta}}{k^2-i\epsilon} + c_4\delta^{\beta}_{\mu}\tfrac{k_{\nu}k^{\alpha}}{k^2-i\epsilon} \\ & - \delta^{\alpha}_{\mu}\delta^{\beta}_{\nu} + \delta^{\beta}_{\nu}\tfrac{k_{\mu}k^{\alpha}}{k^2-i\epsilon} + \delta^{\alpha}_{\mu}\tfrac{k_{\nu}k^{\beta}}{k^2-i\epsilon} \\ & + c_5\tfrac{k_{\mu}k_{\nu}k^{\alpha}k^{\beta}}{(k^2-i\epsilon)^2}\Big], \end{aligned} \tag{4.5.74}$$

where parameters c_1, c_2, c_3, c_4 and c_5 satisfy the following restriction

$$c_1 + c_2 = 2, \tag{4.5.75}$$

$$c_3 + c_4 + c_5 = -1. \tag{4.5.76}$$

Set $c_1 = c_2 = c_5 = 1$ and $c_3 = c_4 = -1$, we got the solution (4.5.70).

The interaction Lagrangian $\mathcal{L}_I$ is a function of gravitational gauge field C_μ^α and ghost fields η^α and $\bar{\eta}_\alpha$,

$$\mathcal{L}_I = \mathcal{L}_I(C, \eta, \bar{\eta}). \tag{4.5.77}$$

Then eq.(4.5.52) is changed into,

$$\begin{aligned} W[J, \beta, \bar{\beta}] &= N \int [\mathcal{D}C][\mathcal{D}\eta][\mathcal{D}\bar{\eta}] \; exp \left\{ i \int \mathrm{d}^4x \mathcal{L}_I(C, \eta, \bar{\eta}) \right\} \\ &\quad \cdot exp \left\{ i \int \mathrm{d}^4x (\mathcal{L}_F + J_\alpha^\mu C_\mu^\alpha + \bar{\eta}_\alpha \beta^\alpha + \bar{\beta}_\alpha \eta^\alpha) \right\} \\ &= exp \left\{ i \int \mathrm{d}^4x \mathcal{L}_I \left(\frac{1}{i} \frac{\delta}{\delta J}, \frac{1}{i} \frac{\delta}{\delta \bar{\beta}}, \frac{1}{-i} \frac{\delta}{\delta \beta} \right) \right\} \cdot W_0[J, \beta, \bar{\beta}], \end{aligned} \tag{4.5.78}$$

where

$$\begin{aligned} W_0[J, \beta, \bar{\beta}] &= N \int [\mathcal{D}C][\mathcal{D}\eta][\mathcal{D}\bar{\eta}] exp \left\{ i \int \mathrm{d}^4x \left(\mathcal{L}_F + J_\alpha^\mu C_\mu^\alpha + \bar{\eta}_\alpha \beta^\alpha + \bar{\beta}_\alpha \eta^\alpha \right) \right\} \\ &= N \int [\mathcal{D}C][\mathcal{D}\eta][\mathcal{D}\bar{\eta}] exp \left\{ i \int \mathrm{d}^4x \left(\frac{1}{2} C_\mu^\alpha \mathbb{M}_{\alpha\beta}^{\mu\nu}(x) C_\nu^\beta + \bar{\eta}_\alpha \partial^2 \eta^\alpha \right. \right. \\ &\quad \left. \left. + J_\alpha^\mu C_\mu^\alpha + \bar{\eta}_\alpha \beta^\alpha + \bar{\beta}_\alpha \eta^\alpha \right) \right\} \\ &= exp \left\{ i \int \int \mathrm{d}^4x \mathrm{d}^4y \left[\frac{1}{2} J_\alpha^\mu(x) \Delta_{F\mu\nu}^{\alpha\beta}(x - y) J_\beta^\nu(y) \right. \right. \\ &\quad \left. \left. + \bar{\beta}_\alpha(x) \Delta_{F\beta}^\alpha(x - y) \beta^\beta(y) \right] \right\}. \end{aligned} \tag{4.5.79}$$

In order to obtain the above relation, eq. (4.5.47) is used.

The interaction Feynman rules for interaction vertices can be obtained from the interaction Lagrangian $\mathcal{L}_I$. For example, the interaction Lagrangian between gravitational gauge field and ghost field is

$$+g(\partial^\mu \bar{\eta}_\alpha) C_\mu^\beta (\partial^\beta \eta^\alpha) - g(\partial^\mu \bar{\eta}_\alpha)(\partial_\sigma C_\mu^\alpha) \eta^\sigma. \tag{4.5.80}$$

This vertex belongs to $C_\mu^\alpha(k) \bar{\eta}_\beta(q) \eta^\delta(p)$ three body interactions, its Feynman rule is

$$-ig\delta_\delta^\beta q^\mu p_\alpha + ig\delta_\alpha^\beta q^\mu k_\delta. \tag{4.5.81}$$

Feynman rules for other interaction vertices can be obtained similarly. We will not calculate Feynman rules for all kinds of interaction vertices here. For interaction vertices which involve in gravitational gauge field, we can find that all Feynman rules for interaction vertices are proportional to energy-momenta of one or more particles, which is one of the most important properties of gravitational interactions. In fact, this interaction property is expected for gravitational interactions, for energy-momentum is the the source of gravity.

4.6 Renormalization

In the quantum gauge theory of gravity, the gravitational coupling constant has the dimensionality of negative powers of mass. According to traditional theory of power counting law, it seems that the quantum gauge theory of gravity is a kind of non-renormalizable theory. But this result is not correct. The power counting law does not work here. General speaking, power counting law does not work when a theory has gauge symmetry. If a theory has gauge symmetry, the constraints from gauge symmetry will make some divergence cancel each other[50, 51, 52, 53, 54, 55]. In the quantum gauge theory of gravity, this mechanism works very well. It can be strictly proved that quantum gauge theory of gravity is renormalizable. It is found that the effect of renormalization is just a scale transformation of the original theory. Though there are infinite number of divergent vertexes in the quantum gauge theory of gravity, we need not introduce infinite number of interaction terms that do not exist in the original Lagrangian and infinite number of parameters. All the divergent vertex can find its correspondence in the original Lagrangian. Therefor, in renormalization, what we need to do is not to introduce extra interaction terms to cancel divergent terms, but to redefine the fields, coupling constants and some other parameters of the original theory. Because most of counterterms come from the factor $J(C)$, this factor is key important for renormalization. Without this factor, the theory is non-renormalizable. In a word, the quantum gauge theory of gravity is a renormalizable quantum gauge theory. However, the proof on the renormalizability of the theory is lengthy and complicated. Because of limited space, we will not discuss it. Details can be found in literature [31].

4.7 Classical Tests of Quantum Gauge theory of gravity

In the above chapters, the quantum gravity is formulated in the traditional framework of quantum field theory, i.e., the physical space-time is always flat and the space-time metric is always selected to be the Minkowski metric. In this picture, gravity is treated as physical interactions in flat physical space-time. Our gravitational gauge transformation does not act on physical space-time coordinates, but act on physical fields, so gravitational gauge transformation does not affect the structure of physical space-time. This is the physics picture of gravity.

According to above discussions, there is another picture of gravity which is widely used in Einstein's general relativity, i.e., the geometrical picture of gravity. For gravitational gauge theory, if we treat G^{α}_{μ} (or $G^{-1\mu}_{\alpha}$) as a fundamental physical input of the theory, we can also set up the geometrical picture of gravity[48]. For gravitational gauge theory, the geometrical nature of gravity essentially originates from the geometrical nature of the gravitational gauge transformation. In the geometrical picture of gravity, gravity is not treated as a kind of physical interactions, but is treated as the geometry of space-time. So, there is no physical gravitational interactions in space-time and space-time is curved if gravitational field does not vanish. In this case, the space-time metric is not Minkowski metric, but $g^{\alpha\beta}$ (or $g_{\alpha\beta}$). In other words, Minkowski metric is the space-time metric if we discuss gravity in physics picture of gravity while metric tensor $g^{\alpha\beta}$ (or $g_{\alpha\beta}$) is space-time metric if we discuss gravity in geometrical picture of gravity. So, if we use Minkowski metric in our discussion, it means that we are in physics picture of gravity; if we use metric tensor $g^{\alpha\beta}$ (or $g_{\alpha\beta}$) in our discussion, it means that we are in geometrical picture of gravity.

In one picture, gravity is treated as physical interactions, while in another picture, gravity is treated as geometry of space-time. But we know that there is only one physics for gravity, so two pictures of gravity must be equivalent. This equivalence means that the nature of gravity is physics-geometry duality, i.e., gravity is a kind of physical interactions which has the characteristics of geometry; it is also a geometry of space-time which is essentially a kind of physical interactions.

When we discuss quantization and renormalization of quantum gauge theory of gravity, it is better to perform all discussions in physics picture of gravity; but when we discuss classical effects of gravity, it's convenient for us to perform our discussions in geometrical picture of gravity. In this chapter, we will discuss classical tests of gravity and our discussions are in geometrical picture of gravity.

In quantum gauge theory of gravity, the field equation of gravitational gauge field is the same as that of the Einstein's field equation. It is known that, the Einstein's field equation for empty space has Schwarzschild solution, it is expected that the field equation of gravitational gauge field in empty space should also have Schwarzschild solution. In spherical coordinate system (t, r, θ, ϕ), the simplest form of gravitational gauge field which has spherical symmetry is

$$C_\mu^\alpha = \begin{pmatrix} u(r) & 0 & 0 & 0 \\ 0 & v(r) & 0 & 0 \\ 0 & 0 & 0 & 0 \\ 0 & 0 & 0 & 0 \end{pmatrix}. \tag{4.7.1}$$

We can prove that, when we select

$$gu(r) = 1 - \frac{1}{\sqrt{1 - \frac{2GM}{r}}}, \tag{4.7.2}$$

$$gv(r) = 1 - \sqrt{1 - \frac{2GM}{r}}, \tag{4.7.3}$$

gravitational gauge field C_μ^α given by (4.7.1) is just the solution of field equation (4.3.73) for pure gravitational gauge field. (4.7.1) is just the Schwarzschild solution in quantum gauge theory of gravity. Using relation (4.3.20), we can get

$$g_{\alpha\beta} = \begin{pmatrix} -1 + \frac{2GM}{r} & 0 & 0 & 0 \\ 0 & \frac{1}{1-\frac{2GM}{r}} & 0 & 0 \\ 0 & 0 & r^2 & 0 \\ 0 & 0 & 0 & r^2 \sin^2\theta \end{pmatrix}, \tag{4.7.4}$$

which is the standard Schwarzschild solution in general relativity. We can easily check that (4.7.4) is the solution to the field equation (4.3.96) when $T^{\alpha\beta}$ vanishes.

As we have stated above, in the geometrical picture of gravity, gravity is not treated as physical interactions in space-time. In the geometrical picture of gravity, if there is no other physical interactions in space time, any mass point can not feel any physical forces when it moves in space-time. So, it must move along the curve which has the least length. Suppose that a particle is moving from point A to point B along a curve. Define

$$T_{BA} = \int_A^B \sqrt{-g_{\mu\nu} \frac{dx^\mu}{dp} \frac{dx^\nu}{dp}} dp, \tag{4.7.5}$$

where p is a parameter that describe the orbit. The real curve that the particle moving along corresponds to the minimum of T_{AB}. The minimum of T_{AB} gives out the following geodesic curve equation

$$\frac{d^2 x^\mu}{dp^2} + \Gamma^\mu_{\nu\lambda} \frac{dx^\nu}{dp} \frac{dx^\lambda}{dp} = 0, \tag{4.7.6}$$

where $\Gamma^\mu_{\nu\lambda}$ is the affine connection defined by equation (4.3.33). Eq.(4.7.6) gives out the curve that a free particle moves along in curved space-time if we discuss physics in the geometrical picture of gravity.

Combine Schwarzschild solution (4.7.4) with the geodesic curve equation (4.7.6), theoretical predictions on classical tests of gravity can be obtained. Details on this calculation can be found in literature [45, 46] and many other standard textbook on general relativity. Here, I only list the final results. For the deflection of light by sun, the theoretical prediction is

$$\Delta\varphi = \frac{4GM}{r_0}, \tag{4.7.7}$$

where M is the mass of the sun and r_0 is the distance of closest approach to the sun. For the procession of perihelia, the theoretical prediction is

$$\Delta\varphi = \frac{6\pi GM}{L}, \tag{4.7.8}$$

where L is the semilatus rectum. For the time delay of radar echo, the theoretical prediction is

$$(\Delta t)_{max} \simeq 4GM \left\{ 1 + \ln\left(\frac{4r_1 r_2}{R_0^2} \right) \right\}, \tag{4.7.9}$$

where r_1 and r_2 are the distances from the center of the sun to the radar antenna on earth, and the point on Mercury's surface closest to the earth, and R_0 is the radius of the sun.

For the precession of orbiting gyroscope, quantum gauge theory of gravity should also give out the same theoretical predictions as that of Einstein's general relativity. In fact, quantum gauge theory of gravity and Einstein's general relativity should have the same classical effects of gravitational interactions, and they should give out the same theoretical predictions on any phenomenon which belongs to the classical effects of gravitational interactions, they have the same post-Newtonian approximation.

Robertson-Walker metric is a solution of Einstein's field equation, so it is also the solution of the field equation of gravitational gauge field in quantum gauge theory of gravity. It is known that, in general relativity, the cosmological model is dominated by classical gravitational interactions. Because quantum gauge theory of gravity has the fundamental laws of classical gravitational interactions as those of general relativity, so it will give out the same cosmological model as that of general relativity. All traditional calculations on cosmological model can be copied to the present model without any changes on the final results.

4.8 Discussions

In this paper, a renormalizable quantum gauge theory of gravity is formulated in the framework of quantum gauge theory of gravity, where gravity is treated as a kind of physical interactions and space-time is always kept flat. This treatment satisfies the fundamental spirit of traditional quantum field theory, and go along this way, four different kinds of fundamental interactions can be unified on the same fundamental baseline[33, 34, 35]. The most advantage of this approach is that the renormalizability of the quantum gravity can be proved. Gravitational gauge interactions is determined by gauge symmetry. In other words, the Lagrangian of the system is determined by gauge symmetry. Using the langurage of Cartan tetrad, we set up the geometrical formulation of the quantum gauge theory of gravity and to study its geometrical foundation. So, gravity theory has both physical picture and geometrical picture, which is the reflection of the physics-geometry duality of gravity. Finally, we give a simple summary to the whole theory.

1. In leading order approximation, the quantum gauge theory of gravity gives out classical Newton's theory of gravity.

2. In quantum gauge theory of gravity, the field equation of the gravitational gauge field is just the Einstein's field equation. In classical level, quantum gauge theory of gravity is the same as that of the Einstein's general relativity.

3. For classical tests of gravity, such as the gravitational red-shift, the perihelion procession, deflection of light, time delay of radar echo and the procession of a gyroscope, the quantum gauge theory of gravity gives out the same theoretical predictions as that of the Einstein's general relativity. It also gives out the same cosmological model as that of the Einstein's general relativity.

4. Quantum gauge theory of gravity is a perturbatively renormalizable quantum gravity.

5. Four different kinds of fundamental interactions can be easily unified in a simple and beautiful way in quantum gauge theory of gravity.

6. Quantum gauge theory of gravity can also explain quantum effects of gravitational interactions, such as gravitational phase effects and gravitational shielding effects.

In a word, we can say that quantum gauge theory of gravity is a self-consistent theory that unifies Einstein's general relativity with quantum theory.

References

[1] Isaac Newton, *Mathematical Principles of Natural Philosophy*, (Camgridge University Press, 1934) .

[2] Albert Einstein, *Annalen der Phys.*, **49** (1916) 769 .

[3] Albert Einstein, *Zeits. Math. und Phys.* **62** (1913) 225.

[4] Albert Einstein, *"Naeherungsweise Integration der Feldgleichungen der Gracitation,"* Preussische Akademie der Wissenschaften (Berlin) Sitzungsberichte, pg 688 (1916) .

[5] G.Amelino-Camelia, *Nature*, **408** (2000) 661-664

[6] M.Tegmark and J.A.Wheeler, *Sci. Am.* **284** (2001) 68-75

[7] H.Weyl, in *Raum. Zeit. Materia,* 3rd ed. (Springer-Verlag, Berlin, 1920)

[8] H.Weyl, *Gravitation and Elektrizitat.* Sitzungsber. Akademie der Wissenschaften Berlin, 465-480(1918). Siehe auch die *Gesammelten Abhandlungen.* 6 Vols. Ed. K.Chadrasekharan, Springer-Verlag.

[9] V.Fock, *Zeit. f. Physik*, **39** (1927) 226.

[10] H.Weyl, *Zeit. f. Physik*, **56** (1929) 330.

[11] W.Pauli, *Handbuch der physik, Geiger and scheel, 2, Aufl.*, Vol. 24, Teil **1** pp. 83-272.

[12] C.N.Yang, R.L.Mills, *Phys Rev* **96** (1954) 191 .

[13] S.Glashow, *Nucl Phys* **22**(1961) 579 .

[14] S.Weinberg, *Phys Rev Lett* **19** (1967) 1264 .

[15] A.Salam, in *Elementary Particle Theory*, eds. N. Svartho . Forlag, Stockholm,1968).

[16] R.Utiyama, *Phys. Rev.* **101** (1956) 1597.

[17] A.Brodsky, D.Ivanenko and G. Sokolik, *JETPH* 41 (1961)1307; *Acta Phys. Hung.* **14** (1962) 21.

[18] T.W.Kibble, *J. Math. Phys.* **2** (1961) 212.

[19] D.Ivanenko and G.Sardanashvily, *Phys. Rep.* **94** (1983) 1.

[20] F.W.Hehl, J.D.McCrea, E.W.Mielke and Y.Ne'eman *Phys. Rep.* **258** (1995) 1-171

[21] F.W.Hehl, P. Von Der Heyde, G.D.Kerlick, J.M.Nester *Rev. Mod. Phys.* **48** (1976) 393-416

[22] F.Gronwald and F.W.Hehl, *On the Gauge Aspects of Gravity*, gr-qc/9602013

[23] F.Brandt, *Phys. Rev.* D64 (2001) 065025

[24] Rolando Gaitan D., *A possible gauge formulation for gravity?*, gr-qc/0109079

[25] M.Botta Cantcheef, *General Relativity as a (constrained) Yang-Mills's theory*, gr-qc/0010080

[26] R.Magnani, E.Pessa, G.Resconi, *Gen. Rel. Grav.* **29**: (1997) 1049.

[27] Ning Wu, *"Gauge Theory of Gravity,"* hep-th/0109145

[28] Ning Wu, *Commun. Theor. Phys.* (Beijing, China) **38** (2002): 151-156.

[29] Ning Wu, *"Quantum Gauge Theory of Gravity,"* hep-th/0112062

[30] Ning Wu, *"Quantum Gauge Theory of Gravity,"* talk given at Meeting of the Devision of Particles and Fields of American Physical Society at the College of William & Mery(DPF2002), May 24-28, 2002, Williamsburg, Virgia, USA; hep-th/0207254; Transparancy can be obtained from: http://dpf2002.velopers.net/talks_pdf/33talk.pdf

[31] Ning Wu, *"Renormalizable Quantum Gauge General Relativity"* gr-qc/0309041.

[32] Ning Wu, *Commun. Theor. Phys.* (Beijing, China) **42** (2004): 543-552.

[33] Ning Wu, *Commun. Theor. Phys.* (Beijing, China) **38** (2002): 322-326.

[34] Ning Wu, *Commun. Theor. Phys.* (Beijing, China) **38** (2002): 455-460.

[35] Ning Wu, *Commun. Theor. Phys.* (Beijing, China) **39** (2003): 561-568.

[36] Ning Wu, *Commun.Theor.Phys.*, (Beijing, China) **36**(2001) 169-172.

[37] Ning Wu, *Commun. Theor. Phys.* (Beijing, China) **39** (2003): 671-674.

[38] R.Colella, A.W.Overhauser and S.A.Werner, *Phys. Rev. Lett.* (1975):1472-1474.

[39] S.A.Werner, R.Colella and A.W.Overhauser, *Phys. Rev. Lett.* (1975): 1053-1055.

[40] A.W.Overhauser and R.Colella, *Phys. Rev. Lett.* (1974):1237-1239.

[41] E.Podkletnov and R.Nieminen, *Physica* **C 203** (1992) 271.

[42] E. Podkletnov, *Weak gravitational shielding properties of composite bulk* $YBa_2Cu_3O_{7-x}$ *superconductor below 70 K under electro-magnetic field*, report MSU-chem 95, cond-mat/9701074.

[43] G.Modanese, *Europhys. Lett.* **35** (1996) 413; G.Modanese, *Phys. Rev.* **D 54** (1996) 5002; G.Modanese and J.Schnurer, *Possible quantum gravity effects in a charged Bose condensate under variable e.m. field*, report UTH-391/96, gr-qc/9612022.

[44] Ning Wu, *Commun. Theor. Phys.*, (Beijing, China) **41**(2004) 567-572; hep-th/0307225.

[45] S. Weinberg, *Gravitation and Cosmology: Principles and Applications of the General Theory of Relativity*, (John Wiley, New York, 1972).

[46] N. Straumann, *General Relativity and Relativistic Astrophysics*, (Springer-Verlag, Berlin Heidelberg New York Tokyo, 1984).

[47] Ning Wu, *Problems on Foundations of General Relativity*, hep-th/0303258.

[48] Ning Wu, Dahua Zhang and Tu-nan Ruan, *"Differential Geometrical Formulation of Gravitational Gauge Theory of Gravity"* hep-th/0201131

[49] L.Fadeev and V.N.Popov, *Phys. Letters*, **25B** (1967),29

[50] G. 't-Hooft, *Nucl. Phys.* **B33** (1971) 173.

[51] G. 't-Hooft and M. Veltman, *Nucl. Phys.* **B50** (1972) 318.

[52] B.W.Lee and J. Zinn-Justin, *Phys. Rev.* **D5** (1972) 3121, 3137.

[53] B.W.Lee and J. Zinn-Justin, *Phys. Rev.* **D7** (1973) 1049.

[54] B.W.Lee and J. Zinn-Justin, *Phys. Rev.* **D9** (1974) 933.

[55] J. Zinn-Justin, *Renormalization of Gauge Theories, Lectures at the* 1974 *Bonn int. Summer Inst. for Theoretical physics.*

[56] B.W.Lee *in Methods in Field Theory* (North-Holland, 1976)

[57] G. 't-Hooft and M. Veltman, *Nucl. Phys.* **B44** (1972) 189.

[58] G. 't-Hooft and M. Veltman, *Diagram* (CERN Theoretical Studies Division 1973).

In: Focus on Quantum Gravity Research
Editor: David C. Moore, pp. 171-214

ISBN 1-59454-660-6

Chapter 5

Long-Range Behavior in Quantum Gravity

Kirill A. Kazakov
Department of Theoretical Physics, Physics Faculty,
Moscow State University, 119899, Moscow, Russian Federation
E-mail: kirill@phys.msu.ru

Quantum gravity effects of zeroth order in the Planck constant are investigated in the framework of the low-energy effective theory. A special emphasis is placed on establishing the correspondence between classical and quantum theories, for which purpose transformation properties of the $\hbar^0$-order radiative contributions to the effective gravitational field under deformations of a reference frame are determined. Using the formalism of canonical transformations, it is shown that the one-loop contributions violate the principle of general covariance, in the sense that the quantities which are classically invariant under such deformations take generally different values in different reference frames. In particular, variation of the scalar curvature under transitions between different reference frames is calculated explicitly. Furthermore, the long-range properties of the two-point correlation function of the gravitational field are examined. Using the Schwinger-Keldysh formalism it is proved that this function is finite in the coincidence limit outside the region of particle localization. In this limit, the leading term in the long-range expansion of the correlation function is calculated explicitly. It shown also that in the case of a macroscopic gravitating body, the terms violating general covariance, and the field fluctuation are both suppressed by a factor $1/N$, where N is the number of particles in the body. This leads naturally to a macroscopic formulation of the correspondence between classical and quantum theories of gravitation. As an application of the obtained results, the secular precession of a test particle orbit in the fluctuating gravitational field is determined.

PACS numbers: 04.60.Ds, 11.15.Kc, 11.10.Lm
Keywords: General covariance, correspondence principle, long-range expansion, quantum fluctuations, correlation function, gauge dependence, Slavnov identities, secular precession

5.1 Introduction

In this Chapter, quantum properties of the gravitational fields produced by quantized matter will be considered in the framework of the low-energy effective theory. This means that our interest will be in quantum aspects of the gravitational interaction at energy scales much lower than the Planck energy

$$E_{\rm P} = \sqrt{\frac{\hbar c^5}{G}}\,.$$

From the practical point of view, this does not present a limitation, since all phenomena observed in the Universe so far fall perfectly into this category. In view of the extreme smallness of the Newton constant G, these phenomena are well described by the lowest-order Einstein theory. On the other hand, the results obtained within the lowest-order approximation are model-independent, i.e., they play the role of low-energy theorems. This universality allows one to draw a number of important conclusions concerning the synthesis of quantum theory and gravitation.

An essential part of investigation of this synthesis consists in establishing a correspondence between classical and quantum theories. The formal rules of this correspondence are contained in the Bohr correspondence principle which gives a general recipe for the construction of operators for various physical quantities. Informally, the problem of correspondence is in elucidating actual conditions to be imposed on a system to allow its classical consideration. Identification of these quasi-classical conditions constitutes an integral part of interpretative basis of quantum theory. Right from this point of view the low-energy behavior in quantum gravity will be considered below.

Two characteristics of the gravitational field produced by a quantized system will be of our main concern in this Chapter – the mean value of the gravitational field, and its correlation function. The results of independent investigation of these quantities will lead us to one and the same macroscopic formulation of the quasi-classical conditions in quantum gravity.

Turning to a more detailed formulation of our approach, let us note, first of all, that the low-energy condition stated above implies a similar condition on the characteristic length scale L of the process under consideration, $L \gg l_{\rm P}$, where $l_{\rm P}$ is the Planck length

$$l_{\rm P} = \sqrt{\frac{G\hbar}{c^3}}\,. \tag{5.1.1}$$

In other words, our interest is in the *long-range* properties of the gravitational field produced by a given system. As far as this system can be treated classically, $l_{\rm P}$ *is* the length scale characterizing quantum properties of its gravitational interactions, because $l_{\rm P}^2$ is the only parameter entering the theory of quantized gravitational field through the Einstein action

$$S_{\rm g} = -\frac{c^3}{16\pi G}\int d^4x\sqrt{-g}R\,. \tag{5.1.2}$$

In the presence of quantized matter, however, another parameter with dimension of length comes into play, namely the gravitational radius

$$r_{\rm g} = \frac{2Gm}{c^2}\,.$$

This parameter appears, of course, already in classical theory. The question of crucial importance is whether $r_{\rm g}$ has an independent meaning in quantum domain, representing a scale of specifically quantum phenomena. This question may seem strange at first sight, as $r_{\rm g}$ does not contain the Planck constant $\hbar$, an inalienable attribute of quantum theory. However, well known is the fact that gravitational radiative corrections do contain pieces independent of $\hbar$. In the framework of the effective theory, they appear as a power series in $r_{\rm g}/r$, just like post-Newtonian corrections in classical general relativity. This fact was first clearly stated by Iwasaki [2]. The reason for the appearance of $\hbar^0$ terms through the loop contributions is that in the case of gravitational interaction, the mass and "kinetic" terms in the matter Lagrangian determine not only the properties of matter quanta propagation, but also their couplings. Thus the mass term of, e.g., scalar field Lagrangian generates the vertices proportional to

$$\left(\frac{mc}{\hbar}\right)^2$$

containing inverse powers of $\hbar$. Naively, one expects these be cancelled by $\hbar$'s coming from the propagators when combining an amplitude. One should remember, however, that such a counting of powers of $\hbar$ in Feynman diagrams is a bad helpmate in the presence of massless particles. Virtual propagation of gravitons interacting with matter field quanta near their mass shells results in a *root* non-analyticity of the massive particle form factors at zero momentum transfer (p). For instance, the low-energy expansion of the one-loop diagram in Fig. 5.1.1(a) begins with terms proportional to $(-p^2)^{-1/2}$, rather than integer powers (or logarithms) of p^2. It is this singularity that is responsible for the appearance of powers of $r_{\rm g}/r$ in the long-range expansion of the loop contributions. The question we ask is of what nature, classical or quantum, these pieces are. This is precisely the question of correspondence in quantum gravity.

Suppose that the matter producing gravitational field satisfies the usual quantum mechanical quasi-classical conditions, e.g., consider sufficiently heavy particles.

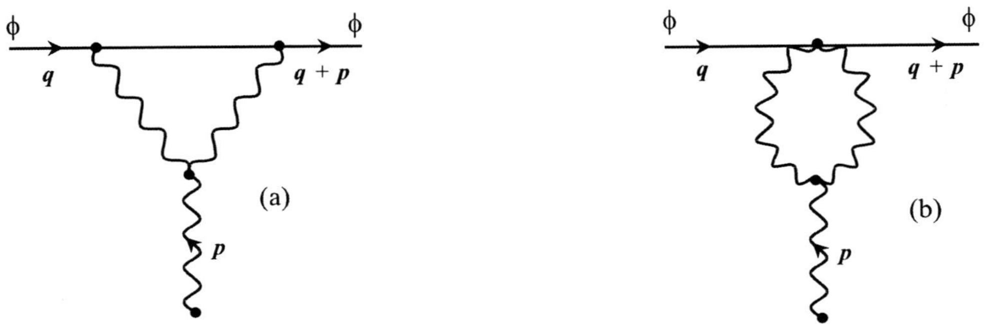

Figure 5.1.1: Some of the one-loop diagrams contributing to the effective gravitational field. (a) The only diagram giving rise to the root singularity with respect to the momentum transfer p. (b) An example of a diagram free of the root singularity. Wavy lines represent gravitons, solid lines massive particles. q is the particle 4-momentum.

Then the quasi-classical conditions for the gravitational field can be inferred from the requirement that the mean value of the spacetime metric, $\langle g_{\mu\nu}\rangle$, coincides with the classical solution of Einstein equations, corresponding to the same distribution of gravitating matter. Practically, the most convenient way of looking for these conditions is to compare *transformation properties* of the quantities involved under deformations of the reference frame, thus avoiding explicit calculation of the expectation values. The latter point of view takes advantage of the fact that the transformation law of classical solutions is known in advance. Hence, we have to check whether $\langle g_{\mu\nu}\rangle$ transforms covariantly with respect to transitions between different reference frames. In other words, we have to consider the question of *general covariance in quantum gravity*. This is the way we follow in Sec. 5.3. Alternatively, one can infer the quasi-classical conditions from the requirement of *vanishing of field fluctuations*. We take this rout in Sec. 5.4. Some results of the Schwinger-Keldysh and anti-canonical formalisms, used in our investigation, are summarized in Sec. 5.2. Section 5.5 contains concluding remarks.

Condensed notations of DeWitt [1] are in force throughout this Chapter. Also, right and left derivatives with respect to the fields and the sources, respectively, are used. The dimensional regularization of all divergent quantities is assumed.

5.2 Preliminaries

Before going into detailed discussion of the question of general covariance in quantum gravity, let us describe the general setting we will be working in.

5.2.1 Frame of reference and interacting fields.

First of all, we should set a frame of reference, *i.e.*, a system of idealized reference bodies with respect to which the 4-position in spacetime can be fixed. Let us assume, for definiteness, that the frame of reference is realized by means of an appropriate distribution of electrically charged matter. For simplicity, the energy-momentum of matter, as well as of the electromagnetic field it produces, will be assumed sufficiently small so as not to alter the gravitational field under consideration. The 4-position in spacetime can be determined by exchanging electromagnetic signals with a number of charged matter species. The electric charge distributions σ_a of the latter are thus supposed to be in a one-to-one correspondence with the spacetime coordinates x_μ,

$$\sigma_a \leftrightarrow x_\mu \,,$$

where index a enumerates the species. The $\sigma_a(x)$ will be assumed smooth scalar functions. Physical properties of the reference frame are determined by the action S_σ which specific form is of no importance for us.

Next, let us consider a system of interacting gravitational and matter fields. The latter are arbitrary species, bosons or fermions, self-interacting or not, denoted collectively by ϕ_i, $i = 1, 2, ..., k$. Dynamical variables of the gravitational field are $h_{\mu\nu} = g_{\mu\nu} - \eta_{\mu\nu}$. Dynamics of the system is described by the action $S_{\rm g} + S_\phi$, where S_ϕ is the matter action, and $S_{\rm g}$ is given[1] by Eq. (5.1.2).

The total action $S = S_{\rm g} + S_\phi + S_\sigma$ is invariant under the gauge transformations

$$\begin{aligned} \delta h_{\mu\nu} &= \xi^\alpha \partial_\alpha h_{\mu\nu} + (\eta_{\mu\alpha} + h_{\mu\alpha})\partial_\nu \xi^\alpha + (\eta_{\nu\alpha} + h_{\nu\alpha})\partial_\mu \xi^\alpha \equiv G^\alpha_{\mu\nu}\xi_\alpha \,, && (5.2.3) \\ \delta\phi_i &= G^\alpha_i \xi_\alpha \,, && (5.2.4) \\ \delta\sigma_a &= \sigma_{a,\alpha}\xi^\alpha && (5.2.5) \end{aligned}$$

The generators $G_{\mu\nu}, G_i$ span the closed algebra

$$\begin{aligned} G^{\alpha,\sigma\lambda}_{\mu\nu} G^\beta_{\sigma\lambda} - G^{\beta,\sigma\lambda}_{\mu\nu} G^\alpha_{\sigma\lambda} &= f^{\alpha\beta}{}_\gamma G^\gamma_{\mu\nu} \,, \\ G^{\alpha,k}_i G^\beta_k - G^{\beta,k}_i G^\alpha_k &= f^{\alpha\beta}{}_\gamma G^\gamma_i \,, \end{aligned} \qquad (5.2.6)$$

where the "structure constants" $f^{\alpha\beta}{}_\gamma$ are defined by

$$f^{\alpha\beta}{}_\gamma \xi_\alpha \eta_\beta = \xi_\alpha \partial^\alpha \eta_\gamma - \eta_\alpha \partial^\alpha \xi_\gamma \,. \qquad (5.2.7)$$

Let the gauge-fixing action be written in the form

$$S_{\rm gf} = \left(F_\alpha - \frac{1}{2}\pi^\beta \zeta_{\beta\alpha}\right)\pi^\alpha \,, \qquad (5.2.8)$$

[1] Our notation is $R_{\mu\nu} \equiv R^\alpha{}_{\mu\alpha\nu} = \partial_\alpha \Gamma^\alpha_{\mu\nu} - \cdots$, $R \equiv R_{\mu\nu} g^{\mu\nu}$, $g \equiv \det g_{\mu\nu}$, $g_{\mu\nu} = {\rm sgn}(+,-,-,-)$, $\eta_{\mu\nu} = {\rm diag}\{+1,-1,-1,-1\}$. The Minkowski tensor η is used to raise and lower tensor indices. The units in which $\hbar = c = 16\pi G = 1$ are chosen in what follows.

where F_α is a set of functions of the fields $h_{\mu\nu}$, fixing general invariance, π^α auxiliary fields introducing the gauge, and $\zeta_{\alpha\beta}$ a non-degenerate symmetric matrix weighting the functions F_α; the particular choice $\zeta_{\alpha\beta} = \xi\eta_{\alpha\beta}$ corresponds to the well-known Feynman weighting of the gauge conditions. Introducing the ghost fields c_α, $\bar{c}^\alpha$, and denoting all the fields collectively by Φ, we write the Faddeev-Popov action [3]

$$S_{\rm FP}[\Phi] = S + S_{\rm gf} + \bar{c}^\beta F_\beta^{,\mu\nu} G^\alpha_{\mu\nu} c_\alpha \,. \tag{5.2.9}$$

$S_{\rm FP}[\Phi]$ is invariant under the following Becchi-Rouet-Stora-Tyutin (BRST) transformations [4]

$$\begin{aligned} \delta h_{\mu\nu} &= G^\alpha_{\mu\nu} c_\alpha \lambda \,, \\ \delta \phi_i &= G^\alpha_i c_\alpha \lambda \,, \\ \delta \sigma_a &= \sigma_{a,\alpha} c^\alpha \lambda \,, \\ \delta c_\gamma &= -\frac{1}{2} f^{\alpha\beta}{}_\gamma c_\alpha c_\beta \lambda \,, \\ \delta \bar{c}^\alpha &= \pi^\alpha \lambda \,, \\ \delta \pi^\alpha &= 0 \,, \end{aligned} \tag{5.2.10}$$

where λ is a constant anticommuting parameter. Finally, the generating functional of Green functions has the form

$$Z[J, K] = \int \mathcal{D}\Phi \exp\{i(\Sigma[\Phi, K] + \bar{\beta}^\alpha c_\alpha + \beta_\alpha \bar{c}^\alpha + t^{\mu\nu} h_{\mu\nu} + j^i \phi_i + s^a \sigma_a)\} \tag{5.2.11}$$

where

$$\Sigma[\Phi, K] = S_{\rm FP}[\Phi] + k^{\mu\nu} G^\alpha_{\mu\nu} c_\alpha + q^i G^\alpha_i c_\alpha + r^a \sigma_{a,\alpha} c^\alpha - \frac{l^\gamma}{2} f^{\alpha\beta}{}_\gamma c_\alpha c_\beta + n_\alpha \pi^\alpha \,,$$

$\{t,\ j,\ s,\ \bar{\beta},\ \beta,\ 0\} \equiv J$ ordinary sources, and $\{k,\ q,\ r,\ l,\ n,\ 0\} \equiv K$ the BRST-transformation sources [5] for the fields $\{h,\ \phi,\ \sigma,\ c,\ \bar{c},\ \pi\} \equiv \Phi$, respectively, and integration is carried over all field configurations satisfying

$$\Phi^\pm \to 0\,, \qquad \text{for} \quad t \to \mp\infty\,,$$

where the superscripts + and − denote the positive- and negative-frequency parts of the fields, respectively. The K-sources are introduced into Z for the purpose of studying gauge dependence of observable quantities.

5.2.2 Effective fields and correlation functions

The central objects of our investigation in the sections below will be the expectation value of the gravitational field produced by elementary particles, and its correlation function. They can be obtained from the one- and two-point Green functions of the gravitational field, respectively,

$$h^{\text{eff}}_{\mu\nu}(j) = \left.\frac{\delta W[J,K]}{\delta t^{\mu\nu}}\right|_{\substack{J\setminus j=0\\ K=0}}, \qquad (5.2.12)$$

$$C_{\mu\nu\alpha\beta}(j) = \left.\frac{\delta^2 W[J,K]}{\delta t^{\mu\nu}\delta t^{\alpha\beta}}\right|_{\substack{J\setminus j=0\\ K=0}}, \qquad (5.2.13)$$

with the help of the well-known reduction formula applied to the external matter lines to go over onto the mass shell of the particles. In Eqs. (5.2.12), (5.2.13), W is the generating functional of connected Green functions,

$$W[J,K] = -i\ln Z[J,K],$$

and the symbol $J \setminus j$ means that the source j is excluded from the set J.

Under the Feynman asymptotic conditions specified above, Eqs. (5.2.12), (5.2.13) give the *in-out* matrix elements of operators, rather than the *in-in* expectation values we are interested in. As is well known, in order to find the latter, the usual Feynman rules for constructing the matrix elements must be modified. According to the so-called closed time path formalism of Schwinger and Keldysh [7, 6] (modern reviews of this formalism can be found in Refs. [8, 9]), this amounts to using in Eqs. (5.2.12), (5.2.13) the generating functional

$$Z_{\text{SK}}[J_\pm, K_\pm] = \int \mathcal{D}\Phi_- \int \mathcal{D}\Phi_+ \exp\{i(\Sigma[\Phi_\pm, K_\pm] + \bar{\beta}^\alpha c_\alpha + \beta_\alpha \bar{c}^\alpha + t^{\mu\nu}h_{\mu\nu} + j^i\phi_i + s^a\sigma_a)\}, \qquad (5.2.14)$$

where

$$\Sigma[\Phi_\pm, K_\pm] = S_{\text{FP}}[\Phi_+] - S_{\text{FP}}[\Phi_-] + k^{\mu\nu}G^\alpha_{\mu\nu}c_\alpha + q^i G^\alpha_i c_\alpha + r^a\sigma_{a,\alpha}c^\alpha - \frac{l^\gamma}{2}f^{\alpha\beta}{}_\gamma c_\alpha c_\beta + n_\alpha \pi^\alpha\,,$$

instead of (5.2.11). Here the subscript + (−) shows that the time argument of the integration variable runs from $-\infty$ to $+\infty$ (from $+\infty$ to $-\infty$). Integration is over all fields satisfying

$$\Phi^+_\pm \to 0\,, \quad \text{for } t \to -\infty\,,$$
$$\Phi_+ = \Phi_- \quad \text{for } t \to +\infty\,. \qquad (5.2.15)$$

We do not distinguish the plus and minus field components in the source terms explicitly, implying that summation over repeated Greek indices includes summation over $\pm$ as well as spacetime integration. For instance,

$$t^{\mu\nu}h_{\mu\nu} \equiv \int d^4x\, (t^{\mu\nu}_+ h_{\mu\nu+} + t^{\mu\nu}_- h_{\mu\nu-})\,,$$

$$l^{\gamma} f^{\alpha\beta}{}_{\gamma} c_{\alpha} c_{\beta} \equiv \int d^4x \left(l^{\gamma}_{+} f^{\alpha\beta}{}_{\gamma} c_{\alpha +} c_{\beta +} + l^{\gamma}_{-} f^{\alpha\beta}{}_{\gamma} c_{\alpha -} c_{\beta -} \right), \qquad \text{etc.}$$

With the help of the new generating functional, the in-in expectation value of, e.g., the product $\hat{h}_{\mu\nu}(x)\hat{h}_{\alpha\beta}(x')$ can be written as

$$\langle \mathrm{in}|\hat{h}_{\mu\nu}(x)\hat{h}_{\alpha\beta}(x')|\mathrm{in}\rangle = -Z^{-1}_{\mathrm{SK}} \left. \frac{\delta^2 Z_{\mathrm{SK}}[J_{\pm}, K_{\pm}]}{\delta t^{\mu\nu}_{-}(x)\delta t^{\alpha\beta}_{+}(x')} \right|_{\substack{J\backslash j=0 \\ K=0}} . \tag{5.2.16}$$

It is seen that $\langle \mathrm{in}|\hat{h}^{\mu\nu}(x)\hat{h}^{\alpha\beta}(x')|\mathrm{in}\rangle$ is given by the ordinary functional integral but with the number of fields doubled, and unusual boundary conditions specified above. Accordingly, diagrammatics generated upon expanding this integral in perturbation theory consists of the following elements. There are four types of pairings for each component of the set Φ, corresponding to the four different ways of placing two field operators on the two branches of the time path. They are conveniently combined into 2×2 matrices according to[2]

$$\mathfrak{D} = \begin{pmatrix} D_{++} & D_{+-} \\ D_{-+} & D_{--} \end{pmatrix}.$$

For instance, pairings matrices for the gravitational and scalar matter fields read

$$\mathfrak{D}_{ik}(x,y) = \begin{pmatrix} i\langle T\hat{\phi}_i(x)\hat{\phi}_k(y)\rangle_0 & i\langle \hat{\phi}_k(y)\hat{\phi}_i(x)\rangle_0 \\ i\langle \hat{\phi}_i(x)\hat{\phi}_k(y)\rangle_0 & i\langle \tilde{T}\hat{\phi}_i(x)\hat{\phi}_k(y)\rangle_0 \end{pmatrix},$$

$$\mathfrak{D}_{\mu\nu\alpha\beta}(x,y) = \begin{pmatrix} i\langle T\hat{h}_{\mu\nu}(x)\hat{h}_{\alpha\beta}(y)\rangle_0 & i\langle \hat{h}_{\alpha\beta}(y)\hat{h}_{\mu\nu}(x)\rangle_0 \\ i\langle \hat{h}_{\mu\nu}(x)\hat{h}_{\alpha\beta}(y)\rangle_0 & i\langle \tilde{T}\hat{h}_{\mu\nu}(x)\hat{h}_{\alpha\beta}(y)\rangle_0 \end{pmatrix},$$

where the operation of time ordering T ($\tilde{T}$) arranges the factors so that the time arguments decrease (increase) from left to right, and $\langle\cdot\rangle_0$ denotes vacuum averaging. The "propagators" $\mathfrak{D}_{\mu\nu\alpha\beta}$, $\mathfrak{D}_{ik}$ satisfy the following matrix equations

$$\mathfrak{G}^{\mu\nu\alpha\beta}\mathfrak{D}_{\alpha\beta\sigma\lambda} = -\mathfrak{e}\delta^{\mu\nu}_{\sigma\lambda}, \quad \mathfrak{G}^{\mu\nu\alpha\beta} = \mathfrak{i}\,\frac{\delta^2 S^{(2)}}{\delta h_{\mu\nu}\delta h_{\alpha\beta}}, \quad \delta^{\mu\nu}_{\sigma\lambda} = \frac{1}{2}\left(\delta^{\mu}_{\sigma}\delta^{\nu}_{\lambda} + \delta^{\nu}_{\sigma}\delta^{\mu}_{\lambda}\right) \tag{5.2.17}$$

$$\mathfrak{G}^{ik}\mathfrak{D}_{kl} = -\mathfrak{e}\delta^{i}_{l}, \quad \mathfrak{G}^{ik} = \mathfrak{i}\,\frac{\delta^2 S^{(2)}}{\delta\phi_i\delta\phi_k}, \tag{5.2.18}$$

where $S^{(2)}$ denotes the free field part of the gauge-fixed action after the gauge introducing fields π^{α} have been integrated out, and $\mathfrak{e}$, $\mathfrak{i}$ are 2×2 matrices with respect to indices $+, -$:

$$\mathfrak{e} = \begin{pmatrix} 1 & 0 \\ 0 & 1 \end{pmatrix}, \quad \mathfrak{i} = \begin{pmatrix} 1 & 0 \\ 0 & -1 \end{pmatrix}.$$

[2]Gothic letters are used to distinguish quantities representing columns, matrices etc. with respect to indices $+, -$.

As in the ordinary Feynman diagrammatics of the S-matrix theory, the propagators are contracted with the vertex factors generated by the interaction part of the action, $S^{\rm int}[\Phi]$, with subsequent summation over $(+,-)$ in the vertices, each "$-$" vertex coming with an extra factor (-1). This can be represented as the matrix multiplication of $\mathfrak{D}_{\mu\nu\alpha\beta}, \mathfrak{D}_{ik}$ with suitable matrix vertices. For instance, the $\phi^2 h$ part of the action generates the matrix vertex $\mathfrak{V}^{\mu\nu,ik}$ which in components has the form

$$V^{\mu\nu,ik}_{stu}(x,y,z) = e_{stu}\,\frac{\delta^3 S}{\delta h_{\mu\nu}(x)\delta\phi_i(y)\delta\phi_k(z)}\Big|_{\Phi=0},$$

where indices s,t,u take the values $+,-$, and e_{stu} is defined by $e_{+++} = -e_{---} = 1$, and zero otherwise. An external ϕ line is represented in this notation by a column

$$\mathfrak{r}_i = \begin{pmatrix} \bar{\phi}_i \\ \bar{\phi}_i \end{pmatrix},$$

satisfying

$$\mathfrak{G}^{ik}\mathfrak{r}_k = \begin{pmatrix} 0 \\ 0 \end{pmatrix}. \tag{5.2.19}$$

For future references, we give explicit expressions for various pairings of a single real scalar field

$$\begin{aligned} D_{++}(x,y) &= \int\frac{d^4k}{(2\pi)^4}\frac{e^{-ik(x-y)}}{m^2-k^2-i0}, \quad D_{--}(x,y) = \int\frac{d^4k}{(2\pi)^4}\frac{e^{-ik(x-y)}}{k^2-m^2-i0}, \\ D_{-+}(x,y) &= i\int\frac{d^4k}{(2\pi)^3}\theta(k^0)\delta(k^2-m^2)e^{-ik(x-y)}, \, D_{+-}(x,y) = D_{-+}(y,x). \end{aligned} \tag{5.2.20}$$

5.2.3 Generating functionals and Slavnov identities.

The functional $\Sigma[\Phi_\pm, K_\pm]$ can be written as [10]

$$\Sigma[\Phi_\pm, K_\pm] = \Sigma^r\left[\Phi_\pm, \frac{\delta\Psi[\Phi_\pm, K_\pm]}{\delta\Phi_\pm}\right],$$

where the so-called reduced action Σ^r and the gauge fermion Ψ are defined by

$$\Sigma^r[\Phi_\pm, K_\pm] = S + k^{\mu\nu}G^\alpha_{\mu\nu}c_\alpha + q^i G^\alpha_i c_\alpha + r^a\sigma_{a,\alpha}c^\alpha - \frac{l\gamma}{2}f^{\alpha\beta}{}_\gamma c_\alpha c_\beta + n_\alpha\pi^\alpha,$$

$$\Psi[\Phi_\pm, K_\pm] = k^{\mu\nu}h_{\mu\nu} + q^i\phi_i + r^a\sigma_a + l^\alpha c_\alpha + n_\alpha\bar{c}^\alpha + \bar{c}^\alpha\left(F_\alpha - \frac{1}{2}\pi^\beta\zeta_{\beta\alpha}\right).$$

Let us further simplify notation abbreviating $k^{\mu\nu}h_{\mu\nu} + q^i\phi_i + r^a\sigma_a + l^\alpha c_\alpha + n_\alpha\bar{c}^\alpha \equiv K\Phi$, and similarly for other sums of products of fields and sources or derivatives with respect to Φ, J, K. Omitting also the subscript $\pm$, the variation of Σ under infinitesimal variation of the gauge conditions takes the form

$$\delta\Sigma[\Phi, K] = \frac{\delta\Delta\Psi[\Phi, K]}{\delta\Phi}\frac{\delta\Sigma[\Phi, K]}{\delta K}.$$

The corresponding variation of the generating functional

$$\delta Z_{\rm SK}[J, K] = i\int \mathcal{D}\Phi_- \int \mathcal{D}\Phi_+ \exp\{i(\Sigma + J\Phi)\}\frac{\delta\Delta\Psi(\Phi, K)}{\delta\Phi}\frac{\delta\Sigma(\Phi, K)}{\delta K}.$$

Integrating by parts and omitting $\delta^2\Sigma/\delta K\delta\Phi \sim \delta^{(4)}(0)$ in the latter equation gives

$$\delta Z_{\rm SK}[J, K] = iJ\frac{\delta}{\delta K}\int \mathcal{D}\Phi_- \int \mathcal{D}\Phi_+ \Delta\Psi \exp\{i(\Sigma + J\Phi)\}. \qquad (5.2.21)$$

Since Σ is invariant under the BRST transformation (5.2.10), a BRST change of integration variables in Eq. (5.2.14) gives the Slavnov identity for the generating functional

$$J\frac{\delta Z_{\rm SK}}{\delta K} = 0, \qquad (5.2.22)$$

which allows one to rewrite Eq. (5.2.21) in terms of the generating functional of connected Green functions as

$$\delta W_{\rm SK}[J, K] = J\frac{\delta}{\delta K}\langle\Delta\Psi\rangle, \qquad (5.2.23)$$

where $\langle X\rangle$ denotes the functional averaging of X [11].

5.3 Transformation properties of effective gravitational field

In this section, we will investigate the structure of the $\hbar^0$-order contributions to the effective gravitational field, concentrating mainly on their transformation properties under deformations of the reference frame. The appearance of such contributions through the radiative corrections has its roots in the field properties of quantized matter sources. To make the field-theoretic aspect of the problem clearer, it is convenient to get rid of the purely quantum mechanical issues related to the quantum particle kinematics assuming matter quanta sufficiently heavy to neglect the position-velocity indeterminacy following from the Heisenberg principle. This assumption justifies the use of the term "particle at rest," and allows one to introduce a fixed distance r between the particle and the point of observation.

The mean gravitational field produced by a massive particle is a function of five dimensional parameters – the fundamental constants $\hbar, G, c$, the particle's mass m,

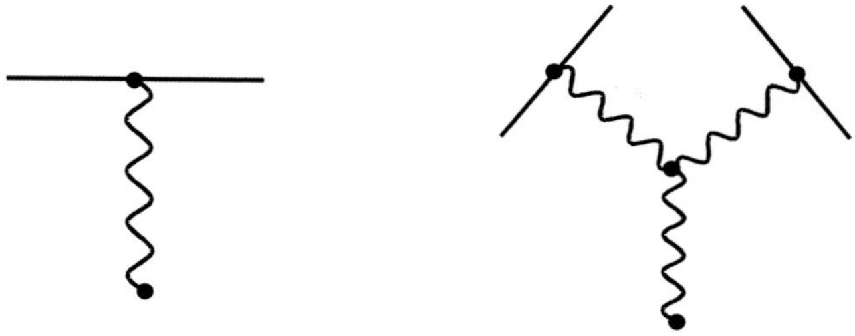

(a) (b)

Figure 5.3.2: Lower-order tree diagrams representing solution of the Einstein equations. (a) The first order (Newtonian) term. (b) One of the second-order diagrams describing the first post-Newtonian correction. Wavy lines represent gravitons, solid lines the auxiliary source $t^{\mu\nu}$, or expectation value of the energy-momentum tensor of matter, $\langle\hat{T}^{\mu\nu}\rangle$.

and the distance r. Of these only two independent dimensionless combinations can be constructed, which we choose to be $\chi = l_{\rm P}/r$ and $\varkappa = l_{\rm C}/r$, where $l_{\rm C} = \hbar/mc$ is the Compton length of the particle. The above assumption means that the gravitational field is considered in the limit $\varkappa \to 0$, leading naturally to the long-range expansion of the mean gravitational field. Accordingly, we will use where appropriate the terminology and general ideas of effective field theories [12, 13, 14, 15]. Equivalently, the $\hbar^0$-contribution can be identified as the leading term of the large mass expansion. It should be stressed that this identification is only a convenient way to extract the leading contribution. In particular, it does not mean that the results derived below are supposed to be applied only to heavy particles.

5.3.1 General covariance at the tree level.

From the point of view of the general formalism outlined in the preceding sections, the classical Einstein theory corresponds to the tree approximation of the full quantum theory. The sum of the tree contributions to the expectation value of a field operator coincides with the corresponding classical field, and the effective equations of motion

$$\left\langle \frac{\delta \Sigma}{\delta h_{\mu\nu}} \right\rangle + t^{\mu\nu} = 0\,,$$

expressing the translation invariance of the functional integral measure, go over into the classical Einstein equations. Some of the tree diagrams are shown in Fig. 5.3.2. The results of the preceding section allow one, in particular, to reestablish the general covariance of these equations.

In quantum theory, symmetry transformations must be formulated in terms of fields, rather than transformations of coordinate systems. In particular, transformations of the reference frame are represented by variations of the fields σ_i, induced by appropriate variations of the gauge conditions. Namely, it follows from Eq. (5.2.23) at the tree level that a gauge variation ΔF_α induces the following variations of the metric and reference fields

$$\delta g_{\mu\nu} = g_{\mu\nu,\alpha}\Xi^\alpha + g_{\mu\alpha}\Xi^\alpha_{,\nu} + g_{\nu\alpha}\Xi^\alpha_{,\mu}\,, \tag{5.3.24}$$

$$\delta\sigma_a = \sigma_{a,\alpha}\Xi^\alpha\,, \qquad \Xi_\alpha = \langle c_\alpha \Delta\Psi\rangle\,, \tag{5.3.25}$$

where $\Delta\Psi$ is the corresponding variation of the gauge fermion.

Let us consider any quantity entering the Einstein equations, for instance, the scalar curvature R measured in a given point σ^0_a of the reference frame. Under the above change of the gauge, σ^0_a goes to $\sigma^0_a + (\sigma_{a,\alpha})^0\,\Xi^\alpha$. In order to return to the initial point σ^0_a of the reference frame, one has to go from the point with coordinates x^μ to the point $x^\mu + \delta x^\mu$, where $\delta x^\mu = -\Xi^\mu$. As a result of these transformations, the tree value of R measured in σ^0_a remains unchanged,

$$\delta R[g(x(\sigma^0))] = \frac{\delta R}{\delta g_{\mu\nu}}\delta g_{\mu\nu} + \frac{\partial R}{\partial x^\mu}\delta x^\mu = \frac{\partial R}{\partial x^\alpha}\Xi^\alpha - \frac{\partial R}{\partial x^\mu}\Xi^\mu = 0\,. \tag{5.3.26}$$

Analogously, the tree contribution to any tensor quantity $O_{\mu\nu\dots}$ (or $O^{\mu\nu\dots}$), for instance, the metric $g_{\mu\nu}$ itself, calculated at a fixed reference point σ^0, transforms covariantly (contravariantly), as prescribed by the position of the tensor indices of the corresponding operator. This is the manifestation of general covariance of the classical Einstein theory in terms of quantum field theory.

5.3.2 General covariance at the one-loop order.

Let us now consider transformation properties of the one-loop contributions. As in Sec. 5.3.1, we have to determine the effect of an arbitrary gauge variation on the value of the effective metric, and also on the functions σ_a, *i.e.*, on the structure of the reference frame. After that, the transformation law of observables, defined generally as the diffeomorphism-invariant functions of the metric and reference fields, can be determined in the way followed in Sec. 5.3.1. For definiteness, we will deal below with the scalar curvature R. Since we are interested in the one-loop contribution to the first post-Newtonian correction, we can linearize R in $h_{\mu\nu}$:

$$R = \partial^\mu\partial^\nu h_{\mu\nu} - \Box h\,, \qquad h \equiv \eta^{\mu\nu}h_{\mu\nu}\,. \tag{5.3.27}$$

The transformation properties of R under variations of the weighting matrix $\zeta_{\alpha\beta}$ are considered in Sec. 5.3.2, and under variations of the gauge functions F_α themselves in Sec. 5.3.2.

Dependence of effective metric on weighting parameters.

Dependence of the effective fields on the weighting matrix $\zeta_{\alpha\beta}$ can be determined in a quite general way without specifying either the gauge functions F_α, or the properties of gravitating matter fields. We begin with the classical theory in Sec. 5.3.2, and then consider the one-loop order in Sec. 5.3.2.

The tree level. As we saw in Sec. 5.3.1, arbitrary gauge variations lead to the transformations of the classical fields, equivalent to the spacetime diffeomorphisms, and thus do not affect the values of R. In the particular case of variations of the weighting matrix, however, not only R, but also the effective fields themselves remain unchanged. This means that the structure of reference frames in classical theory is determined by the functions F_α only. As this differs in the full quantum theory, a somewhat more detailed discussion of this issue will be given in this Section.

To demonstrate $\zeta_{\alpha\beta}$-independence of the classical metric, let us first integrate the auxiliary fields π^α out of the gauge-fixing action (5.2.8),

$$S_{\rm gf} \to S^{\xi}_{\rm gf} = \frac{1}{2} F_\alpha \xi^{\alpha\beta} F_\beta \,, \qquad \zeta_{\alpha\beta}\xi^{\beta\gamma} = \delta^{\gamma}_{\alpha} \,. \tag{5.3.28}$$

The classical equations of motion thus become

$$\frac{\delta(S_{\rm g} + S^{\xi}_{\rm gf})}{\delta g_{\mu\nu}} = -T^{\mu\nu}, \tag{5.3.29}$$

where $T^{\mu\nu}$ is the energy-momentum tensor of matter. Using invariance of the action S under the gauge transformations

$$\delta g_{\mu\nu} = \xi^\alpha \partial_\alpha g_{\mu\nu} + g_{\mu\alpha}\partial_\nu \xi^\alpha + g_{\nu\alpha}\partial_\mu \xi^\alpha \equiv \nabla_\mu \xi_\nu \,, \tag{5.3.30}$$

and taking into account the "conservation law" $\nabla_\mu T^{\mu\nu} = 0$, one has from Eq. (5.3.29)

$$F_\alpha \xi^{\alpha\beta} \frac{\delta F_\beta}{\delta g_{\mu\nu}(x)} \nabla^x_\mu \delta(x-y) = 0 \,. \tag{5.3.31}$$

The matrix $M^\nu_\beta(x,y) = \delta F_\beta / \delta g_{\mu\nu}(x) \nabla^x_\mu \delta(x-y)$ is non-degenerate; its determinant $\Delta \equiv \det M^\nu_\beta(x,y)$ is just the Faddeev-Popov determinant, and therefore $\Delta \neq 0$. Hence, Eq. (5.3.31) gives $F_\alpha \xi^{\alpha\beta} = 0$, and, in view of non-degeneracy of $\xi^{\alpha\beta}$, $F_\alpha = 0$. The classical metric is thus independent of the choice of the matrix $\xi^{\alpha\beta}$, and in particular, of the replacements $F_\alpha \to A^\beta_\alpha F_\beta$. One can put this in another way by saying that the weighting matrix has no geometrical meaning in classical theory.

This differs, however, in quantum domain. The classical equations (5.3.29) are replaced in quantum theory by the effective equations

$$\frac{\delta \Gamma}{\delta g^{\rm eff}_{\mu\nu}} = -T^{\mu\nu}_{\rm eff} \,,$$

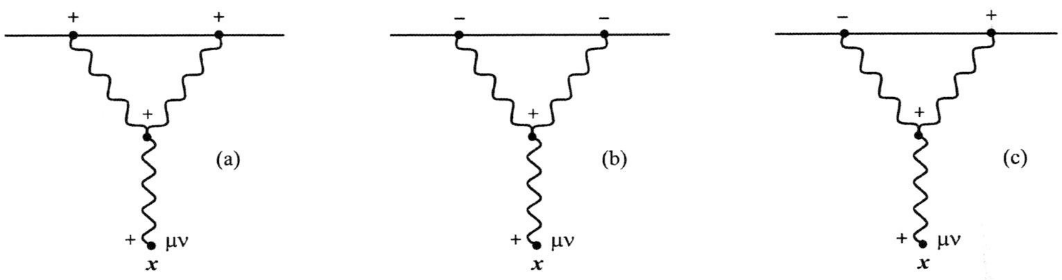

Figure 5.3.3: Some of diagrams arising upon writing out the matrix diagram in Fig. 5.1.1(a) according to the Schwinger-Keldysh rules. (a) The only diagram giving rise to a nonzero contribution in the long-range limit. (b) Diagram vanishing in the long-range limit because of the condition $p^0 = 0$. (c) An example of a diagram which is zero identically because of the vanishing of one of its vertices (the left $\phi^2 h$ vertex).

where Γ, $g^{\rm eff}_{\mu\nu}$, and $T^{\mu\nu}_{\rm eff}$ are the effective action, metric, and energy-momentum tensor of matter, respectively. In general, the fields $g^{\rm eff}_{\mu\nu}$ do not satisfy the gauge conditions $F_\alpha = 0$, and moreover, depend on the choice of the weighting matrix $\xi^{\alpha\beta}$; $\xi^{\alpha\beta}$-independence is inherited only by the tree contribution.

Dependence on the choice of the weighting matrix generally represents an excess of the gauge arbitrariness over the arbitrariness in the choice of reference frame; it is therefore a potential source of ambiguity in the values of observables. This dependence causes no gauge ambiguity of observables only if it reduces to a symmetry transformation. In other words, under (infinitesimal) variations of the matrix $\xi^{\alpha\beta}$, the fields $g^{\rm eff}_{\mu\nu}$ must transform as in Eq. (5.3.30)

$$\delta g^{\rm eff}_{\mu\nu} = \Xi^\alpha g^{\rm eff}_{\mu\nu,\alpha} + g^{\rm eff}_{\mu\alpha}\Xi^\alpha_{,\nu} + g^{\rm eff}_{\nu\alpha}\Xi^\alpha_{,\mu}\,, \tag{5.3.32}$$

with some functions Ξ^α. It will be shown in the following Section that this is the case indeed.

The one-loop level. Let us now turn to the examination of the gauge dependence of $\hbar^0$ loop contribution to the effective gravitational field. This contribution comes from diagrams in which virtual propagation of matter fields is near their mass shells, and is represented by terms containing the root singularity with respect to the momentum transfer between gravitational and matter fields. In the first post-Newtonian approximation, the only diagram we need to consider is the one-loop diagram depicted in Fig. 5.1.1(a). As a simple analysis shows, other one-loop diagrams do not contain the root singularities, while many-loop diagrams are of higher orders in the Newton constant. For instance, the loop in the diagram of Fig. 5.1.1(b) does not contain massive particle propagators, and therefore, expands in integer powers (or logarithms) of the ratio $\alpha = -p^2/m^2$.

Let us show, first of all, that the Schwinger-Keldysh rules applied to the diagram in Fig. 5.1.1(a) reduce to the ordinary Feynman rules of the standard S-matrix theory. According to the former, there are eight diagrams corresponding to the eight different ways of placing the three internal vertices of this diagram into the plus or minus branches of the time path (see Fig. 5.3.3). But all diagrams in Fig. 5.3.3 except the first two diagrams [2(a) and 2(b)] are zeros identically because of the energy-momentum conservation in the vertices, and the mass shell condition for external matter lines (massive particle cannot emit a real massless graviton). As to the diagram 2(b), its $\hbar^0$ order contribution is also zero. Indeed, the time component of the momentum transfer

$$p^0 = \sqrt{m^2 + (\boldsymbol{p} + \boldsymbol{q})^2} - \sqrt{m^2 + \boldsymbol{q}^2} = O\left(\frac{\boldsymbol{pq}}{m}\right).$$

To the leading order, therefore, p^0 is to be set zero, and the product of the two internal graviton propagators vanishes because the arguments of the corresponding θ-functions are opposite:

$$D_{-+}(k)D_{+-}(k+p) \sim \theta(k^0)\theta(-k^0) = 0\,.$$

The remaining diagram 2(a) contains only Feynman (causal) propagators, and therefore so do all other diagrams arising in the derivations of this section. Accordingly, we will omit the subscripts SK, $\pm$, and denote the D_{++} functions simply by D.

It follows from Eq. (5.2.23) that under a variation $\Delta\Psi$ of the gauge fermion, variation of the effective gravitational field

$$h^{\text{eff}}_{\mu\nu} = \frac{\delta W}{\delta t^{\mu\nu}}$$

has the form

$$\delta h^{\text{eff}}_{\mu\nu} = \left(\frac{\delta}{\delta k^{\mu\nu}}\langle\Delta\Psi\rangle + j^i\frac{\delta^2}{\delta t^{\mu\nu}\delta q^i}\langle\Delta\Psi\rangle\right)_{\substack{J\backslash j=0\\ K=0}}\,. \qquad (5.3.33)$$

We are interested presently in variations of the weighting matrix $\xi^{\mu\nu}$, therefore,

$$\Delta\Psi(\Phi, K) = -\frac{\bar{c}^\alpha}{2}\pi^\beta\Delta\zeta_{\beta\alpha}\,,$$

or, integrating π^α out,

$$\Delta\Psi(\Phi, K) = \frac{\bar{c}^\alpha}{2}\zeta_{\alpha\beta}\Delta\xi^{\beta\gamma}F_\gamma\,. \qquad (5.3.34)$$

According to general rules, in order to find the contribution of a diagram with n external ϕ-lines, one has to take the nth derivative of the right hand side of Eq. (5.3.33) with respect to j^i, multiply the result by the product of n factors $e_i(q^2 - m^2)$, where q, e_i are the 4-momentum and polarization of the external ϕ-field quanta, and set

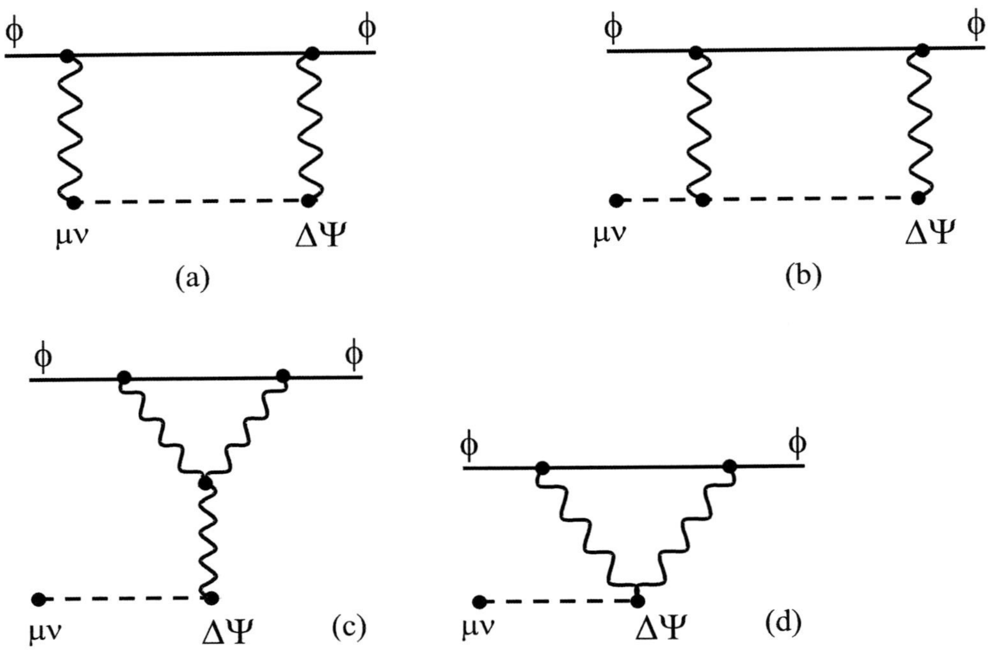

Figure 5.3.4: The one-loop diagrams giving rise to the root singularity in the right hand side of Eq. (5.3.35). Dashed lines represent the Faddeev-Popov ghosts.

$q^2 = m^2$ afterwards. The second term on the right of Eq. (5.3.33) is proportional to the source j^i contracted with the vertex $D_i^\alpha c_\alpha$. This term represents contribution of the graviton propagators ending on the external matter lines. Multiplied by $(q^2 - m^2)$, it gives rise to a non-zero value as $q^2 \to m^2$ only if the corresponding diagram is one-particle-reducible with respect to the ϕ-line, in which case it describes the variation of $h_{\mu\nu}^{\rm eff}$ under the gauge variation of external matter lines. It is well-known, however, that ϕ-operators must be renormalized[3] so as to cancel all the radiative corrections to the external lines[4]. Therefore, this term can be omitted, and Eq. (5.3.33) rewritten finally as

$$\delta h_{\mu\nu}^{\rm eff} = \frac{1}{2}\zeta_{\alpha\beta}\Delta\xi^{\beta\gamma}\,\frac{\delta\langle\bar{c}^\alpha F_\gamma\rangle}{\delta k^{\mu\nu}}\bigg|_{\substack{J\backslash j=0\\ K=0}}\,. \tag{5.3.35}$$

The one-loop diagrams representing the right hand side of Eq. (5.3.35), which give rise to the root singularity, are shown in Fig. 5.3.4. Let us consider the diagram

---

[3]One might think that the gauge dependence of the renormalization constants could spoil the above derivation of Eq. (5.2.23). In fact, this equation holds true for renormalized as well as unrenormalized quantities [11].

[4]The above discussion is nothing but the well-known reasoning underlying the proof of gauge-independence of the S-matrix [16].

of Fig. 5.3.4(a) first. It turns out that this diagram is actually free of the root singularity despite the presence of the internal ϕ-line. The rightmost vertex in this diagram is generated by $\bar{c}^{\alpha}F^{(1)}_{\gamma}$, where $F^{(1)}_{\gamma}$ denotes the linear part of F_{γ}. The graviton propagator connecting this vertex to the ϕ-line can be expressed through the ghost propagator with the help of the equation

$$\xi^{\alpha\beta}F^{(1),\sigma\lambda}_{\beta}D_{\sigma\lambda\mu\nu} = G^{(0)\beta}_{\mu\nu}\tilde{D}^{\alpha}_{\beta}\,, \qquad G^{(0)\beta}_{\mu\nu} \equiv G^{\beta}_{\mu\nu}\Big|_{h=0}\,, \tag{5.3.36}$$

where $\tilde{D}^{\alpha}_{\beta}$ is the ghost propagator defined by

$$F^{(1),\mu\nu}_{\alpha}G^{(0)\beta}_{\mu\nu}\tilde{D}^{\gamma}_{\beta} = -\delta^{\gamma}_{\alpha}\,, \tag{5.3.37}$$

respectively. Equation (5.3.36) is the Slavnov identity (5.2.22) at the tree level, differentiated twice with respect to $t^{\mu\nu}$, β_{α}. Using this identity in the diagram Fig. 5.3.4(a) we see that the ghost propagator is attached to the matter line through the generator $G^{(0)\alpha}_{\mu\nu}$. On the other hand, the action S_{ϕ} is invariant under the gauge transformations (5.2.3) – (5.2.5),

$$\frac{\delta S_{\phi}}{\delta\phi_i}G^{\alpha}_{i} + \frac{\delta S_{\phi}}{\delta h_{\mu\nu}}G^{\alpha}_{\mu\nu} = 0. \tag{5.3.38}$$

Differentiating Eq. (5.3.38) twice with respect to ϕ, and setting $h_{\mu\nu} = 0$ yields

$$\frac{\delta^2 S^{(2)}_{\phi}}{\delta\phi_i\delta\phi_k}\frac{\delta G^{\alpha}_{i}}{\delta\phi_l} + \frac{\delta^2 S^{(2)}_{\phi}}{\delta\phi_i\delta\phi_l}\frac{\delta G^{\alpha}_{i}}{\delta\phi_k} + \frac{\delta^3 S_{\phi}}{\delta h_{\mu\nu}\delta\phi_l\delta\phi_k}\Bigg|_{h=0} G^{(0)\alpha}_{\mu\nu} = 0\,. \tag{5.3.39}$$

Taking into account also the mass shell condition

$$\frac{\delta^2 S^{(2)}_{\phi}}{\delta\phi_i\delta\phi_k}\bar{\phi}_k = 0,$$

we see that under contraction of the vertex factor with the external and internal matter lines, the ϕ-particle propagator, D_{ik}, satisfying

$$\frac{\delta^2 S^{(2)}_{\phi}}{\delta\phi_i\delta\phi_k}D_{kl} = -\delta^{i}_{l}\,,$$

is cancelled by the vertex factor

$$\frac{\delta^3 S_{\phi}}{\delta\phi_l\delta\phi_k\delta h_{\mu\nu}}\Bigg|_{h=0} G^{(0)\alpha}_{\mu\nu}D_{ki}\bar{\phi}_l = \frac{\delta G^{\alpha}_{i}}{\delta\phi_l}\bar{\phi}_l\,. \tag{5.3.40}$$

We conclude that the $\hbar^0$ contribution of the diagram Fig. 5.3.4(a) is zero. As to the rest of diagrams, they are all proportional to the generator $G^{(0)\alpha}_{\mu\nu}$. Thus, the right hand side of Eq. (5.3.35) can be written

$$\delta h^{\rm eff}_{\mu\nu} = G^{(0)\alpha}_{\mu\nu}\Xi_{\alpha} + O(\hbar)\,, \qquad \Xi_{\alpha} = \frac{1}{2}\tilde{D}^{\beta}_{\alpha}\zeta_{\beta\gamma}\Delta\xi^{\gamma\delta}\langle F_{\delta}\rangle\,.$$

Since Ξ_α are of the order G^2, one can also write, within the accuracy of the first post-Newtonian approximation,

$$\delta h^{\rm eff}_{\mu\nu} = G^{\alpha}_{\mu\nu}\Xi_\alpha\,, \tag{5.3.41}$$

where $G^{\alpha}_{\mu\nu}$ are defined by Eq. (5.2.3) with $h_{\mu\nu} \to h^{\rm eff}_{\mu\nu}$.

We thus see that under variations of the weighting matrix, the effective metric does transform according to Eq. (5.3.32). To determine the effect of these variations on the values of observables, one has to find also the induced transformation of the reference frame, *i.e.*, of the functions σ_a. Obviously, the gauge variation of σ_a's is represented by the same set of diagrams pictured in Fig. 5.3.4(b,c,d),[5] with the only difference that the leftmost vertex $(\mu\nu)$ in these diagrams is now generated by $\sigma_{a,\alpha}c^\alpha$ instead of $D^{\alpha}_{\mu\nu}c_\alpha$. Thus, under variations of the weighting matrix, the functions $\sigma_a(x)$ transform according to

$$\delta\sigma_a = \sigma_{a,\alpha}\Xi^\alpha\,, \tag{5.3.42}$$

where Ξ^α's are the same as in Eq. (5.3.41).

Equations (5.3.41) and (5.3.42) are of the same form as Eqs. (5.3.24) and (5.3.25), respectively, which implies that the value of any observable O, measured in a given point of the reference frame, is invariant under variations of the weighting matrix,

$$\delta O[h^{\rm eff}(x(\sigma^0))] = \frac{\delta O}{\delta h^{\rm eff}_{\mu\nu}}\delta h^{\rm eff}_{\mu\nu} + \frac{\partial O}{\partial x^\mu}\delta x^\mu = \frac{\partial O}{\partial x^\alpha}\Xi^\alpha - \frac{\partial O}{\partial x^\mu}\Xi^\mu = 0\,. \tag{5.3.43}$$

In particular,

$$\delta R[g^{\rm eff}(x(\sigma^0))] = 0.$$

Furthermore, any tensor quantity $O_{\alpha\beta\dots}$ (or $O^{\mu\nu\dots}$), calculated at a fixed reference point σ^0, transforms covariantly (contravariantly), as prescribed by the position of the tensor indices of the corresponding operator. This is in accord with the principle of general covariance.

Dependence of effective metric on the form of F_α

Having established the general law of the effective metric transformation under variations of the weighting matrix, let us turn to investigation of the variations of the functions F_α themselves.

According to the general Eq. (5.3.33), a variation ΔF_α induces the following variation of the effective metric

$$\delta h^{\rm eff}_{\mu\nu} = \left.\frac{\delta\langle\bar{c}^\alpha\Delta F_\alpha\rangle}{\delta k^{\mu\nu}}\right|_{\substack{J\setminus j=0\\ K=0}}. \tag{5.3.44}$$

[5]The diagram of Fig. 5.3.4(a) does not contribute in this case.

The general structure of diagrams representing the one-loop contribution to the right hand side of this equation is the same as before, and given by Fig. 5.3.4. Contribution of diagrams (b), (c), and (d) is again a spacetime diffeomorphism. In the present case, however, diagram (a) gives rise to a non-zero contribution already in the order $\hbar^0$. Namely, it is not difficult to show that the combination $\Delta F_{\alpha}^{(1),\mu\nu} D_{\mu\nu\sigma\lambda}$ cannot be brought to the form proportional to the generator $G^{(0)}$. Note, first of all, that the variation of $\Delta F_{\alpha}^{(1),\mu\nu} D_{\mu\nu\sigma\lambda}$ with respect to $\xi^{\alpha\beta}$ *is* proportional to $G^{(0)}$; in the highly condensed DeWitt's notation,

$$\delta(\Delta F_1^{(1)} D) = \Delta F_1^{(1)} D (F_1^{(1)} \delta\xi F_1^{(1)}) D = \Delta F_1^{(1)} D F_1^{(1)} \delta\xi\zeta\tilde{D} G^{(0)},$$

where the Slavnov identity (5.3.36) has been used. Hence, without changing the $\hbar^0$ part of diagram (a), $\zeta_{\alpha\beta}$ can be set zero, in which case Eq. (5.3.36) gives $F_1^{(1)} D = 0$. Suppose that $\Delta F_{\alpha}^{(1),\mu\nu} D_{\mu\nu\sigma\lambda} = X_{\alpha\beta} G_{\sigma\lambda}^{(0)\beta}$, or shorter, $\Delta F_1^{(1)} D = X G^{(0)}$, with some X. Then one has $0 = \Delta F_1^{(1)} D F_1^{(1)} = X F_1^{(1)} G^{(0)} \equiv X M(h = 0)$. Since the Faddeev-Popov determinant $\det M \neq 0$, it follows that $X = 0$, which is impossible, because $\Delta F_{\alpha}^{(1),\mu\nu} D_{\mu\nu\sigma\lambda}$ is in any case nonzero. Thus, the argument used in the preceding section does not work, and the question is whether contribution of the diagram (a) can be actually represented in the form $G_{\mu\nu}^{\alpha} \Xi_{\alpha}$.

The answer to this question is negative, as an explicit calculation shows. This will be demonstrated below in the simplest case of scalar matter described by the action

$$S_{\phi} = \frac{1}{2} \int d^4x \sqrt{-g} \left\{ g^{\mu\nu} \partial_{\mu}\phi \partial_{\nu}\phi - m^2 \phi^2 \right\}, \tag{5.3.45}$$

and linear gauge conditions

$$F_{\gamma} = \eta^{\mu\nu} \partial_{\mu} h_{\nu\gamma} - \left(\frac{\varrho - 1}{\varrho - 2} \right) \partial_{\gamma} h, \qquad h \equiv \eta^{\mu\nu} h_{\mu\nu}, \qquad \zeta_{\alpha\beta} = 0, \tag{5.3.46}$$

where ϱ is an arbitrary parameter. According to Eq. (5.3.44), the ϱ-derivative of the effective metric is given by

$$\frac{\partial h_{\mu\nu}^{\text{eff}}}{\partial \varrho} = \left. \frac{\delta \langle \bar{c}^{\alpha} \partial F_{\alpha} / \partial \varrho \rangle}{\delta k^{\mu\nu}} \right|_{\substack{J \setminus j = 0 \\ K = 0}}. \tag{5.3.47}$$

There are two diagrams with the structure of Fig. 5.3.4(a), in which the scalar particle propagates in opposite directions. They are represented in Fig. 5.3.5. In fact, it is sufficient to evaluate either of them. Indeed, these diagrams have the following tensor structure

$$a_1 q_{\mu} q_{\nu} + a_2 (p_{\mu} q_{\nu} + p_{\nu} q_{\mu}) + a_3 p_{\mu} p_{\nu} + a_4 \eta_{\mu\nu},$$

Figure 5.3.5: Diagrams responsible for the nontrivial contribution to the right hand side of Eq. (5.3.47).

where $a_i, i = 1, ..., 4$, are some functions of p^2. When transformed to the coordinate space, the second and third terms become spacetime gradients, hence, they can be written in the form $G^{(0)\alpha}_{\mu\nu}\Xi_\alpha$. As was discussed in the preceding sections, the terms of this type respect general covariance, therefore, we can restrict ourselves to the calculation of a_1 and a_4 only. On the other hand, diagrams of Fig. 5.3.5 go over one into another under the substitution $q \to -q - p$ which leaves a_1, a_4 unchanged. Thus, we have

$$\begin{aligned} I_{\mu\nu}(x) &= \iint \frac{d^3\boldsymbol{q}}{(2\pi)^3}\frac{d^3\boldsymbol{p}}{(2\pi)^3}\frac{a(\boldsymbol{q})a^*(\boldsymbol{q}+\boldsymbol{p})}{\sqrt{2\varepsilon_{\boldsymbol{q}}2\varepsilon_{\boldsymbol{q}+\boldsymbol{p}}}} e^{ipx}\tilde{I}_{\mu\nu}(p,q), \ p_0 = \varepsilon_{\boldsymbol{q}+\boldsymbol{p}} - \varepsilon_{\boldsymbol{q}}, \quad (5.3.48) \\ \tilde{I}_{\mu\nu}(p,q) &= \tilde{I}^{3(a)}_{\mu\nu}(p,q) + \tilde{I}^{3(b)}_{\mu\nu}(p,q)\,, \\ \tilde{I}^{3(a)}_{\mu\nu}(p,q) &= -i\mu^\epsilon \int \frac{d^{4-\epsilon}k}{(2\pi)^4}\left\{\frac{1}{2}W^{\alpha\beta\gamma\delta}(q_\gamma + p_\gamma)(q_\delta + k_\delta) - m^2\frac{\eta^{\alpha\beta}}{2}\right\} \\ &\times D(q+k)\left\{\frac{1}{2}W^{\rho\tau\sigma\lambda}q_\sigma(q_\lambda + k_\lambda) - m^2\frac{\eta^{\rho\tau}}{2}\right\} D_{\rho\tau\pi\omega}(k)\frac{\partial F^{\xi,\pi\omega}}{\partial\varrho} \\ &\times \tilde{D}^\zeta_\xi(k)\left\{-(k_\zeta - p_\zeta)\delta^{\chi\theta}_{\mu\nu} + \delta^{\chi\theta}_{\zeta\mu}k_\nu + \delta^{\chi\theta}_{\zeta\nu}k_\mu\right\} D_{\chi\theta\alpha\beta}(k-p)\,, \\ \tilde{I}^{3(b)}_{\mu\nu}(p,q) &= \tilde{I}^{3(a)}_{\mu\nu}(p,-q-p)\,, \end{aligned}$$

where $a(\boldsymbol{q})$ is the momentum space amplitude for the scalar particle, normalized by

$$\int \frac{d^3\boldsymbol{q}}{(2\pi)^3}|a(\boldsymbol{q})|^2 = 1\,,$$

μ is an arbitrary mass scale, $\varepsilon_{\boldsymbol{q}} = \sqrt{\boldsymbol{q}^2 + m^2}$, and $\epsilon = 4 - d$, d being the dimensionality of spacetime. Explicit expressions for the propagators

$$
\begin{aligned}
D_{\mu\nu\sigma\lambda}(k) &= \frac{W_{\mu\nu\sigma\lambda}}{k^2} - \varrho(\eta_{\mu\nu}k_\sigma k_\lambda + \eta_{\sigma\lambda}k_\mu k_\nu)\frac{1}{k^4} \\
&+ (\eta_{\mu\sigma}k_\nu k_\lambda + \eta_{\mu\lambda}k_\nu k_\sigma + \eta_{\nu\sigma}k_\mu k_\lambda + \eta_{\nu\lambda}k_\mu k_\sigma)\frac{1}{k^4} \\
&+ (3\varrho^2 - 4\varrho)k_\mu k_\nu k_\sigma k_\lambda \frac{1}{k^6}\,, \qquad (5.3.49) \\
\tilde{D}^\alpha_\beta(k) &= \frac{\delta^\alpha_\beta}{k^2} - \frac{\varrho}{2}\frac{k^\alpha k_\beta}{k^4}\,, \\
D(k) &= \frac{1}{m^2 - k^2}\,,
\end{aligned}
$$

Calculation of (5.3.48) can be further simplified using the relation

$$\frac{\partial F_1}{\partial \varrho}D = -F_1\frac{\partial D}{\partial \varrho}\,,$$

which follows from $F_1 D = 0$, and noting that all gradient terms in the graviton propagators, contracted with the $\phi^2 h$ vertices, can be omitted (see Sec. 5.3.2), i.e., only the first line in Eq. (5.3.49) actually contributes.

Upon extraction of the leading contribution, Eq. (5.3.48) considerably simplifies. Note, first of all, that to the leading order in the long-range expansion, one has $\varepsilon_{\boldsymbol{q}+\boldsymbol{p}} \approx \varepsilon_{\boldsymbol{q}} \approx m$, and hence, $p_0 \approx 0$. Next, take into account that the function $a(\boldsymbol{q})$ is generally of the form

$$a(\boldsymbol{q}) = b(\boldsymbol{q})e^{-i\boldsymbol{q}\boldsymbol{x}_0}\,,$$

where $\boldsymbol{x}_0$ is the mean particle position, and $b(\boldsymbol{q})$ is such that

$$\int d^3\boldsymbol{q}\; b(\boldsymbol{q})e^{i\boldsymbol{q}\boldsymbol{x}} = 0 \qquad (5.3.50)$$

for $\boldsymbol{x}$ outside of some finite region W around $\boldsymbol{x} = 0$. In the long-range limit, $b(\boldsymbol{q}+\boldsymbol{p})$ may be substituted by $b(\boldsymbol{q})$: This implies that we disregard spatial spreading of the wave packet, neglecting the multipole moments of the particle mass distribution. Hence, Eq. (5.3.48) can be written as

$$I_{\mu\nu}(x) = \frac{1}{2m}\iint \frac{d^3\boldsymbol{q}}{(2\pi)^3}\frac{d^3\boldsymbol{p}}{(2\pi)^3}|b(\boldsymbol{q})|^2 e^{-i\boldsymbol{p}(\boldsymbol{x}-\boldsymbol{x}_0)}\tilde{I}_{\mu\nu}(p,q)\,, \quad p^0 = 0\,. (5.3.51)$$

Let the equality of two functions up to a spacetime diffeomorphism be denoted by "$\sim$". Then, performing tensor multiplications in Eq. (5.3.48), and omitting terms proportional to p_μ, one obtains

$$\tilde{I}^{3(a)}_{\mu\nu}(p,q) \sim -i\mu^\epsilon \int \frac{d^{4-\epsilon}k}{(2\pi)^4} \frac{1}{k^4} \frac{1}{(k+p)^2} \frac{1}{m^2-(k+q)^2}$$
$$\Big\{\eta_{\mu\nu}\Big[\frac{\varrho}{2}(k^4+2PQ)(P-Q)(Q-m^2)$$
$$+\frac{\varrho}{2}k^2P(P+Q)(Q-m^2)) - \varrho k^2Q^2(Q-m^2)$$
$$+\left(\frac{\varrho}{4}p^2(k^2+2Q)+(k+p)^2m^2\right)(k^2+P)(Q-m^2)\Big]$$
$$+k_\mu k_\nu\left[\varrho(P^2+k^2m^2)(Q-m^2)+4\varrho PQm^2-3\varrho PQ^2\right.$$
$$+\frac{\varrho}{2}p^2(P-2Q)(Q-m^2)+2\varrho Q^2(Q-m^2)-\varrho Pm^4$$
$$\left.+2(k+p)^2(Q(Q-2m^2)-P(Q-m^2)+m^4)\right]$$
$$+2(k_\mu q_\nu+k_\nu q_\mu)(k+p)^2(Q-P)(Q-m^2)$$
$$-2q_\mu q_\nu(k+p)^2(k^2+P)(Q-m^2)\Big\}\,, \qquad Q\equiv(kq), \quad P\equiv(kp). \quad (5.3.52)$$

Introducing the Schwinger parameterization of denominators

$$\frac{1}{k^2} = -\int_0^\infty dy\,\exp\{yk^2\}\,, \qquad \frac{1}{(k+p)^2} = -\int_0^\infty dx\,\exp\{x(k+p)^2\}\,,$$
$$\frac{1}{k^2+2(kq)} = -\int_0^\infty dz\,\exp\{z[k^2+2(kq)]\}\,,$$

one evaluates the loop integrals using

$$\int d^dk\,\exp\{k^2(x+y+z)+2k^\mu(xp_\mu+zq_\mu)+p^2x\}$$
$$= i\left(\frac{\pi}{x+y+z}\right)^{d/2}\exp\left\{\frac{p^2xy-m^2z^2}{x+y+z}\right\},$$

$$\int d^dk\;k_\alpha\exp\{k^2(x+y+z)+2k^\mu(xp_\mu+zq_\mu)+p^2x\} =$$
$$= i\left(\frac{\pi}{x+y+z}\right)^{d/2}\exp\left\{\frac{p^2xy-m^2z^2}{x+y+z}\right\}\left[-\frac{xp_\alpha+zq_\alpha}{x+y+z}\right],$$

etc., up to six k-factors in the integrand. This calculation can be automated to a considerable extent with the help of the tensor package [17] for the REDUCE system. Changing the integration variables (x,y,z) to (t,u,v) via

$$x = \frac{t(1+t+u)v}{m^2(1+\alpha tu)}\,, \qquad y = \frac{u(1+t+u)v}{m^2(1+\alpha tu)}\,, \qquad z = \frac{(1+t+u)v}{m^2(1+\alpha tu)}\,, \qquad \alpha \equiv -\frac{p^2}{m^2}\,,$$

integrating v out, subtracting the ultraviolet divergence[6]

$$\tilde{I}^{3(a)\mathrm{div}}_{\mu\nu}(p,q) = -\frac{1}{16\pi^2\epsilon}\left(\frac{\mu}{m}\right)^{\epsilon}\left[\frac{1}{3}q_\mu q_\nu + \eta_{\mu\nu}(p^2 - 2m^2)\frac{3\varrho - 2}{24}\right],$$

setting $\epsilon = 0$, omitting gradient terms, and retaining only the $\hbar^0$-contribution, we obtain

$$\begin{aligned}
\tilde{I}^{3(a)\mathrm{ren}}_{\mu\nu}(p,q) &\equiv (\tilde{I}^{3(a)}_{\mu\nu} - \tilde{I}^{3(a)\mathrm{div}}_{\mu\nu})_{\epsilon\to 0} \\
&\sim \frac{m^2}{16\pi^2}\int_0^\infty\int_0^\infty dudt\left\{\frac{\eta_{\mu\nu}\varrho}{H^3\alpha}\left(\frac{1}{D^2} - \frac{1}{2D}\right)\right. \\
&+ \left.\frac{q_\mu q_\nu}{H^3 m^2}\left[\frac{2H^2}{D^2}\left(1 - \frac{1}{D}\right) + \frac{1}{\alpha}\left(\frac{4\varrho}{D^3} - \frac{11\varrho + 4}{D^2} + \frac{8\varrho + 4}{D}\right)\right]\right\}, \\
& D \equiv 1 + \alpha ut\,, \qquad H \equiv 1 + u + t\,.
\end{aligned} \tag{5.3.53}$$

The root singularity in the right hand side of Eq. (5.3.53) can be extracted using Eqs. (5.5.97) derived in the Appendix. Denoting

$$\int_0^\infty\int_0^\infty dudt\; H^{-n}D^{-m} \equiv J_{nm}\,,$$

one has

$$\begin{aligned}
J^{\mathrm{root}}_{12} &= \frac{\pi^2}{4\sqrt{\alpha}}\,, \quad J^{\mathrm{root}}_{13} = \frac{3\pi^2}{16\sqrt{\alpha}}\,, \\
J^{\mathrm{root}}_{31} &= -\frac{\pi^2}{16}\sqrt{\alpha}\,, \quad J^{\mathrm{root}}_{32} = -\frac{3\pi^2}{32}\sqrt{\alpha}\,, \quad J^{\mathrm{root}}_{33} = -\frac{15\pi^2}{128}\sqrt{\alpha}\,.
\end{aligned}$$

Substituting these into Eq. (5.3.53) gives

$$\tilde{I}^{\mathrm{ren}}_{\mu\nu}(p,q) \sim \frac{1}{128\sqrt{\alpha}}[q_\mu q_\nu(\varrho + 1) - \eta_{\mu\nu}m^2\varrho]. \tag{5.3.54}$$

Finally, restoring ordinary units, substituting Eq. (5.3.54) into Eq. (5.3.51), and using the formula

$$\int\frac{d^3\boldsymbol{p}}{(2\pi)^3}\frac{e^{i\boldsymbol{p}\boldsymbol{x}}}{|\boldsymbol{p}|} = \frac{1}{2\pi^2 r^2}\,, \quad r^2 \equiv \delta_{ik}x^i x^k\,, \tag{5.3.55}$$

[6]A technicality must be mentioned here. By itself, the diagram of Fig. 5.3.3(a) is free of infrared divergences. As a result of the BRST-operating with this diagram, however, some fictitious infrared divergences are brought into individual diagrams representing the right hand side of Eq. (5.3.33). This is because the vertex $D^{\alpha}_{\mu\nu}C_\alpha$ contains the term $C^\alpha\partial_\alpha h_{\mu\nu}$ in which the spacetime derivatives act on the gravitational, rather than the ghost field. These divergences occur as $u, t \to \infty$. They are proportional to integer powers of p^2, and therefore do not interfere with the part containing the root singularity. Since these divergences must eventually cancel in the total sum in Eq. (5.3.33), they will be simply omitted in what follows.

we obtain the following expression for the ϱ-derivative of the $G^2\hbar^0$-order contribution to the effective metric [18]

$$\frac{\partial h^{\rm eff}_{\mu\nu}}{\partial\varrho} = \frac{\partial h^{\rm tree}_{\mu\nu}}{\partial\varrho} + \frac{\partial h^{\rm loop}_{\mu\nu}}{\partial\varrho} \sim \frac{\partial h^{\rm loop}_{\mu\nu}}{\partial\varrho} = I^{\rm ren}_{\mu\nu}(x) \sim \frac{G^2m^2}{2c^2r^2}\left[\frac{\delta^0_\mu\delta^0_\nu}{c^2}(\varrho+1) - \eta_{\mu\nu}\varrho\right]. \qquad (5.3.56)$$

The right hand side of this equation cannot be represented in the form (5.3.32). This result can be made more expressive by calculating the ϱ-variation of the scalar curvature R in a given point σ^0 of the reference frame

$$\begin{aligned}\delta R[h^{\rm eff}(x(\sigma^0))] &= \frac{\delta R}{\delta h^{\rm eff}_{\mu\nu}}\left(\frac{\partial h^{\rm tree}_{\mu\nu}}{\partial\varrho} + \frac{\partial h^{\rm loop}_{\mu\nu}}{\partial\varrho}\right)\delta\varrho + \frac{\partial R}{\partial x^\mu}\delta x^\mu \\ &= \frac{\partial R}{\partial x^\alpha}\Xi^\alpha_{\rm tree} + (\partial^\mu\partial^\nu - \eta^{\mu\nu}\Box)\frac{h^{\rm loop}_{\mu\nu}}{\partial\varrho}\delta\varrho - \frac{\partial R}{\partial x^\mu}\Xi^\mu_{\rm tree} \\ &= \partial_i\partial_k\frac{\partial h^{\rm loop}_{ik}}{\partial\varrho}\delta\varrho + \Delta\frac{\partial h^{\rm loop}}{\partial\varrho}\delta\varrho = \frac{G^2m^2}{c^4r^4}(1-2\varrho)\delta\varrho\,,\end{aligned}$$

or[7]

$$\frac{\partial R[h^{\rm eff}(x(\sigma^0))]}{\partial\varrho} = \frac{G^2m^2}{c^4r^4}(1-2\varrho)\,. \qquad (5.3.58)$$

Equations (5.3.56), (5.3.58) express violation of general covariance by the loop corrections.

Thus, despite their independence of the Planck constant, the post-Newtonian loop contributions turn out to be of a purely quantum nature.

We are now in a position to ask for conditions to be imposed on a system in order to allow classical consideration of its gravitational field. Such a condition providing vanishing of the $\hbar^0$ loop contributions can easily be found out by examining their dependence on the number of particles in the system. Let us consider a body with mass M, consisting of a large number $N = M/m$ of elementary particles with mass

[7]Another way to obtain this result is to introduce the sources tR and $kR_{,\alpha}c^\alpha$ for the scalar curvature and its BRST-variation, respectively, into the generating functional (5.2.11), instead of the corresponding sources for the metric. Then Eq. (5.3.47) is replaced by

$$\frac{\partial R^{\rm eff}}{\partial\varrho} = \left.\frac{\delta\langle\bar{c}^\alpha\partial F_\alpha/\partial\varrho\rangle}{\delta k}\right|_{\substack{J\backslash j=0\\K=0}}. \qquad (5.3.57)$$

At the second order in G, the nontrivial contribution comes again from the diagram of Fig. 5.3.4(a) in which the lower left vertex is now generated by $R_{,\alpha}c^\alpha$. Thus, only the linear part of R gives rise to a non-zero contribution to the right hand side of Eq. (5.3.57). In other words, $\delta R[h^{\rm eff}] = \delta R^{\rm eff}$, though generally $R[h^{\rm eff}] \neq R^{\rm eff}$.

m. Then it is readily seen that the n-loop contribution to the effective gravitational field of the body turns out to be suppressed by a factor $1/N^n$ in comparison with the tree contribution. For instance, at the first post-Newtonian order, the tree diagram in Fig. 5.3.2(b) is *bilinear* in the energy-momentum tensor $\langle \hat{T}^{\mu\nu} \rangle$ of the particles, and therefore proportional to $(m \cdot N) \cdot (m \cdot N) = M^2$. On the other hand, the post-Newtonian contribution of the diagram in Fig. 5.1.1(a) is proportional to $m^2 \cdot N = M^2/N$, since it has only two external matter lines.

Thus, we are led to the following macroscopic formulation of the correspondence principle in quantum gravity: *the effective gravitational field produced by a macroscopic body of mass M consisting of N particles turns into corresponding classical solution of the Einstein equations in the limit $N \to \infty$* [19]. In particular, the principle of general covariance is to be considered as approximate, valid only for the description of macroscopic phenomena.

The $\hbar^0$-order loop contributions are normally highly suppressed. For the solar gravitational field, for instance, their relative value is $m_{\text{proton}}/M_{\odot} \approx 10^{-57}$. However, they are the larger the more gravitating body resembles an elementary particle, and can become noticeable for a sufficiently massive compact body.

In connection with the macroscopic formulation of the correspondence principle, given above, the following important circumstance must be emphasized. For any finite number of particles N, the tree diagrams representing the mean value $\langle N|g_{\mu\nu}|N\rangle$ constitute only a finite subset of the infinite set of tree diagrams generated by the corresponding classical solution of the Einstein equations (these are connected diagrams with no more than $2N$ external matter lines). Therefore, in investigating the correspondence between classical and quantum theories, it is more natural to work with eigenstates of the matter fields, $|\phi(\boldsymbol{x})\rangle$, i.e., coherent states, rather than the states $|N\rangle$ with definite number of particles. In such states, the quantities N, M, etc. are substituted by the corresponding averaged values $\langle N\rangle$, $\langle M\rangle$, while the amplitude $a(\boldsymbol{q})$ represents the Fourier transform of the given field $\phi(\boldsymbol{x})$. Of course, the two pictures are completely equivalent in the classical limit. Since coherent states are of special interest in the light of the correspondence between classical and quantum theories, below we consider an example of calculation of the effective gravitational field produced by a coherent field configuration.

5.3.3 Effective gravitational field of a coherent field configuration

As an application of the obtained results, we will calculate the effective gravitational field of a coherent scalar field state in which $\langle N\rangle = 1$, in the first post-Newtonian approximation.

The complete expression of the order $\hbar^0$ for the spacetime metric is the sum of two pieces. The first is the tree contribution corresponding to the classical Schwarzschild solution

$$ds^2_{\rm cl} = \left(1 - \frac{r_g}{r}\right) c^2 dt^2 - \frac{dr^2}{1 - \frac{r_g}{r}} - r^2(d\theta^2 + \sin^2\theta\, d\varphi^2), \tag{5.3.59}$$

where θ, φ are the standard spherical angles, r is the radial coordinate, and $r_g = 2Gm/c^2$ is the averaged gravitational radius of the given (spherically-symmetric) field configuration. The second is the one-loop post-Newtonian correction contained in the diagram (a) of Fig. 5.1.1. Restoring the ordinary units, and using expressions (55), (65) of Ref. [15] for the vertex formfactors, and (5.3.49) with $\varrho = 0$ for the graviton propagator,[8] we obtain

$$h^{\rm loop}_{\mu\nu}(p) = -\frac{\pi^2 G^2}{c^2\sqrt{-p^2}}\left(3m^2\eta_{\mu\nu} + \frac{q_\mu q_\nu}{c^2} + 7m^2\frac{p_\mu p_\nu}{p^2}\right). \tag{5.3.60}$$

Written down in the coordinate space with the help of the formulas (5.3.55) and

$$\int \frac{d^3\boldsymbol{p}}{(2\pi)^3}\frac{p_i p_k}{|\boldsymbol{p}|^3} e^{i\boldsymbol{p}\boldsymbol{x}} = \frac{1}{2\pi^2 r^2}\left(\delta_{ik} - \frac{2x_i x_k}{r^2}\right),$$

equation (5.3.60) gives, in the static case,

$$h^{\rm loop}_{00} = -\frac{2G^2m^2}{c^2r^2}, \quad h^{\rm loop}_{ik} = \frac{G^2m^2}{c^2r^2}\left(-2\delta_{ik} + \frac{7x_i x_k}{r^2}\right). \tag{5.3.61}$$

Before adding the two contributions, however, one has to transform Eq. (5.3.59) which form is fixed by the requirements $g_{ti} = 0,\ i = r, \theta, \varphi;\ g_{\theta\theta} = -r^2$, to the DeWitt gauge

$$F_\gamma = \eta^{\mu\nu}\partial_\mu h_{\nu\gamma} - \frac{1}{2}\partial_\gamma h\,, \qquad \zeta_{\alpha\beta} = \eta_{\alpha\beta}\,, \tag{5.3.62}$$

under which Eq. (5.3.60) was derived. According to the general theorems about $\zeta_{\alpha\beta}$-independence of the $\hbar^0$-order contributions, proved in Sec. 5.3.2, we can transform our expressions to the singular case $\zeta_{\alpha\beta} = 0$ instead of $\zeta_{\alpha\beta} = \eta_{\alpha\beta}$. Then the effective gravitational field will satisfy

$$\eta^{\mu\nu}\partial_\mu h^{\rm eff}_{\nu\gamma} - \frac{1}{2}\partial_\gamma h^{\rm eff} = 0\,. \tag{5.3.63}$$

The t, θ, φ-components of Eq. (5.3.63) are already satisfied by the classical solution (5.3.59). To meet the remaining condition, let us substitute $r \to f(r)$, where f is a function of r only. Then the t, θ, φ-components of Eq. (5.3.46) are still satisfied, while its r-component gives the following equation for the function $f(r)$:

$$\frac{1}{r^2}\frac{\partial}{\partial r}\left(\frac{r^2 f'^2}{1 - r_g/f}\right) - \frac{2f^2}{r^2} - \frac{1}{2}\frac{\partial}{\partial r}\left(1 - \frac{r_g}{f} + \frac{f'^2}{1 - r_g/f} + \frac{2f^2}{r^2}\right) = 0, \tag{5.3.64}$$

[8]Note the notation differences between Ref. [15] and this Chapter.

where $f' \equiv \partial f(r)/\partial r$.

Since we are interested only in the long-distance corrections to the Newton law, one may expand $f(r)/r$ in powers of r_g/r keeping only the first few terms:

$$f(r) = r\left[1 + c_1\frac{r_g}{r} + c_2\left(\frac{r_g}{r}\right)^2 + \cdots\right].$$

Substituting this into Eq. (7.13.4), one obtains successively $c_1 = 1/2$, $c_2 = 1/2$, etc. Therefore, up to terms of the order r_g^2/r^2, the Schwarzschild solution takes the following form

$$ds^2_{\rm cl} = \left(1 - \frac{r_g}{r} + \frac{r_g^2}{2r^2}\right)c^2dt^2 - \left(1 + \frac{r_g}{r} - \frac{r_g^2}{2r^2}\right)dr^2$$
$$-r^2\left(1 + \frac{r_g}{r} + \frac{5r_g^2}{4r^2}\right)(d\theta^2 + \sin^2\theta\, d\varphi^2)\,. \tag{5.3.65}$$

Rewriting Eq. (5.3.61) in spherical coordinates and adding the tree contribution, we finally obtain the following expression for the interval [20]

$$ds^2_{\rm eff} \equiv g^{\rm eff}_{\mu\nu}dx^\mu dx^\nu = \left(1 - \frac{r_g}{r}\right)c^2dt^2 - \left(1 + \frac{r_g}{r} - \frac{7r_g^2}{4r^2}\right)dr^2$$
$$-r^2\left(1 + \frac{r_g}{r} + \frac{7r_g^2}{4r^2}\right)(d\theta^2 + \sin^2\theta\, d\varphi^2)\,. \tag{5.3.66}$$

In connection with this result it should be noted the following. Taking into account higher-order radiative post-Newtonian corrections will result in a further modification of the Schwarzschild solution. Since quantum contributions are of the same order of magnitude as those given by general relativity, this modification can lead to a significant shift of the horizon. In particular, the metric $g^{\rm eff}_{\mu\nu}(r)$ may well turn out to be a regular function of r when all the $\hbar^0$ loop corrections are taken into account.

5.4 Quantum fluctuations of gravitational field

We turn now to another aspect of the long-range behavior in quantum gravity – quantum fluctuations of the gravitational fields produced by matter fields. On various occasions, the issue of fluctuations has been the subject of a number of investigations (see Refs. [21, 22, 23, 24, 25], where references to early works can be found). It should be mentioned, however, that despite extensive literature in the area, only vacua contributions of quantized matter fields to the metric fluctuations have been studied in detail. At the same time, it is effects produced by real matter that are of special interest concerning the structure of field fluctuation.

Although our general considerations apply to systems in any quantum state, fluctuations exhibited by systems in coherent states will be of our primary concern below. Before we proceed to calculation of the correlation function, we shall examine its general properties in more detail. Namely, the structure of the long-range expansion of the correlation function, and the question of its gauge dependence will be considered.

5.4.1 Properties of correlation function

As defined by Eq. (5.2.13), the correlation function $C_{\mu\nu\alpha\beta}$ is a function of two spacetime arguments (x, x'). Of special interest is its "diagonal element" corresponding to coinciding arguments and Lorentz indices ($x = x'$, $\mu\nu = \alpha\beta$), which describes dispersion of the spacetime metric around its mean value in a given spacetime point. However, it is well-known that this element is not generally well-defined because of the singular behavior of the product of field operators in the coincidence limit $x \to x'$. This difficulty is naturally resolved when one takes into account the fact (realized long ago, see, e.g., Ref. [26]) that in any field measurement in a given spacetime point, one deals actually with the field averaged over a small but finite spacetime domain surrounding this point. Thus, the physically sensible expression for the field operator is the following

$$\hat{\mathcal{H}}_{\mu\nu} = \frac{1}{VT}\int_T dt \int_V d^3\boldsymbol{x}\; \hat{h}_{\mu\nu}(\boldsymbol{x}, t)\,. \tag{5.4.67}$$

Respectively, the product of two fields in a given point is understood as the limit of

$$\hat{\mathcal{B}}_{\mu\nu\alpha\beta} = \frac{1}{(VT)^2}\int_T dt \int_T dt' \int_V d^3\boldsymbol{x} \int_V d^3\boldsymbol{x}'\; \hat{h}_{\mu\nu}(\boldsymbol{x}, t)\hat{h}_{\alpha\beta}(\boldsymbol{x}', t') \tag{5.4.68}$$

when the size of the domain tends to zero. Finally, correlation function of the spacetime metric in this domain is

$$\mathcal{C}_{\mu\nu\alpha\beta} = \langle\hat{\mathcal{B}}_{\mu\nu\alpha\beta}\rangle - \langle\hat{\mathcal{H}}_{\mu\nu}\rangle\langle\hat{\mathcal{H}}_{\alpha\beta}\rangle \equiv \frac{1}{(VT)^2}\iint d^4x d^4x' C_{\mu\nu\alpha\beta}(x, x')\,. \tag{5.4.69}$$

Figure 5.4.6 depicts the tree diagrams contributing to the right hand side of Eq. (5.2.16) in the case of a coherent matter field state. The disconnected part shown in Fig. 5.4.6(a) cancels in the expression for the correlation function, Eq. (5.4.69), which is thus represented by the diagrams (b)–(h).

Correlation function in the long-range limit

As we mentioned in the beginning of Sec. 5.3, the mean gravitational field produced by the given field configuration is a function of five independent dimensional parameters. This applies also to the correlation function (in the coincidence limit),

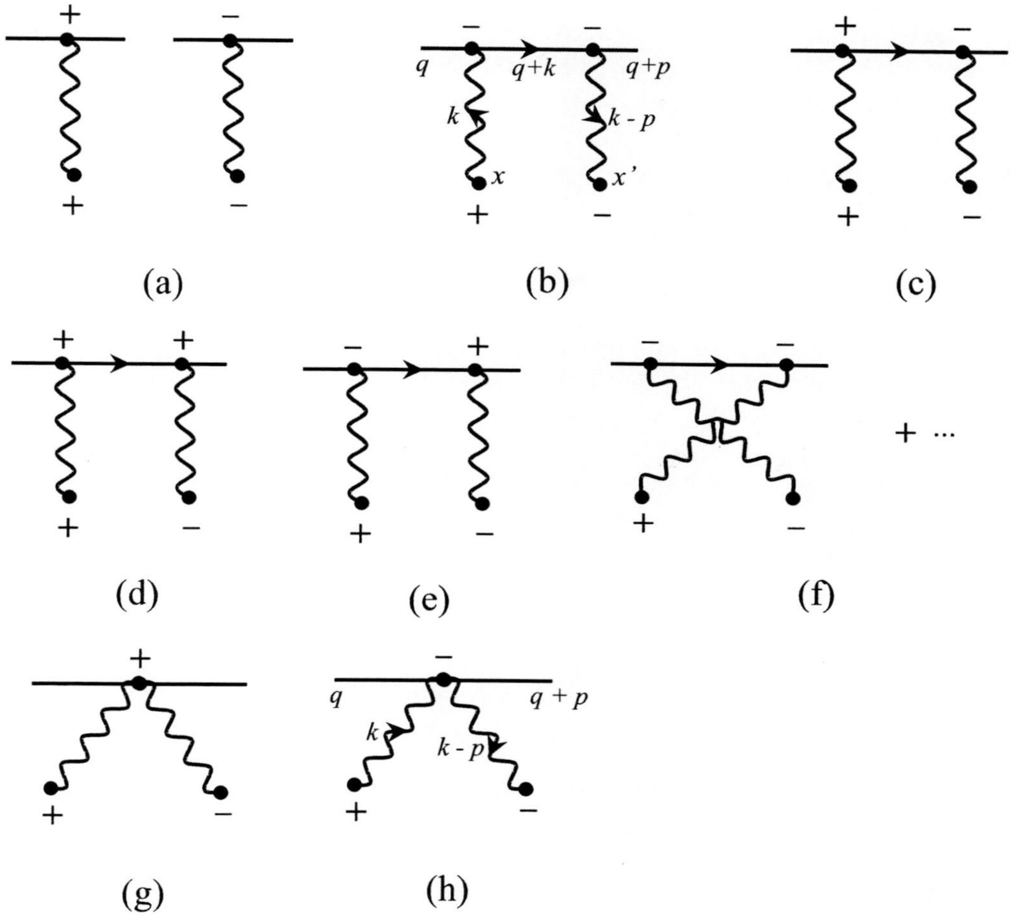

Figure 5.4.6: Tree contribution to the right hand side of Eq. (5.2.16). Part (f) of the figure represents the "transposition" of diagrams (b)–(e) (see Sec. 5.4.2).

but with one important specification. The point is that in addition to the five parameters mentioned above, there are also parameters characterizing spatial mass distribution. These parameters determine the multipole components of the quantities under consideration. As long as the mean field is considered, this circumstance does not matter, because the multipoles of the spherically-symmetric Newton field vanish. However, this is not the case for the correlation function. Thus, within the long-range expansion, this function is represented by a series in powers of D/r, where D is the characteristic length of spatial spreading of the particle wave function. To avoid this complication when extracting the $\hbar^0$-contribution, one has to speak about the integral kernel of the correlation function before it is contracted with the amplitudes $a(\boldsymbol{q})$, rather than the correlation function itself. Hence, extraction of the leading long-range contribution to the correlation function consists of two steps: the first and most laborious is the extraction of the leading term from the integral kernel within the large-mass expansion (where parameters characterizing spatial charge distribution do not appear), and the second is the extraction of the leading term in the multipole decomposition of the correlation function.

Next, let us discuss the role of the spacetime domain used in the definition of physical gravitational field operators, Eq. (5.4.67), in more detail. In any actual field measurement, the size of this domain must be small in comparison with the characteristic length at which the mean field changes significantly. In particular, we expect that the correlation function is well-defined in the limit of $V, T \to 0$. However, one cannot set $x = x'$ directly in the expression $C_{\mu\nu\alpha\beta}(x, x')$ appearing in the right hand side of Eq. (5.4.69). It is not difficult to see that the formal expression $\langle \hat{h}_{\mu\nu}(x)\hat{h}_{\alpha\beta}(x)\rangle$ does not exist. Consider, for instance, the diagram 5.4.6(b). It is proportional to the integral

$$I_{\mu\nu\alpha\beta}(x - x', p) = \int d^4k \frac{V_{\mu\nu\alpha\beta}\theta(k^0)\delta(k^2)e^{ik(x-x')}}{[(k+q)^2 - m^2](k-p)^2},$$

where q_μ is the 4-momentum of the particle, p_μ the momentum transfer, and $V_{\mu\nu\alpha\beta}$ a vertex factor combined of $\eta_{\mu\nu}$'s and momenta q, p, k. For small but nonzero $(x - x')$ this integral is effectively cut-off at large k's by the oscillating exponent, but for $x = x'$ it is divergent. This divergence arises from integration over large values of virtual photon momenta, and therefore has nothing to do with the long-range behavior of the correlation function, because this behavior is determined by the low-energy properties of the theory. Evidently, the singularity of $I_{\mu\nu\alpha\beta}(x - x', p)$ for $x \to x'$ is not worse than $\ln(x^\mu - x'^\mu)^2/(x^\mu - x'^\mu)^2$. Therefore, $\langle \text{in}|\hat{h}_{\mu\nu}(x)\hat{h}_{\alpha\beta}(x')|\text{in}\rangle$ is integrable, and $\mathcal{C}_{\mu\nu}$ given by Eqs. (5.4.68), (5.4.69) is well defined.

Our aim below will be to show that this singularity can be consistently isolated and removed from the expression for $I_{\mu\nu\alpha\beta}(x - x', p)$ and similar integrals for the rest of diagrams in Fig. 5.4.6, without changing the long-range properties of the correlation function. After this removal, it is safe to set $x = x'$ in the finite remainder, and to consider $\mathcal{C}_{\mu\nu\alpha\beta}$ as a function of the single variable – the distance

r. An essential point of this procedure is that the singularity turns out to be *local*, and hence does not interfere with terms describing the long-range behavior, which guaranties unambiguity of the whole procedure.

It should be emphasized that in contrast to what takes place in the scattering theory, the ultraviolet divergences appearing in calculating the coincidence limit of individual diagrams within the Schwinger-Keldysh formalism are generally non-polynomial with respect to the momentum transfer. The reason for this is the different analytic structure of various elements in the matrix propagators $\mathfrak{D}_{\mu\nu\alpha\beta}$, $\mathfrak{D}$, which spoils the simple ultraviolet properties exhibited by the ordinary Feynman amplitudes.[9] Take the above integral as an example. Because of the delta function in the integrand, differentiation of $I_{\mu\nu\alpha\beta}(x - x', p)$ with respect to the momentum transfer does not remove the ultraviolet divergence of $I_{\mu\nu\alpha\beta}(0, p)$. However, we will show in Sec. 5.4.2 that in the coincidence limit, the sum of diagrams in Fig. 5.4.6 can be transformed to the form which involves only the usual Feynman propagators. This means that the non-polynomial parts of divergent contributions eventually cancel each other, and the overall divergence turns out to be completely local.

Next, let us establish general form of the leading term in the long-range expansion of the correlation function. In momentum representation, an expression of lowest order in the momentum transfer with suitable dimensionality is the following

$$\mathcal{C}^{(0)}_{\mu\nu\alpha\beta}(p) \sim G^2 m^2 \frac{\mathcal{F}_{\mu\nu\alpha\beta}}{\sqrt{-p^2}} \,, \tag{5.4.70}$$

where $\mathcal{F}_{\mu\nu\alpha\beta}$ is a dimensionless positive definite tensor combined from $\eta_{\mu\nu}$ and q_μ.

Not all of diagrams in Fig. 5.4.6 contain contributions of this type. It is not difficult to identify those which do not. Consider, for instance, the diagram (h). It is proportional to the integral

$$\int d^4k \theta(k^0) \delta(k^2) \frac{e^{ik(x-x')}}{(k-p)^2} \,,$$

which does not involve the particle mass at all. Taking into account m^2 coming from the vertex factor, and $(2\varepsilon_{\boldsymbol{q}})^{-1/2} \sim \sqrt{m}$ from each external matter line, we see that the contribution of the diagram (h) is proportional to m, not m^2. The same is true of all other diagrams without internal matter lines. As to diagrams involving such lines, it will be shown in Sec. 5.4.2 by direct calculation that they do contain contributions of the type Eq. (5.4.70). But prior to this the question of their dependence on the gauge will be considered.

[9] As is well known, the proof of locality of the S-matrix divergences relies substantially on the causality of the pole structure of Feynman propagators. This property allows Wick rotation of the energy contours, thus revealing the essentially Euclidean nature of the ultraviolet divergences.

Gauge independence of the leading contribution

The function $\mathfrak{C}^{(0)}_{\mu\nu\alpha\beta}$ contains information about fluctuations in a quantity of direct physical meaning – the static potential energy of interacting particles. As such, it is expected to be independent of the choice of gauge conditions used to fix general covariance. More precisely, we have to verify that under variations of the gauge conditions, $\mathfrak{C}^{(0)}_{\mu\nu\alpha\beta}$ varies in a way that does not affect the values of observables built from it.

As we saw in Sec. 5.4.1, the only diagrams contributing in the long-range limit are those containing internal massive particles lines. We will show presently that the gauge-dependent part of these diagrams can be reduced to the form without such lines. First, it follows form Eq. (5.2.17) that the gauge variation of the graviton propagator satisfies

$$\delta\mathfrak{D}_{\mu\nu\sigma\lambda} = \mathfrak{D}_{\mu\nu\alpha\beta}\delta\mathfrak{G}^{\alpha\beta\gamma\delta}\mathfrak{D}_{\gamma\delta\sigma\lambda}\,. \tag{5.4.71}$$

On the other hand, contracting Eq. (5.2.17) with $G^{(0)\rho}_{\mu\nu}$ yields

$$\mathrm{i}\,F^{,\mu\nu}_{\alpha}G^{(0)\rho}_{\mu\nu}\xi^{\alpha\beta}F^{,\gamma\delta}_{\beta}\mathfrak{D}_{\gamma\delta\sigma\lambda} = -\mathfrak{e}G^{(0)\rho}_{\sigma\lambda}.$$

Defining the matrix ghost propagator $\tilde{\mathfrak{D}}^{\alpha}_{\beta}$ according to Eqs. (5.2.17), (5.3.37) by

$$\mathrm{i}F^{,\mu\nu}_{\alpha}G^{(0)\beta}_{\mu\nu}\tilde{\mathfrak{D}}^{\gamma}_{\beta} = -\mathfrak{e}\delta^{\gamma}_{\alpha}$$

one finds

$$F^{,\mu\nu}_{\alpha}\mathfrak{D}_{\mu\nu\sigma\lambda} = \zeta_{\alpha\gamma}G^{(0)\beta}_{\sigma\lambda}\tilde{\mathfrak{D}}^{\gamma}_{\beta}\,, \tag{5.4.72}$$

Taking into account that

$$\delta\mathfrak{G}^{\alpha\beta\gamma\delta} = \mathrm{i}\delta F^{,\alpha\beta}_{\rho}\xi^{\rho\tau}F^{,\gamma\delta}_{\tau} + \mathrm{i}F^{,\alpha\beta}_{\rho}\xi^{\rho\tau}\delta F^{,\gamma\delta}_{\tau}\,,$$

and using Eq. (5.4.72), we get

$$\delta\mathfrak{D}_{\mu\nu\sigma\lambda} = \left(\delta F^{,\alpha\beta}_{\delta}\mathfrak{D}_{\mu\nu\alpha\beta}\,\mathrm{i}\,\tilde{\mathfrak{D}}^{\delta}_{\gamma}\right)G^{(0)\gamma}_{\sigma\lambda} + G^{(0)\beta}_{\mu\nu}\left(\tilde{\mathfrak{D}}^{\alpha}_{\beta}\,\mathrm{i}\,\delta F^{,\gamma\delta}_{\alpha}\mathfrak{D}_{\gamma\delta\sigma\lambda}\right)\,. \tag{5.4.73}$$

Let us assume for definiteness that the pair of indices $\mu\nu$ refers to the point of observation, and consider the first term in Eq. (5.4.73). This part of the gauge variation of the graviton propagator is attached to the matter line through the generator $G^{(0)\gamma}_{\sigma\lambda}$. Suppressing all Lorentz and matrix indices except those referring to the $\phi^2 h$-vertex, it can be written as

$$\mathfrak{b}_{\sigma\lambda} = \begin{pmatrix} a_{\gamma+} \\ a_{\gamma-} \end{pmatrix} G^{(0)\gamma}_{\sigma\lambda}\,, \qquad \mathfrak{a}_{\gamma} = \delta F^{,\alpha\beta}_{\delta}\mathfrak{D}_{\mu\nu\alpha\beta}\,\mathrm{i}\,\tilde{\mathfrak{D}}^{\delta}_{\gamma}\,.$$

The matrix vertex $\mathfrak{V}^{\sigma\lambda,ik}$ is obtained multiplying $\delta^3 S_\phi/\delta\phi_i\delta\phi_k\delta h_{\sigma\lambda}$ by the matrix e_{stu}. Equation (5.3.39) then shows that the combination $\mathfrak{b}_{\sigma\lambda}\mathfrak{V}^{\sigma\lambda,ik}$ can be written as

$$\mathfrak{b}_{\sigma\lambda}\mathfrak{V}^{\sigma\lambda,ik} = -\mathfrak{f}_\gamma \left\{ \frac{\delta^2 S^{(2)}_\phi}{\delta\phi_l\delta\phi_k}\frac{\delta G^\gamma_l}{\delta\phi_i} + \frac{\delta^2 S^{(2)}_\phi}{\delta\phi_l\delta\phi_i}\frac{\delta G^\gamma_l}{\delta\phi_k} \right\},$$

where

$$\mathfrak{f}_\gamma = \begin{pmatrix} a_{\gamma+} & 0 \\ 0 & -a_{\gamma-} \end{pmatrix}.$$

Finally, contracting $\mathfrak{b}_{\sigma\lambda}\mathfrak{V}^{\sigma\lambda,ik}$ with the matrix $\mathfrak{D}_{km}$ and the vector $\tilde{\mathfrak{r}}_i = \left(\bar{\phi}_i, \bar{\phi}_i\right)$, and using Eqs. (5.2.18), (5.2.19) gives

$$\tilde{\mathfrak{r}}_i \mathfrak{b}_{\sigma\lambda}\mathfrak{V}^{\sigma\lambda,ik}\mathfrak{D}_{km} = \tilde{\mathfrak{a}}_\gamma \frac{\delta G^\gamma_m}{\delta\phi_i}\bar{\phi}_i .$$

Similarly to what we have found in Sec. 5.3.2 considering ζ-dependence of the one-loop contribution to the effective metric, the matrix matter propagator is cancelled upon contraction with the vertex factor. As in the case of the one-loop post-Newtonian contributions, the $\hbar^0$ terms in the correlation function are associated with the virtual matter quanta propagating near their mass shells. Hence, the first term in the gauge variation of the propagator, Eq. (5.4.73), gives rise to terms of higher order in the Planck constant. As to the second term, it gives rise to a $\hbar^0$ order variation of the correlation function, but its structure with respect to the indices $\mu\nu$ is that of an ordinary gauge transformation, so

$$\delta\mathfrak{C}^{(0)}_{\mu\nu\alpha\beta} \sim G^\gamma_{\mu\nu}\Omega_{\gamma,\alpha\beta} + G^\gamma_{\alpha\beta}\Omega_{\gamma,\mu\nu}$$

with some infinitesimal $\Omega_{\gamma,\alpha\beta}$, which proves gauge independence of gauge-invariant functionals built from $\mathfrak{C}^{(0)}_{\mu\nu\alpha\beta}$. Note also that as far as variations of the weighting matrix $\xi^{\alpha\beta}$ are considered, not only these functionals, but also $\mathfrak{C}^{(0)}_{\mu\nu\alpha\beta}$ itself turns out to be gauge independent. Indeed, a variation $\delta\xi^{\alpha\beta}$ of the weighting matrix is equivalent to the variation of the gauge condition F_α :

$$\delta F_\alpha = \theta^\beta_\alpha F_\beta ,$$

provided that

$$\delta\xi^{\alpha\beta} = \xi^{\alpha\gamma}\theta^\beta_\gamma + \xi^{\beta\gamma}\theta^\alpha_\gamma .$$

Then using Eq. (5.4.72) in Eq. (5.4.73) gives

$$\delta\mathfrak{D}_{\mu\nu\sigma\lambda} = -G^{(0)\beta}_{\mu\nu}\tilde{\mathfrak{D}}^\alpha_\beta\, \mathfrak{i}\, \delta\zeta_{\alpha\delta}\tilde{\mathfrak{D}}^\delta_\gamma G^{(0)\gamma}_{\sigma\lambda},$$

which implies that $\mathfrak{C}^{(0)}_{\mu\nu\alpha\beta}$ is independent of the choice of the matrix $\zeta_{\alpha\beta}$.

5.4.2 Evaluation of the leading contribution

Let us proceed to the calculation of the leading contribution to the correlation function. We assume that the gravitational field is produced by a scalar matter field configuration $\bar{\phi}$. The leading contribution is contained in the sum of diagrams (b)–(f) in Fig. 5.4.6, which has the symbolic form

$$B_{\mu\nu\alpha\beta} = I_{\mu\nu\alpha\beta} + I^{\rm tr}_{\mu\nu\alpha\beta}\,, \qquad I_{\mu\nu\alpha\beta} = \frac{1}{i}\left\{\mathfrak{D}_{\mu\nu\sigma\lambda}\left[\mathfrak{r}^{\dagger}\mathfrak{V}^{\sigma\lambda}\mathfrak{D}\mathfrak{V}^{\tau\rho}\mathfrak{r}\right]\mathfrak{D}_{\tau\rho\alpha\beta}\right\}_{+-}\,,$$

where the superscript "tr" means transposition of the indices and spacetime arguments referring to the points of observation: $\mu\nu \leftrightarrow \alpha\beta$, $+ \leftrightarrow -$, $x \leftrightarrow x'$ [the transposed contribution is represented by the diagrams collected in part (f) of Fig. 5.4.6].

In the DeWitt gauge

$$\mathfrak{D}_{\mu\nu\alpha\beta} = -W_{\mu\nu\alpha\beta}\mathfrak{D}^0\,, \qquad \mathfrak{D}^0 \equiv \mathfrak{D}|_{m=0}\,,$$

$I_{\mu\nu\alpha\beta}$ reads

$$\begin{aligned} I_{\mu\nu\alpha\beta}(x,x') &= \frac{1}{i}\iint d^4z d^4z' \\ &\times\Big\{ + D^0_{++}(x,z)\left[\bar{\phi}(z)\overleftrightarrow{V}_{\mu\nu}\, D_{++}(z,z')\,\overleftrightarrow{V'}_{\alpha\beta}\,\bar{\phi}(z')\right] D^0_{+-}(z',x') \\ &\quad - D^0_{++}(x,z)\left[\bar{\phi}(z)\overleftrightarrow{V}_{\mu\nu}\, D_{+-}(z,z')\,\overleftrightarrow{V'}_{\alpha\beta}\,\bar{\phi}(z')\right] D^0_{--}(z',x') \\ &\quad - D^0_{+-}(x,z)\left[\bar{\phi}(z)\overleftrightarrow{V}_{\mu\nu}\, D_{-+}(z,z')\,\overleftrightarrow{V'}_{\alpha\beta}\,\bar{\phi}(z')\right] D^0_{+-}(z',x') \\ &\quad + D^0_{+-}(x,z)\left[\bar{\phi}(z)\overleftrightarrow{V}_{\mu\nu}\, D_{--}(z,z')\,\overleftrightarrow{V'}_{\alpha\beta}\,\bar{\phi}(z')\right] D^0_{--}(z',x')\Big\}\,, \end{aligned} \tag{5.4.74}$$

where

$$\varphi\overleftrightarrow{V}_{\mu\nu}\,\psi = \frac{1}{2}W_{\mu\nu\alpha\beta}\left\{W^{\alpha\beta\gamma\delta}\varphi\,\overleftarrow{\partial}_{\gamma}\overrightarrow{\partial}_{\delta}\,\psi - m^2\eta^{\alpha\beta}\varphi\psi\right\}.$$

Contribution of the third term in the right hand side of Eq. (5.4.74) is zero identically. Indeed, using Eq. (5.2.20), and performing spacetime integrations we see that the three lines coming, say, into z-vertex are all on the mass shell, which is inconsistent with the momentum conservation in the vertex. The remaining terms in Eq. (5.4.74) take the form

$$\begin{aligned} I_{\mu\nu\alpha\beta}(x,x') &= \iint \frac{d^3\boldsymbol{q}}{(2\pi)^3}\frac{d^3\boldsymbol{p}}{(2\pi)^3}\frac{a(\boldsymbol{q})a^*(\boldsymbol{q}+\boldsymbol{p})}{\sqrt{2\varepsilon_{\boldsymbol{q}}2\varepsilon_{\boldsymbol{q}+\boldsymbol{p}}}}e^{ipx'}\tilde{I}_{\mu\nu\alpha\beta}(p,q)\,, \\ & p_0 = \varepsilon_{\boldsymbol{q}+\boldsymbol{p}} - \varepsilon_{\boldsymbol{q}}\,, \end{aligned} \tag{5.4.75}$$

where

$$\begin{aligned}
\tilde{I}_{\mu\nu\alpha\beta}(p,q) = & - i\int\frac{d^4k}{(2\pi)^4}e^{ik(x-x')}\{m^4\eta_{\mu\nu}\eta_{\alpha\beta} - 2m^2(\eta_{\mu\nu}q_\alpha q_\beta + \eta_{\alpha\beta}q_\mu q_\nu) \\
& - 4m^2\eta_{\mu\nu}q_{(\alpha}k_{\beta)} \\
& - 2m^2\eta_{\alpha\beta}(p_{(\mu}q_{\nu)} + p_{(\mu}k_{\nu)} + q_{(\mu}k_{\nu)}) + 4p_{(\mu}q_{\nu)}q_{(\alpha}k_{\beta)} \\
& + 4q_\alpha q_\beta(p_{(\mu}q_{\nu)} + p_{(\mu}k_{\nu)}) + 4p_{(\mu}k_{\nu)}q_{(\alpha}k_{\beta)} \\
& + 8q_{(\mu}q_\nu q_\alpha k_{\beta)} + 4q_{(\mu}k_{\nu)}q_{(\alpha}k_{\beta)} + 4q_\mu q_\nu q_\alpha q_\beta\} \\
& \times \Big\{D^0_{++}(k)D_{++}(q+k)D^0_{+-}(k-p) \\
& - D^0_{++}(k)D_{+-}(q+k)D^0_{--}(k-p) \\
& + D^0_{+-}(k)D_{--}(q+k)D^0_{--}(k-p)\Big\}\,, \qquad (5.4.76)
\end{aligned}$$

$(\mu_1\mu_2\cdots\mu_n)$ denoting symmetrization over indices enclosed in the parentheses,

$$(\mu_1\mu_2\cdots\mu_n) = \frac{1}{n!}\sum_{\substack{\{i_1 i_2\cdots i_n\}=\\ \text{perm}\{12\cdots n\}}}\mu_{i_1}\mu_{i_2}\cdots\mu_{i_n}\,.$$

In the long-range limit, Eq. (5.4.75) simplifies to

$$I_{\mu\nu\alpha\beta}(x,x') = \frac{1}{2m}\iint\frac{d^3\boldsymbol{q}}{(2\pi)^3}\frac{d^3\boldsymbol{p}}{(2\pi)^3}|a(\boldsymbol{q})|^2 e^{-i\boldsymbol{p}(\boldsymbol{x}'-\boldsymbol{x}_0)}\tilde{I}_{\mu\nu\alpha\beta}(p,q)\,,\, p_0 = (0.4.77)$$

The leading term in $\tilde{I}_{\mu\nu}(p,q)$ has the form [Cf. Eq. (5.4.70)]

$$\tilde{I}^{(0)}_{\mu\nu\alpha\beta}(p,q) \sim G^2m^2\frac{\mathcal{F}_{\mu\nu\alpha\beta}}{\sqrt{-p^2}}\,. \qquad (5.4.78)$$

This singular at $p \to 0$ contribution comes from integration over small k in Eq. (5.4.76). Therefore, to the leading order, the momenta k, p in the vertex factors can be neglected in comparison with q. Thus,

$$\begin{aligned}
& \tilde{I}_{\mu\nu\alpha\beta}(p,q) = \\
-iQ_{\mu\nu}Q_{\alpha\beta}\int\frac{d^4k}{(2\pi)^4}e^{ik(x-x')}\Big\{ & \quad D^0_{++}(k)D_{++}(q+k)D^0_{+-}(k-p) \\
& - D^0_{++}(k)D_{+-}(q+k)D^0_{--}(k-p) \\
& + D^0_{+-}(k)D_{--}(q+k)D^0_{--}(k-p)\Big\} \quad (5.4.79)
\end{aligned}$$

where

$$Q_{\mu\nu} \equiv 2q_\mu q_\nu - m^2\eta_{\mu\nu}\,.$$

Furthermore, it is convenient to combine various terms in this expression with the corresponding terms in the transposed contribution. Noting that the right hand side of Eq. (5.4.79) is explicitly symmetric with respect to $\{\mu\nu\} \leftrightarrow \{\alpha\beta\}$ and that the variables x, x' in the exponent can be freely interchanged because they appear symmetrically in Eq. (5.4.81), we may write

$$\tilde{B}_{\mu\nu\alpha\beta}(p,q) \equiv \tilde{I}_{\mu\nu\alpha\beta}(p,q) + \tilde{I}^{\rm tr}_{\mu\nu\alpha\beta}(p,q) = -iQ_{\mu\nu}Q_{\alpha\beta}\int \frac{d^4k}{(2\pi)^4}e^{ik(x-x')} \times$$
$$\Big\{ \left[D^0_{++}(k)D_{++}(q+k)D^0_{+-}(k-p) + D^0_{--}(k)D_{--}(q+k)D^0_{-+}(k-p)\right]$$
$$+ \left[D^0_{+-}(k)D_{--}(q+k)D^0_{--}(k-p) + D^0_{-+}(k)D_{++}(q+k)D^0_{++}(k-p)\right]$$
$$- \left[D^0_{++}(k)D_{+-}(q+k)D^0_{--}(k-p) + D^0_{--}(k)D_{-+}(q+k)D^0_{++}(k-p)\right]\Big\} \qquad (5.4.80)$$

For future reference, we mention also that as it follows from the considerations of Sec. 5.4.1, $\mathfrak{C}^{(0)}_{\mu\nu\alpha\beta}$ can be expressed through $\tilde{B}_{\mu\nu\alpha\beta}$ as

$$\mathfrak{C}^{(0)}_{\mu\nu\alpha\beta} = \lim_{V,T\to 0}\left\{\frac{1}{(VT)^2}\iint\limits_{(V,T)} d^4x d^4x' \right. \qquad (5.4.81)$$
$$\left. \times \iint \frac{d^3\boldsymbol{q}}{(2\pi)^3}\frac{d^3\boldsymbol{p}}{(2\pi)^3}\frac{|a(\boldsymbol{q})|^2}{2m}e^{-i\boldsymbol{p}(\boldsymbol{x}'-\boldsymbol{x}_0)}\lim_{m\to\infty}\tilde{B}_{\mu\nu\alpha\beta}(p,q,x-x')\right\}.$$

With the help of the relation

$$D_{--} + D_{++} = D_{+-} + D_{-+}\,, \qquad (5.4.82)$$

which is a consequence of the identity

$$T\hat{\phi}(x)\hat{\phi}(y) + \tilde{T}\hat{\phi}(x)\hat{\phi}(y) - \hat{\phi}(x)\hat{\phi}(y) - \hat{\phi}(y)\hat{\phi}(x) = 0\,,$$

the first term in the integrand of Eq. (5.4.80) can be transformed as

$$D^0_{++}(k)D_{++}(q+k)D^0_{+-}(k-p) + D^0_{--}(k)D_{--}(q+k)D^0_{-+}(k-p)$$
$$= D^0_{++}(k)D_{++}(q+k)D^0(k-p) + D^0(k)D_{--}(q+k)D^0_{-+}(k-p)\,,$$

where

$$D(k) \equiv D_{+-}(k) + D_{-+}(k) = 2\pi i\delta(k^2-m^2)\,, \quad D^0(k) = D(k)|_{m=0}\,.$$

Here we used the already mentioned fact that $D^0_{+-}(k)D_{-+}(k+q) \equiv 0$ for q on the mass shell. Analogously, the second term becomes

$$D^0(k)D_{--}(q+k)D^0_{--}(k-p) - D^0_{-+}(k)D_{--}(q+k)D^0(k-p)\,.$$

The contributions involving the functions D_{-+} in the obtained expressions cancel each other in the long-range limit. Indeed,

$$D^0(k)D_{--}(q+k)D^0_{-+}(k-p) - D^0_{-+}(k)D_{--}(q+k)D^0(k-p)$$
$$= D_{--}(q+k)(2\pi i)^2\delta(k^2)\delta((k-p)^2)(\theta(k^0-p^0) - \theta(k^0)) \to 0$$
$$\text{as} \quad m\to\infty\,. \qquad (5.4.83)$$

Finally, we rewrite the rest of the integrand in Eq. (5.4.80) as

$$\begin{aligned}&D^0_{++}(k)D_{+-}(q+k)D^0_{--}(k-p) + D^0_{--}(k)D_{-+}(q+k)D^0_{++}(k-p)\\&= -D^0_{++}(k)D(q+k)D^0_{++}(k-p)\,.\end{aligned}$$

Using the identity (5.4.82), it is now straightforward to verify that

$$\begin{aligned}&D^0_{++}(k)D_{++}(q+k)D^0(k-p) + D^0(k)D_{--}(q+k)D^0_{--}(k-p)\\&+D^0_{++}(k)D(q+k)D^0_{++}(k-p) =\\&D^0_{++}(k)D_{++}(q+k)D^0_{++}(k-p) + D^0_{--}(k)D_{--}(q+k)D^0_{--}(k-p)\,.\end{aligned}$$

Thus, Eq. (5.4.80) takes the form

$$\begin{aligned}\tilde{B}_{\mu\nu\alpha\beta}(p,q,x-x') &= -iQ_{\mu\nu}Q_{\alpha\beta}\int\frac{d^4k}{(2\pi)^4}e^{-ik(x-x')}\\&\times\left\{D^0_{++}(k)D_{++}(q+k)D^0_{++}(k-p) + D^0_{--}(k)D_{--}(q+k)D^0_{--}(k-p)\right\}\end{aligned}\qquad(5.4.84)$$

As was discussed in Sec. 5.4.1, the exponent in the integrand in Eq. (5.4.84) plays the role of an ultraviolet cutoff, ensuring convergence of the integral at large k's. On the other hand, the leading contribution is determined by integrating over $k \sim p$ where it is safe to take the limit $x \to x'$. Since $(x-x')$ is eventually set equal to zero, one can further simplify the k-integral by using the dimensional regulator instead of the oscillating exponent. Namely, introducing the dimensional regularization of the k-integral, one may set $x = x'$ afterwards to obtain

$$\begin{aligned}&\tilde{B}_{\mu\nu\alpha\beta}(p,q,x-x') \to \tilde{B}^{\epsilon}_{\mu\nu\alpha\beta}(p,q)\\&= 2Q_{\mu\nu}Q_{\alpha\beta}\mu^{\epsilon}\,\mathrm{Im}\int\frac{d^{4-\epsilon}k}{(2\pi)^4}D^0_{++}(k)D_{++}(q+k)D^0_{++}(k-p)\,,\end{aligned}\qquad(5.4.85)$$

where it is taken into account that $D_{--} = -D^*_{++}$. We thus see that the regularized value of $\tilde{B}_{\mu\nu\alpha\beta}$ in the coincidence limit is given by the expression that would have been obtained if the ordinary Feynman rules were used instead of the Schwinger-Keldysh technique, except the "Im" in front of the k-integral (using Feynman rules would replace "Im" by the factor $(-i)$). This implies, in particular, that the integral in Eq. (5.4.85) is ultraviolet finite. Indeed, the power counting shows that this integral converges as k^{-2} for $k \to \infty$, which ensures finiteness of the integral according to the general theory of the S-matrix.[10] Thus, we may set $\epsilon = 0$ in Eq. (5.4.85), and conclude that the $\hbar^0$-contribution to the two-point correlation function has a well-defined coincidence limit everywhere.

---

[10]This finiteness is the result of neglecting k's in the vertex factors in the integrand of Eq. (5.4.76), which are irrelevant in the long-range limit, but give rise to ultraviolet divergences. In this respect, the operations "Im" and multiplication by $(-i)$ are generally not equivalent: as an explicit calculation shows, the integral in Eq. (5.4.76) has a non-vanishing real part.

Evaluation of the integral (5.4.85) proceeds exactly as in Sec. 5.3.2, and gives for the leading long-range term

$$\tilde{B}_{\mu\nu\alpha\beta}(p,q) \equiv \tilde{B}^{\epsilon=0}(p,q) = \frac{mQ_{\mu\nu}Q_{\alpha\beta}}{16\sqrt{-p^2}}\,. \tag{5.4.86}$$

Substituting this into Eq. (5.4.81), using Eq. (5.3.55), and restoring ordinary units yields

$$\mathcal{C}^{(0)}_{\mu\nu\alpha\beta} = \frac{4G^2m^2}{r^2c^4}\delta_{\mu\nu}\delta_{\alpha\beta}\,. \tag{5.4.87}$$

In particular, the root mean square fluctuation of $\mu\nu$-component of the metric

$$\sqrt{\left\langle (\Delta h_{\mu\nu})^2 \right\rangle} = \delta_{\mu\nu}\frac{2Gm}{rc^2}\,. \tag{5.4.88}$$

The nonrelativistic gravitational potential $\Phi^{\rm g}$ is related to the 00-component of metric via $\Phi^{\rm g} = h_{00}c^2/2$, thus,

$$\sqrt{\left\langle (\Delta \Phi^{\rm g})^2 \right\rangle} = \frac{Gm}{r}\,. \tag{5.4.89}$$

We can now ask for conditions to be imposed on a system in order to justify neglecting quantum fluctuations of its gravitational field. As before, this condition can easily be found out by examining dependence of the $\hbar^0$ contribution to the correlation function on the number of field producing particles. The structure of ϕ-lines in the diagram of Fig. 5.4.6(b) is the same as that in Fig. 5.1.1(a). Therefore, in the case when the gravitational field is produced by an N-particle body, this diagram is proportional to m^2N, where m is the mass of a constituent particle. Correspondingly, the root mean square fluctuation of the potential is proportional to $m\sqrt{N} = M/\sqrt{N}$ (M is the total mass of the body), while its relative value, to $1/\sqrt{N}$. We thus arrive to the macroscopic condition found already in Sec. 5.3.2 in connection with the post-Newtonian loop contributions. In the present context, it coincides with the well-known criterion for neglecting quantum gravity fluctuations, formulated in a somewhat different manner in Refs. [27, 28, 21].

5.4.3 Orbit precession in the fluctuating gravitational field

Here we turn back to the example given in Sec. 5.3.3, and apply the results obtained in the preceding sections to the investigation of dynamics of a classical particle in a fluctuating gravitational field. The particle will be considered as *testing*, *i.e.*, it will be assumed sufficiently light to neglect its contribution to the gravitational field, and sufficiently small compared with $r_{\rm g} = 2Gm/c^2$ to allow considering it as a pointlike object. The latter assumption, in particular, justifies the use of the expression (5.4.87) for the correlation function, obtained in the coincidence limit.

In order to find equations of motion of the particle, we have to calculate its effective action. The action functional for a pointlike particle has the form

$$S_{\rm t} = -m_{\rm t}\int\sqrt{g_{\mu\nu}dx^{\mu}dx^{\nu}} = -m_{\rm t}\int d\tau\sqrt{1+h_{\mu\nu}\dot{x}^{\mu}\dot{x}^{\nu}}\,, \qquad \dot{x}^{\mu}\equiv\frac{dx^{\mu}}{d\tau}$$

where $m_{\rm t}$ is the particle mass, and $d\tau$ is the particle proper time defined with respect to the flat metric, $d\tau^2 = \eta_{\mu\nu}dx^{\mu}dx^{\nu}$. Inserting this expression into the generating functional integral of Green functions, one can write, in view of the assumed smallness of the test particle contribution,

$$\begin{aligned}
&\int\mathcal{D}\Phi_{-}\int\mathcal{D}\Phi_{+}\exp\{i(S+S_{\rm t}+j\phi)\}\\
&= Z_{\rm SK}|_{\substack{J\backslash j=0\\ K=0}} + \int\mathcal{D}\Phi_{-}\int\mathcal{D}\Phi_{+}\,iS_{\rm t}\,\exp\{i(S+j\phi)\}\\
&= Z_{\rm SK}\left(1+i\langle{\rm in}|S_{\rm t}|{\rm in}\rangle\right)|_{\substack{J\backslash j=0\\ K=0}}\,. \qquad (5.4.90)
\end{aligned}$$

The generating functional of connected Green functions takes the form

$$W = -i\ln\{Z_{\rm SK}\left(1+i\langle{\rm in}|S_{\rm t}|{\rm in}\rangle\right)\} = W_{\rm SK} + \langle{\rm in}|S_{\rm t}|{\rm in}\rangle\,.$$

Taking its Legendre transform, we see that the effective particle action is just

$$\Gamma_{\rm t} = \langle{\rm in}|S_{\rm t}|{\rm in}\rangle.$$

Expanding $S_{\rm t}$ in powers of $h_{\mu\nu}$, the right hand side of this equation can be evaluated in the first post-Newtonian approximation as

$$\begin{aligned}
\Gamma_{\rm t} &= -m_{\rm t}\int d\tau\left\{1+\frac{1}{2}\langle h_{\mu\nu}\rangle\dot{x}^{\mu}\dot{x}^{\nu} - \frac{1}{8}\left[\langle h_{\mu\nu}\rangle\langle h_{\alpha\beta}\rangle + \mathcal{C}^{(0)}_{\mu\nu\alpha\beta}\right]\dot{x}^{\mu}\dot{x}^{\nu}\dot{x}^{\alpha}\dot{x}^{\beta}\right\}\\
&= -m_{\rm t}\int\sqrt{(\eta_{\mu\nu}+\langle h_{\mu\nu}\rangle)dx^{\mu}dx^{\nu} - \frac{1}{4}\mathcal{C}^{(0)}_{0000}(dx^0)^2}\,. \qquad (5.4.91)
\end{aligned}$$

Substituting Eqs. (5.3.66), (5.4.87) for the mean gravitational field and its fluctuation, we see from Eq. (5.4.91) that the test particle motion in the fluctuating field is effectively the same as in a non-fluctuating gravitational field described by the following spacetime metric

$$\begin{aligned}
d\bar{s}^2 \equiv \bar{g}^{\rm eff}_{\mu\nu}dx^{\mu}dx^{\nu} &= \left(1-\frac{r_g}{r}-\frac{r_g^2}{4r^2}\right)dt^2 - \left(1+\frac{r_g}{r}-\frac{7r_g^2}{4r^2}\right)dr^2\\
&-r^2\left(1+\frac{r_g}{r}+\frac{7r_g^2}{4r^2}\right)(d\theta^2+\sin^2\theta\, d\varphi^2), \qquad (5.4.92)
\end{aligned}$$

Now, it is not difficult to calculate the secular precession of the particle's orbit. Let the test particle move in the equatorial plane ($\theta = \pi/2$). We have the following Hamilton-Jacobi equation for the action $\Gamma_{\rm t}$

$$\bar{g}^{\mu\nu}_{\rm eff}\frac{\partial\Gamma_{\rm t}}{\partial x^{\mu}}\frac{\partial\Gamma_{\rm t}}{\partial x^{\nu}} - m_{\rm t}^2 = 0,$$

where $\bar{g}^{\mu\nu}_{\rm eff}$ is the reciprocal of $\bar{g}^{\rm eff}_{\mu\nu}$. A simple calculation gives, to the leading order,

$$\begin{aligned} S_{\rm b} &= -Et + L\varphi \\ &+ \int dr \left[\left(E'^2 + 2m_{\rm t}E'\right) + \frac{r_g}{r}\left(m_{\rm t}^2 + 4m_{\rm t}E'\right) - \frac{1}{r^2}\left(L^2 - \frac{9}{4}r_g^2 m_{\rm t}^2\right)\right]^{1/2} \end{aligned} \quad (5.4.93)$$

where E, L are the energy and angular momentum of the particle, respectively, and $E' = E - m_{\rm t}$ its non-relativistic energy. The first two terms in the integrand in Eq. (5.4.93) coincide with the corresponding terms of classical theory, while the third does not, leading to the angular shift of the orbit

$$\delta\varphi = \frac{9\pi G m}{c^2 a(1 - e^2)} \quad (5.4.94)$$

per period (a and e are the major semiaxis and the eccentricity of the orbit, respectively). This result is to be compared with the well-known result of classical general relativity

$$\delta\varphi = \frac{6\pi G m}{c^2 a(1 - e^2)}\,.$$

On the other hand, the angular shift calculated in Ref. [20] neglecting fluctuations of the gravitational field differs from the classical result by the factor 4/3.

5.5 Concluding remarks

In this Chapter, we have determined the long-range behavior of the gravitational fields produced by quantized matter in the first post-Newtonian approximation. From the point of view of establishing the correct correspondence between classical and quantum theories of gravitation, Eq. (5.3.56) demonstrating violation of general covariance by the $\hbar^0$-order loop contributions, and Eq. (5.4.87) describing $\hbar^0$-fluctuations of the spacetime metric turned out to be of special importance, and led us independently to a macroscopic formulation of the correspondence principle. In turn, this formulation endows $\hbar^0$ loop contributions with direct *physical meaning as describing deviations of the spacetime metric from classical solutions of the Einstein equations in the case of finite number of elementary particles constituting the gravitating body.*

In connection with these results, it is worth also to make the following general comment.

As we have seen, an essentially quantum character of elementary particle interactions makes classical consideration inapplicable to systems whose dynamics is governed by interactions of a finite number of constituent particles. On the other hand, there is a deep-rooted belief in the literature that the quantum field description of interacting remote systems, each of which consists of many particles, is equivalent to that in which these systems are replaced by elementary particles with masses and charges equal to the total masses and charges of the systems. In other words, without calling it into question, the familiar notion of a point particle is carried over from classical mechanics to quantum field theory. This point of view is adhered, for instance, in the classic paper by Iwasaki [2] where it is applied to the solar system to calculate the secular precession of Mercury's orbit, considered as a "Lamb shift". The Sun and Mercury are regarded in Ref. [2] as scalar particles. However, as we have seen, the gravitational fields of elementary particles have properties which are quite different from those predicted by the classical general relativity.

Finally, we mention that investigation of quantum fluctuations of the Coulomb potential similar to that carried out in Sec. 5.4 shows that the macroscopic formulation of the correspondence principle given above extends also to the case of electromagnetic interaction [29].

A. Root singularities of Feynman integrals

The integrals

$$J_{nm} \equiv \int_0^\infty \int_0^\infty \frac{dudt}{(1+t+u)^n(1+\alpha tu)^m},$$

encountered in Sec. 5.3.2, can be evaluated as follows. Consider the auxiliary quantity

$$J(A,B) = \int_0^\infty \int_0^\infty \frac{dudt}{(A+t+u)(B+\alpha tu)},$$

where $A, B > 0$ are some numbers eventually set equal to 1. Performing an elementary integration over u, we get

$$J(A,B) = \int_0^\infty dt\, \frac{\ln B - \ln\{\alpha t(A+t)\}}{B-\alpha t(A+t)}.$$

Now consider the integral

$$\tilde{J}(A,B) = \int_C dw f(w,A,B), \quad f(w,A,B) = \frac{\ln B - \ln\{\alpha w(A+w)\}}{B-\alpha w(A+w)} \tag{5.5.95}$$

taken over the contour C shown in Fig. 5.5.7. $\tilde{J}(A,B)$ is zero identically. On the other hand,

$$\tilde{J}(A,B) = \int_{-\infty}^{-A} dz\, \frac{\ln B - \ln\{\alpha z(A+z)\}}{B-\alpha z(A+z)} + \int_{-A}^{0} dz\, \frac{\ln B - \ln\{-\alpha z(A+z)\} + i\pi}{B-\alpha z(A+z)}$$
$$+ \int_0^{+\infty} dz\, \frac{\ln B - \ln\{\alpha z(A+z)\} + 2i\pi}{B-\alpha z(A+z)} - i\pi \sum_{w_+,w_-} \mathrm{Res} f(w,A,B),$$

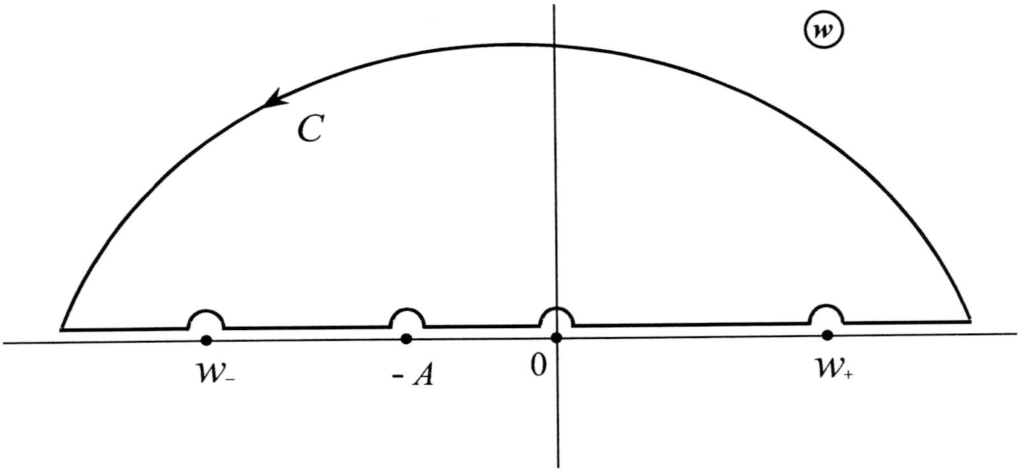

Figure 5.5.7: Contour of integration in Eq. (5.5.95).

$w_\pm$ denoting the poles of the function $f(w, A, B)$,

$$w_\pm = -\frac{A}{2} \pm \sqrt{\frac{B}{\alpha} + \frac{A^2}{4}}.$$

Change $z \to -A - z$ in the first integral. A simple calculation then gives

$$\begin{aligned} J(A, B) &= \frac{\pi^2}{2\sqrt{\alpha}} B^{-1/2} \left(1 + \frac{\alpha A^2}{4B}\right)^{-1/2} \\ &- \frac{1}{2}\int_0^A dt\, \frac{\ln B - \ln\{\alpha t(A - t)\}}{B + \alpha t(A - t)}. \end{aligned} \tag{5.5.96}$$

The roots are contained entirely in the first term on the right of Eq. (5.5.96). The integrals J_{nm} are found by repeated differentiation of Eq. (5.5.96) with respect to A, B. Expanding $(1 + \alpha A^2/4B)^{-1/2}$ in powers of α, we find the leading terms needed in Sec. 5.3.2

$$\begin{aligned} J_{12}^{\rm root} &= \frac{\pi^2}{4\sqrt{\alpha}}, \quad J_{13}^{\rm root} = \frac{3\pi^2}{16\sqrt{\alpha}}, \\ J_{31}^{\rm root} &= -\frac{\pi^2}{16}\sqrt{\alpha}, \quad J_{32}^{\rm root} = -\frac{3\pi^2}{32}\sqrt{\alpha}, \quad J_{33}^{\rm root} = -\frac{15\pi^2}{128}\sqrt{\alpha}. \end{aligned} \tag{5.5.97}$$

References

[1] B. S. DeWitt, *Phys. Rev.* **162**, 1195 (1967).

[2] Y. Iwasaki, *Prog. Theor. Phys.* **46**, 1587 (1971).

[3] L. D. Faddeev and V. N. Popov, *Phys. Lett.* **25B**, 29 (1967).

[4] C. Becchi, A. Rouet, and R. Stora, *Ann. of Phys.* **98**, 287 (1976); *Commun. Math. Phys.* **42**, 127 (1975); I. V. Tyutin, *Report FIAN* 39 (1975). The BRST method was extended to the case of gravity independently by J. Dixon, Ph.D. thesis, Oxford, 1975 (unpublished), R. Delbourgo and M. Ramon-Medrano, *Nucl. Phys.* **B110**, 467 (1976), and K. S. Stelle, *Phys. Rev.* D**16**, 953 (1977).

[5] J. Zinn-Justin, in *Trends in Elementary Particle Physics*, edited by H. Rollnik and K. Dietz, Springer-Verlag, Berlin (1975), p. 2.

[6] J. Schwinger, *J. Math. Phys.* **2**, 407 (1961); *Particles, Sources and Fields* (Addison-Wesley, Reading, Mass., 1970).

[7] L. V. Keldysh, *Zh. Eksp. Teor. Fiz.* **47**, 1515 (1964) [*Sov. Phys. JETP* **20**, 1018 (1965)].

[8] R. D. Jordan, *Phys. Rev.* D**33**, 444 (1986); **36**, 3593 (1987).

[9] J. P. Paz, *Phys. Rev.* D**41**, 1054 (1990).

[10] I. A. Batalin and G. A. Vilkovisky, *Phys. Lett.*, **102B**, 27 (1981).

[11] P. M. Lavrov, I. V. Tyutin, and B. L. Voronov, *Yad. Fiz.* **36**, 498 (1982); *Sov. J. Nucl. Phys.*

[12] S. Weinberg, *Physica* (Amsterdam) **96A**, 327 (1979).

[13] H. Leutwyler and S. Weinberg, in: *Proc. of the XXVIth Int. Conf. on High Energy Physics*, ed. J. Sanford (Dallas, TX, 1992), *AIP Conf. Proc.* No. 272 (AIP, NY, 1993) pp. 185, 346.

[14] J. F. Donoghue, in: *Effective Field Theories of the Standard Model*, ed. U.-G. Meissner (World Scientific, Singapore, 1994).

[15] J. F. Donoghue, *Phys. Rev. Lett.* **72**, 2996 (1994); *Phys. Rev.* D**50**, 3874 (1994).

[16] S. Mandelstam, *Phys. Rev.* **175**, 1580 (1968); R. Kallosh and I. V. Tyutin, *Yad. Fiz.*, **17**, 190 (1973) [*Sov. J. Nucl. Phys.* **17**, 98 (1973)].

[17] P. I. Pronin and K. V. Stepanyantz, in *New Computing Technick in Physics Research. IV.*, edited by B. Denby and D. Perred-Gallix, World Scientific, Singapure (1995), p. 187.

[18] K. A. Kazakov, *Class. Quantum Grav.* **19**, 3017 (2002).

[19] K. A. Kazakov, *Class. Quantum Grav.* **18**, 1039 (2001); *Nucl. Phys. Proc. Suppl.* **104**, 232 (2002); *Int. J. Mod. Phys.* **D12**, 1715 (2003).

[20] K. A. Kazakov, *Phys. Rev.* D**66**, 044003 (2002).

[21] G. Cognola, E. Elizalde, and S. Zerbini, *Phys. Rev.* D**65**, 085031 (2002).

[22] Chung-I Kuo and L. H. Ford, *Phys. Rev.* D**47**, 4510 (1993).

[23] L. H. Ford, *Phys. Rev.* D**51**, 1692 (1995).

[24] B. L. Hu and A. Matacz, *Phys. Rev.* D**51**, 1577 (1995).

[25] N. G. Phillips and B. L. Hu, *Phys. Rev.* D**55**, 6123 (1997).

[26] N. Bohr and L. Rosenfeld, Kgl. Danske Vidensk. Selskab., *Math.-Fys. Medd.* **12**, 3 (1933).

[27] E. Tomboulis, *Phys. Lett.* **B70**, 361 (1977).

[28] C. G. Callan,Jr., S. B. Giddings, J. A. Harvey, and A. Strominger, *Phys. Rev.* D**45**, 1005 (1992).

[29] K. A. Kazakov, *"Quantum fluctuations of a Coulomb potential,"* hep-th/0402178; to be published in *Physical Review D.*

In: Focus on Quantum Gravity Research
Editor: David C. Moore, pp. 215-259
ISBN 1-59454-660-6

Chapter 6

The Cosmological Constant and the Stability of Flat Space

Remo Garattini
Università degli Studi di Bergamo
Facoltà di Ingegneria, Viale Marconi 5
24044 Dalmine (Bergamo) ITALY
and I.N.F.N. - sezione di Milano, Milan, Italy.
E-mail: Remo.Garattini@unibg.it

We extend the Gross-Perry-Yaffe approach of hot flat space instability to Minkowski space. This is done by examining the massless gravitational perturbation in a saddle point approximation of the partition function. In particular, we show that a nonconformal instability appears in the transverse-traceless tensor sector (massless gravitons) if a Schwarzschild wormhole background is considered. The appearance of an instability in the whole manifold is here interpreted as a black hole pair creation. The instability of flat space via black hole pair creation is further explored evaluating the Casimir energy to one loop in a Hamiltonian approach. A variational approach associated to a Sturm-Liouville problem is used to evaluate the Casimir energy by means of trial wave functionals of Gaussian form. The Casimir energy is here interpreted as an induced cosmological constant associated to the Sturm-Liouville eigenvalue. To handle with divergences which appear to one loop approximation, we use a zeta function regularization. A renormalization and a renormalization group equation are involved to handle with ultra-violet divergences.

6.1 Introduction

The issue of stability with respect to quantum fluctuations in the path integral approach to quantum gravity is a fundamental problem as important as quantum gravity itself. Due to the attractive nature of the gravitational field, we expect to find a lot of unstable physical configurations. Therefore one might worry about the stability of the ground state of quantum gravity. One generally accepts to regard Minkowski space as the ground state, or vacuum, of quantum gravity. Classical small perturbations about this vacuum are stable. This is guaranteed by Positivity Theorems of Schoen and Yau[1, 2]and Witten[3]. However one might find that flat space is quantum mechanically unstable. This problem was investigated for the first time by Gross, Perry and Yaffe (GPY)[4] in the Euclidean path integral context. They discovered that hot flat space is unstable and the loss of stability was interpreted as a spontaneous nucleation of a black hole with an inverse temperature $\beta = 8\pi MG$, where M is the black hole mass and G is the Newton constant. In the semiclassical approximation to the Euclidean path integral

$$Z = \int_{\bar{\mathcal{M}}} \mathcal{D}g_{\mu\nu} \exp\left(-I_g\left[g_{\mu\nu}\right]\right), \tag{6.1.1}$$

the decay probability per unit volume and time Γ is defined as

$$\Gamma = A\exp\left(-I_{cl}\right) = \exp\left(-I_g\left[\bar{g}_{\mu\nu}\right]\right)\int \mathcal{D}h_{\mu\nu} \exp\left(-I_g^{(2)}\left[h_{\mu\nu}\right]\right), \tag{6.1.2}$$

where the gravitational field $g_{\mu\nu}$ has been separated into a background field $\bar{g}_{\mu\nu}$ and a quantum fluctuation $h_{\mu\nu}$

$$g_{\mu\nu} = \bar{g}_{\mu\nu} + h_{\mu\nu}, \tag{6.1.3}$$

A is the prefactor coming from the saddle point evaluation of Z and I_{cl} is the classical part of the action. $\bar{\mathcal{M}}$ is an analytic continuation to a Euclidean section of a manifold $\mathcal{M}$, with given boundary conditions. If a single negative eigenvalue appears in the prefactor A, it means that the related bounce shifts the energy of the false ground state[5]. In the Schwarzschild case GPY discovered one single negative mode $\lambda_{neg} = -.19/\left(MG\right)^2$. Allen reconsidered the effect of finite boundaries on the appearance of a negative mode showing that, if the box enclosing the black hole is little enough the instability disappears[6]. Since the pioneering paper of Gross, Perry and Yaffe, many other configurations have been investigated in different contexts. In particular, the de Sitter case involving a positive cosmological constant has been examined in the inflationary context by Bousso and Hawking[7]. In the same context of the de Sitter background, Ginsparg and Perry[8], Young[9] and more recently Volkov and Wipf[10] have computed the partition function to one loop. On the other hand, the Anti-de Sitter case involving a negative cosmological constant has been discussed by Hawking and Page[11] and subsequently by Prestidge[12]. The parallel problem in a higher dimensional context of the GPY-instability was firstly taken under examination by Witten[13], who examined the stability of Kaluza-Klein theories. After a decade Gregory and Laflamme reconsidered the problem of the gravitational stability in the context of branes[14]. This opened an interest in string theory where the stability of black branes was widely examined by different authors[15]. It is interesting to note that the only case of instability involving not a nucleation of a single black hole but a pair, is the de Sitter case. Moreover a deep difference appears between this case and the Anti-de Sitter or Schwarzschild case: temperatures before and after pair creation are different, at least in the neutral de Sitter black hole pair creation. Motivated by this results, we would like to reconsider the stability of flat space without involving temperatures as in the GPY analysis, but looking at a neutral black hole pair creation mediated by a wormhole of the Schwarzschild type. Recall that in the GPY instability is not possible to reach a vanishing temperature, because $T = 1/\beta = 1/(8\pi MG)$. Therefore to investigate the stability at zero temperature, we need to consider a Schwinger-like process of black hole pair creation[1]. Nevertheless, this is not sufficient to understand what will happen to the system once such an instability appears. One way to improve the situation is to compute the "*induced*" cosmological constant by means of Zero Point Energy (ZPE) of gravitons. This kind of computation can give information about the trend of a possible non trivial vacuum. A crude estimate of the ZPE of some field of mass m with a cutoff at the Planck scale gives

$$E_{ZPE} = \frac{1}{2}\int_0^{\Lambda_p} \frac{d^3k}{(2\pi)^3}\sqrt{k^2+m^2} \simeq \frac{\Lambda_p^4}{16\pi^2}, \tag{6.1.4}$$

which formally looks like a cosmological constant. However, one has to note that energy computation in General Relativity is not a trivial issue, because of time reparametrization invariance. Nevertheless, the form of the Hamiltonian in presence of a cosmological constant can be used to compute the cosmological constant itself and therefore the associated ZPE. The rest of the paper is structured as follows, in section 6.2 we report the Schwarzschild solution in both Lorentzian and Euclidean signatures and we write the Kruskal-Szekeres coordinates connected to a Penrose-Carter diagram, in section 6.3 we write the action with boundary for a Schwarzschild metric in real and imaginary time and we compute their contribution, in section 6.4 we report the GPY procedure leading to the existence of a negative mode for the Euclidean Schwarzschild metric, in section 6.5 we extend the GPY procedure to a Euclidean Schwarzschild wormhole, in section 6.6 we compute the Hamiltonian for the Schwarzschild wormhole, in section 6.7 we extract the Wheeler-De Witt operator from the Hamiltonian computed in section 6.6 and we show how to apply the variational approach to the Sturm-Liouville problem associated with the WDW equation, in section 6.8, we analyze the spin-2 operator or the operator acting on transverse traceless tensors, in section 6.9 we use the zeta function to regularize the divergences coming from the evaluation of the ZPE for TT tensors and we write the renormalization group equation. We summarize and conclude in section 6.10. Units in which $\hbar = c = k = 1$ are used throughout the paper.

[1]A recent detailed historical overview about the black hole pair creation process can be found in Ref.[16].

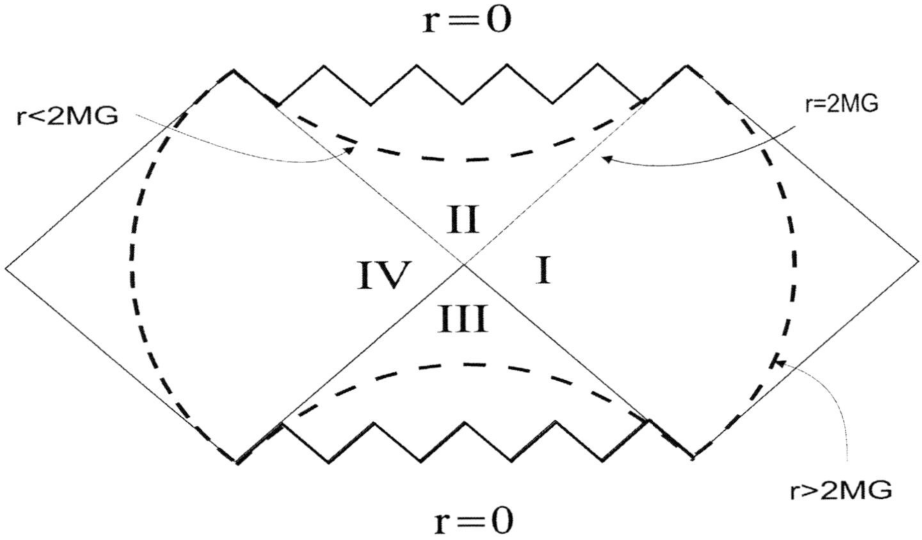

Figure 6.2.1: Kruskal-Szekeres coordinates in a Carter-Penrose diagram.

6.2 Gravitational Action with boundaries for the Schwarzschild wormhole

A classical Lorentzian wormhole with a metric of the spherically symmetric type is defined by a throat of a given radius, connecting two asymptotically flat three-dimensional spaces. An example can be the Schwarzschild metric,

$$ds^2 = -\left(1 - \frac{2MG}{r}\right) dt^2 + \left(1 - \frac{2MG}{r}\right)^{-1} dr^2 + r^2 d\Omega^2, \qquad (6.2.5)$$

describing a four-dimensional Lorentzian wormhole with the horizon located at $2MG$. $d\Omega^2 = d\theta^2 + \sin^2\theta d\phi^2$ is the line element of the unit sphere, G is Newton's constant and M represents the Arnowitt-Deser-Misner (ADM) mass of the wormhole[17]. This metric is asymptotically flat. The apparent singularity located at $r = 2MG$ can be removed by means of the Kruskal-Szekeres coordinates described by[18, 19, 20]

$$\begin{cases} U = -\left(\frac{r}{2MG} - 1\right)^{\frac{1}{2}} \exp\left(\frac{r-t}{4MG}\right) \\ V = \left(\frac{r}{2MG} - 1\right)^{\frac{1}{2}} \exp\left(\frac{r+t}{4MG}\right) \end{cases} \quad U < 0, \quad V > 0 \quad (region\ I), \qquad (6.2.6)$$

$$\begin{cases} U = \left(1 - \frac{r}{2MG}\right)^{\frac{1}{2}} \exp\left(\frac{r-t}{4MG}\right) \\ V = \left(1 - \frac{r}{2MG}\right)^{\frac{1}{2}} \exp\left(\frac{r+t}{4MG}\right) \end{cases} \quad U > 0, \quad V > 0 \quad (region\ II), \qquad (6.2.7)$$

$$\begin{cases} U = \left(\frac{r}{2MG} - 1\right)^{\frac{1}{2}} \exp\left(\frac{r-t}{4MG}\right) \\ V = -\left(\frac{r}{2MG} - 1\right)^{\frac{1}{2}} \exp\left(\frac{r+t}{4MG}\right) \end{cases} \quad U > 0, \quad V < 0 \quad (\textit{region IV}), \tag{6.2.8}$$

and

$$\begin{cases} U = -\left(1 - \frac{r}{2MG}\right)^{\frac{1}{2}} \exp\left(\frac{r-t}{4MG}\right) \\ V = -\left(1 - \frac{r}{2MG}\right)^{\frac{1}{2}} \exp\left(\frac{r+t}{4MG}\right) \end{cases} \quad U < 0, \quad V < 0 \quad (\textit{region III}). \tag{6.2.9}$$

In this way r is implicitly defined by

$$\left(1 - \frac{r}{2MG}\right) \exp\left(-\frac{r}{2MG}\right) = UV \tag{6.2.10}$$

and t is implicitly defined by

$$\exp\left(\frac{t}{2MG}\right) = -\frac{U}{V}. \tag{6.2.11}$$

In terms of these coordinates we have

$$ds^2 = -\frac{32\,(MG)^3}{r} \exp\left(-\frac{r}{2MG}\right) dU\,dV + r^2 d\Omega^2. \tag{6.2.12}$$

The only true singularities are at curves $UV = 1$, where $r = 0$. The region $\{U < 0, V > 0\}$ is the "*outside region*", the only region from which distant observers can obtain any information. The line $U = 0$, where $r = 2MG$, is the "*future horizon*"; the line $V = 0$ where also $r = 2MG$, is the "*past horizon*". An important property of the line element (6.2.12) is its invariance with respect to the discrete symmetries

$$\mathbf{I} : U \to -U;\ V \to -V, \qquad \mathbf{T} : U \to -V;\ V \to -U\ . \tag{6.2.13}$$

We can obtain a Euclidean Schwarzschild wormhole simply by the familiar replacement $t \to i\tau$. The new metric will solve the same (analytic) Einstein equations as the original metric. The form is

$$ds_E^2 = \left(1 - \frac{2MG}{r}\right) d\tau^2 + \left(1 - \frac{2MG}{r}\right)^{-1} dr^2 + r^2 d\Omega^2. \tag{6.2.14}$$

This is singular at $r = 2MG$. To examine the region *near* $r = 2MG$ we set

$$r - 2MG = \frac{x^2}{8MG} \tag{6.2.15}$$

to get

$$ds_E^2 \approx (\kappa x)^2\, d\tau^2 + dx^2 + \frac{1}{4\kappa^2} d\Omega^2. \tag{6.2.16}$$

The metric near $r = 2MG$ is the product of the metric on S^2 and the Euclidean Rindler spacetime

$$ds_E^2 = x^2 d(\kappa\tau)^2 + dx^2. \tag{6.2.17}$$

This is just E^2 in polar coordinates if we make the *periodic identification*

$$\tau \sim \tau + \frac{2\pi}{\kappa} \tag{6.2.18}$$

i.e. the singularity of Euclidean Schwarzschild at $r = 2MG$ (and of Euclidean Rindler at $x = 0$) is just a coordinate singularity provided that imaginary time coordinate τ is periodic with period $2\pi/\kappa$. This means that if we consider the Euclidean functional integral, like in Eq.(6.1.1), this must be taken over fields that are periodic in the imaginary time variable with period $2\pi/\kappa$. If the functional integral is taken over fields that are periodic in imaginary time with period $\hbar\beta$ then it can be written as

$$Z = Tr\exp(-\beta H), \tag{6.2.19}$$

which is the partition function for a quantum mechanical system with Hamiltonian H at temperature T given by $\beta = (k_B T)^{-1}$. k_B is the Boltzmann's constant. But we just saw that $\hbar\beta = 2\pi/\kappa$ for Schwarzschild, so we deduce that a Quantum Field Theory can be in equilibrium with a black hole only at the *Hawking temperature*

$$T_H = \frac{\kappa}{2\pi}\frac{\hbar}{k_B} \tag{6.2.20}$$

i.e. in units for which $\hbar = k_B = 1$

$$T_H = \frac{\kappa}{2\pi}. \tag{6.2.21}$$

Thus

- At any other temperature, Euclidean Schwarzschild has a conical singularity $\longrightarrow$ no equilibrium. The underlying physical reason is that the conical singularity introduces a delta function which corresponds to an involved extra mass.

- Equilibrium at Hawking temperature is unstable since if the black hole absorbs radiation its mass increases and its temperature decreases, i.e. the black hole has a negative specific heat. Nevertheless, as presented in the introduction, if the black hole is enclosed in a box it is possible to show that there exists a critical radius, say it r_c, below which the specific heat turns to be positive and the instability disappears[6].

6.3 Evaluating the action on the boundary for both signatures

To properly analyze the action on the boundary, we begin with the Lorentzian sector of the Schwarzschild-Kruskal manifold and with the inclusion of time-like and space-like boundaries in Fig.(6.2.1). The result appears in Fig.(6.3.2) where we show the corresponding Penrose diagram. We can see that the Schwarzschild-Kruskal spacetime is the union of four regions (wedges) R_+, R_-, T_+, and T_-. The regions R_+ and R_- are asymptotically flat . In the R_+ region, the Kruskal coordinates (U, V) satisfy $U < 0$, $V > 0$, while in R_-, $U > 0$, $V < 0$. Because of the the discrete transformation (6.2.13) in R_+ and R_- we expect the physics be symmetric with respect to the bifurcation surface S_0 defined as the intersection of H^+ and H^-, which are the future (H^+) and past (H^-) horizons, respectively. We denote

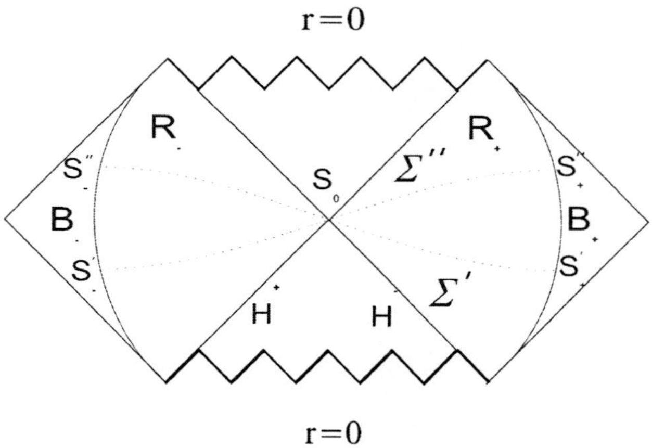

Figure 6.3.2: This is a Penrose diagram for the Schwarzschild wormhole.

with Σ_t the t = *constant* hypersurface. Any of such hypersurfaces is invariant under T-reflections and has Einstein-Rosen bridge topology $R^1 \times S^2$, represented in Fig.(6.3.3), whose intrinsic geometry and time derivatives are chosen to satisfy the gravitational constraint equations. We denote by $\Sigma_\pm$ the parts of Σ lying in $R_\pm$ respectively. The line element (6.2.12) restricted to the surface Σ reads

$$dl^2 = dy^2 + r^2(y)d\Omega^2 \ , \tag{6.3.22}$$

where the quantity y, defined by

$$dy = \pm\frac{dr}{\sqrt{1 - 2MG/r}}, \tag{6.3.23}$$

represents the proper geodesic distance from the "throat" of the bridge located at $r = r(y = 0) = r_+ = 2MG$. We choose y to be positive in Σ_+ and negative

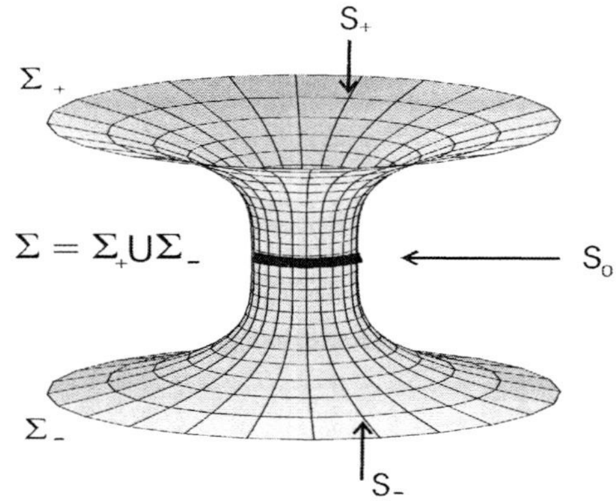

Figure 6.3.3: Wormhole representation for the Schwarzschild metric. The bifurcation surface S_0 divides the hypersurfaces Σ_+ and Σ_-.

in Σ_-. The set (t, y, θ, ϕ) can be used as canonical coordinates everywhere in $R_\pm$ outside the bifurcation surface S_0. These coordinates are right-oriented in R_+ and left-oriented in R_-. The configuration we wish to examine is described by boundary surfaces located in different regions with two different slices $\Sigma' \equiv \Sigma_{t'}$ and $\Sigma'' \equiv \Sigma_{t''}$ intersecting at the same two-dimensional bifurcation sphere S_0. We call this sequence of slices Σ_t (defined by the equation t = constant, with $t' \leq t \leq t''$) a "tilted foliation"[21]. For a "tilted foliation" the spacetime domain M lying between Σ' and Σ'' consists of two wedges $\mathcal{M}_+$ and $\mathcal{M}_-$ located in the right (R_+) and left (R_-) sectors of the Kruskal diagram. The region $\mathcal{M} = \mathcal{M}_+ \cup \mathcal{M}_-$ is bounded by Σ' and Σ'' and by a three-dimensional timelike boundary B that consists of two disconnected parts B_+ and B_-. For a general eternal black hole geometry the boundaries B_+ and B_- are located in $\mathcal{M}_+$ and $\mathcal{M}_-$, respectively. We define $S^2_\pm = \Sigma_\pm \cap B_\pm$. The topology of the slices Σ is therefore $I_\Sigma \times S^2$, where I_Σ is a finite spacelike distance, while the topology of $B_\pm$ is $I_t \times S^2_\pm$, where I_t is a finite timelike distance. In Schwarzschild coordinates the line element of Eq.(6.2.12) can be written as

$$ds^2 = -N^2 dt^2 + dy^2 + r^2(y)\, d\Omega^2, \tag{6.3.24}$$

where y is the proper radial distance from the throat defined in Eq.(6.3.23). The coordinate y covers the entire range $(-\infty, +\infty)$. N is the corresponding lapse function written in terms of the proper distance from the throat

$$N^2 = 1 - \frac{2MG}{r(y)}. \tag{6.3.25}$$

We define the four-velocity vector with $u_\mu = -N\partial_\mu t$, while the timelike unit vector normal to the hypersurface Σ_t is u^μ. This is chosen to be future oriented in $\mathcal{M}$ and normalized by the condition $u \cdot u = -1$. The lapse N is positive at Σ_+, negative at Σ_- and vanishing at the bifurcation surface. The spacelike normal n^μ to the three-dimensional boundaries $B_\pm$, is defined to be outward pointing at B_+, inward pointing at B_-, and normalized so that $n \cdot n = 1$ with the further condition $(u \cdot n)|_{B_\pm} = 0$. Following Ref.[21], greek indices are used for tensors in $\mathcal{M}$ while latin indices are used for tensors defined in either Σ or $B_\pm$. The metric and extrinsic curvature of Σ as a surface embedded in $\mathcal{M}$ are denoted by h_{ij} and $K_{ij} = -h_i^k \nabla_k u_j$, respectively, while the metric and extrinsic curvature of the boundaries $B_\pm$ as surfaces embedded in $\mathcal{M}$ are γ_{ij} and $\Theta_{ij} = -\gamma_i^k \nabla_k n_j$. Covariant differentiation with respect to the metric $g_{\mu\nu}$ and h_{ij} is denoted by ∇ and D, respectively. The induced metric and extrinsic curvature of the boundaries $S^2_\pm$ as surfaces embedded on Σ are denoted by σ_{ab} and $k_{ab} = -\sigma_a^k D_k n_b$, respectively, $(a, b = 2, 3)$. Explicitly, the metric tensors for the different surfaces are

$$h_{\mu\nu} = g_{\mu\nu} + u_\mu u_\nu, \qquad \gamma_{\mu\nu} = g_{\mu\nu} - n_\mu n_\nu, \qquad \sigma_{\mu\nu} = g_{\mu\nu} + u_\mu u_\nu - n_\mu n_\nu. \tag{6.3.26}$$

The determinants of the metric tensors are related by

$$\begin{aligned} \sqrt{-g} &= |N| \sqrt{h}, \\ \sqrt{-\gamma} &= |N| \sqrt{\sigma}. \end{aligned} \tag{6.3.27}$$

The covariant form for the gravitational action generated by a tilted foliation with fixed three-dimensional boundary of $\mathcal{M}$ is[2]

$$\begin{aligned} S = \frac{1}{2\kappa} \int_{\mathcal{M}_+} d^4x \sqrt{-g}\, \Re - \frac{1}{\kappa} \int_{B_+} d^3x \sqrt{-\gamma}\, (\Theta - \Theta_0)_+ \\ -\frac{1}{2\kappa} \int_{\mathcal{M}_-} d^4x \sqrt{-g}\, \Re - \frac{1}{\kappa} \int_{B_-} d^3x \sqrt{-\gamma}\, (\Theta - \Theta_0)_- , \end{aligned} \tag{6.3.29}$$

where $\Re$ denotes the four-dimensional scalar curvature, $\kappa \equiv 8\pi G$. The integrations are taken over the coordinates x^μ which have the same orientation as the canonical coordinates (t, y, θ, ϕ) of the tilted foliation. The differing signs in $\mathcal{M}_+$ and $\mathcal{M}_-$ reflect the fact that the coordinates have different time orientations in $\mathcal{M}_+$ and $\mathcal{M}_-$. The subtraction term

$$S_{0(\pm)} = \frac{1}{\kappa} \int_{B_\pm} d^3x \sqrt{-\gamma}\, \Theta_0 \tag{6.3.30}$$

is the extrinsic curvature evaluated on the reference space, which in this case has been chosen to be flat. The effect of $S_{0(\pm)}$ is to normalize the energy E to zero

[2]The complete action should be considered with additional boundary terms of the type

$$\frac{1}{\kappa} \int_{t', \Sigma_\pm}^{t''} d^3x \sqrt{h} K_\pm . \tag{6.3.28}$$

Nevertheless, since we will look at the Euclideanized version of the action with periodic boundary conditions in the Euclidean time, the previous boundary term disappears.

for the Schwarzschild wormhole with $M = 0$. In the next section we will carefully examine the contribution of the volume terms of the action. Here, we restrict the evaluation on the boundaries $B_{\pm}$ obtaining

$$S = -\frac{1}{\kappa}\int_{B_+} d^3x\,\sqrt{-\gamma}\,(\Theta - \Theta_0)_+ - \frac{1}{\kappa}\int_{B_-} d^3x\,\sqrt{-\gamma}\,(\Theta - \Theta_0)_- . \tag{6.3.31}$$

The trace of the extrinsic curvature is

$$\Theta_{|\pm} = -\nabla_k n^k = -\frac{1}{\sqrt{-g}}\partial_k\left(\sqrt{-g}g^{kj}n_j\right) = -\frac{1}{|N|\,r^2}\partial_y\left(|N|\,r^2\right). \tag{6.3.32}$$

With the help of Eqs.(6.3.23) and (6.3.25), we can write

$$\frac{\partial\,|N|}{\partial y} = \pm\frac{MG}{r^2\,(y)} \qquad \text{in } \Sigma_{\pm}, \tag{6.3.33}$$

where we have used

$$\frac{\partial r}{\partial y} = \pm\sqrt{1 - \frac{2MG}{r\,(y)}} \qquad \text{in } \Sigma_{\pm}. \tag{6.3.34}$$

Thus Eq.(6.3.32) becomes

$$\Theta_{|\pm} = \mp\frac{1}{|N|\,r^2\,(y)}\left[2r\,(y) - 3MG\right], \tag{6.3.35}$$

while the subtraction term can be defined by

$$\Theta_{0|\pm} = \lim_{M\to 0}\Theta_{|\pm} = \mp\frac{2}{r} \tag{6.3.36}$$

and

$$(\Theta - \Theta_0)_{\pm} = \mp\frac{1}{|N|\,r^2\,(y)}\left[2r\,(y) - 3MG - 2\frac{r^2\,(y)}{r}\,|N|\right]. \tag{6.3.37}$$

The proper "*time*" length in Eq.(6.3.31) is

$$\frac{1}{\kappa}\int_{B_{\pm}} dt\,|N| = \frac{1}{\kappa}T\,|N\,(y_{\pm})| \qquad \text{in } \Sigma_{\pm}, \tag{6.3.38}$$

where $y_{\pm}$ is the location of the spatial boundary on $B_{\pm}$ and $T = t'' - t'$. The substitutions $t \to -i\tau$ and $N \to -iN_E$ do not modify the behavior of the boundary action in the Euclidean context. Thus Eq.(6.3.38) becomes

$$\beta_{\pm} = T_{\pm}^{-1} = N_E\,(r_{\pm})\,\beta^* \qquad \text{in } \Sigma_{\pm}, \tag{6.3.39}$$

where $r_{\pm}$ stands for $r\,(y_{\pm})$ and where we have integrated over the period $\beta^* = 8\pi MG$ obtained by imposing regularity condition on the horizon. Note that the inverse temperature at infinity is the same in both wedges. The difference between β_+ and

β_- essentially arises from the Tolman law involving the different location of the boundaries. Then the total action contribution is

$$\begin{aligned} S_{B\pm} &= \pm\frac{4\pi}{\kappa}T\left[2r - 3MG - 2rN_{E\pm}\right]|_{r=r_\pm} \longrightarrow \\ \longrightarrow I_{B\pm} &= \mp\left(8\pi Mr - 12\pi M^2G - \beta r/G\right)|_{r=r_\pm}. \end{aligned} \tag{6.3.40}$$

Like in the case of the Hamiltonian discussed in Ref.[21], the tree level action for the Schwarzschild wormhole can be cast in the form

$$\frac{1}{\kappa}\left(I_+ - I_-\right) = \left(12\pi M^2G - 8\pi Mr_+ + \beta_+r_+/G\right) - \left(12\pi M^2G - 8\pi Mr_- + \beta_-r_-/G\right). \tag{6.3.41}$$

From Eq.(6.3.41), we can compute the energy of the system at the fixed temperature $\beta_\pm$ in each wedge. This is obtained by regarding Eq.(6.3.41) as a function of r and T rather than of r and M. We obtain

$$\begin{aligned} E_+ &= \frac{\partial I_+}{\partial\beta_+} &= \frac{1}{G}\left[r_+ - r_+\sqrt{1 - 2MG/r_+}\right] \\ E_- &= \frac{\partial I_-}{\partial\beta_-} &= \frac{1}{G}\left[r_- - r_-\sqrt{1 - 2MG/r_-}\right] \end{aligned}. \tag{6.3.42}$$

Note that each term tends asymptotically to the Arnowitt-Deser-Misner ($\mathcal{ADM}$) mass[17]. Note also that if the boundary location is different in the respective wedge, implies that $E_+ \neq E_-$. Nevertheless for the entropy we discover

$$\begin{aligned} S_+ &= \beta_+\frac{\partial I_+}{\partial\beta_+} - I_+ &= \beta_+E_+ - I_+ = 4\pi M^2G \\ S_- &= \beta_-\frac{\partial I_-}{\partial\beta_-} - I_- &= \beta_-E_- - I_- = 4\pi M^2G \end{aligned}, \tag{6.3.43}$$

namely it is independent on the wedge. This is a consequence of the definition we have used for computing the entropy. The same result has been obtained by Martinez in Ref.[22] in the context of microcanonical approach to the entropy of an eternal black hole. Note that if one adopts the method used in Ref.[22] for computing the entropy, it is not immediate to arrive at the result of Eq.(6.3.43). By means of Eq.(6.3.43), Eq.(6.3.41) can be explicitly rewritten as

$$\beta_+E_+ - \beta_-E_- = 8\pi MG\left(N_{E+}E_+ - N_{E-}E_-\right) = \beta^*\left(M_+ - M_-\right). \tag{6.3.44}$$

Note that β^* is the periodically identified "asymptotic Euclidean time" which is independent on the considered wedge. We can recognize three different cases

1. $M_+ > M_-$. The limit case is when $r_+ \rightarrow \infty\,(y_+ \rightarrow +\infty)$, while $r_- = 2GM\,(y_- = 0)$. This corresponds to $M_+ = M_{\mathcal{ADM}}$ and $M_- = 0$. This situation is the standard one, where the black hole thermodynamics is investigated in the wedge $\mathcal{M}_+$. Indeed

2. $M_+ < M_-$. The limit case is when $r_- \to \infty\,(y_- \to -\infty)$, while $r_+ = 2GM\,(y_+ = 0)$. This corresponds to $M_- = M_{\mathcal{ADM}}$ and $M_+ = 0$. This situation is "*dual*" to the previous in the sense that the black hole thermodynamics is investigated in the wedge $\mathcal{M}_-$.

3. $M_+ = M_-$.The left boundary y_- and the right boundary y_+ are symmetric with respect to the bifurcation surface S_0^2, which implies that $r_+ = r_-$.

Case 3 corresponds to a vanishing action. Nevertheless, it is well known that a vanishing action with a Euclidean metric describes flat space[2]. Therefore, in this very particular case we have found an alternative way to flat space at the price of having a topology change.

6.4 Instability in 4 dimensions for the Schwarzschild instanton

In this section, we report the result obtained by GPY. Negative modes, if they exist, arise from the analysis of geometric fluctuations about classical Euclidean solutions of the Einstein field equations. However, the analysis to confirm their existence must be performed with care, since the gauge freedom of the Euclidean action will in general introduce a large number of non-physical negative modes associated with conformal deformations of the metric. For pure gravity, the contributions from the conformal and the nonconformal modes decouple if a suitable gauge is chosen. In the path integral approach [23, 24], the partition function Z is given by Eq.(6.1.1). The partition function Z may be approximated using saddle-point techniques, by Taylor expanding about the known stationary points of the Euclidean action – the solutions to the Einstein field equations

$$R_{\mu\nu} = 0. \tag{6.4.45}$$

The expansions are performed by writing the metric $g_{\mu\nu}$ as in Eq.(6.1.3) with $h_{\mu\nu}$ treated as a quantum field on the classical fixed background $\bar{g}_{\mu\nu}$ which vanishes on the boundary $\partial\bar{\mathcal{M}}$. Thus the Euclidean action becomes

$$I[g] = I[\bar{g}] + I_2[h] + \cdots \tag{6.4.46}$$

where the linear term I_1 vanishes precisely because $\bar{g}_{\mu\nu}$ is a classical solution, I_2 is quadratic in the field $h_{\mu\nu}$ in each complex sector, and '$\cdots$' represents terms of higher than quadratic order. The Taylor expansion of the action in $\bar{\mathcal{M}}$ leads to the following one loop approximation of the partition function

$$\log Z = -I\left[\bar{g}\right] + \log \int_{\bar{\mathcal{M}}} D\left[h\right] \exp\left(-I_2\left[h\right]\right). \tag{6.4.47}$$

Since the quadratic action I_2 is rather complicated, we introduce the following decomposition:

$$h_{\mu\nu} = h_{\mu\nu}^{TT} + \frac{1}{4}\bar{g}_{\mu\nu}h + \left(\nabla_\mu \xi_\nu + \nabla_\nu \xi_\mu - \frac{1}{2}\bar{g}_{\mu\nu}\nabla_\mu \xi^\mu\right). \tag{6.4.48}$$

$h_{\mu\nu}^{TT}$ is the transverse traceless $\left(\nabla_\mu h^{\mu\nu TT} = 0,\ \bar{g}^{\mu\nu}h_{\mu\nu}^{TT} = 0\right)$ part of $h_{\mu\nu}$; h is the trace part and ξ_μ is the longitudinal traceless part. It is useful at this stage that it will be used in this part a covariant Lorentz gauge

$$\nabla_\mu \left(h^{\mu\nu} - \frac{1}{2}\bar{g}^{\mu\nu}h\right) = X^\nu, \tag{6.4.49}$$

where X^ν is some vector field defined on $\bar{g}_{\mu\nu}$. This gauge fixing introduces a set of anticommuting Faddeev-Popov vector fields η^μ. Further, we decompose the two vector fields ξ_μ and η_μ using the Hodge-de Rham decomposition

$$\begin{aligned} \eta_\mu &= \eta_\mu^c + \nabla_\mu \chi + \eta_\mu^H, \\ \xi_\mu &= \xi_\mu^c + \nabla_\mu \phi + \xi_\mu^H. \end{aligned} \tag{6.4.50}$$

η_μ^c,ξ_μ^c are the coexact parts of η and ξ and are both divergence-free. χ and ϕ are the exact parts, hence curl-free. η_μ^H and ξ_μ^H are the harmonic pieces. In what follows, we will ignore the harmonic sector, because in this paragraph we are dealing with simply connected manifolds. The action expanded to quadratic order is the sum of the following six terms

$$\textit{tensor} \qquad \frac{1}{16\pi G}\left[\frac{1}{4}h_{\mu\nu}^{TT}G^{\mu\nu\rho\sigma}h_{\rho\sigma}^{TT}\right], \tag{6.4.51}$$

$$\textit{scalar} \qquad \frac{1}{16\pi G}\left[-\frac{1}{2}hFh\right], \tag{6.4.52}$$

$$\textit{scalar} \qquad \frac{1}{16\pi G}\left[2X_{\mu\nu}\left(\psi\right)X^{\mu\nu}\left(F\psi\right)\right], \tag{6.4.53}$$

$$\textit{vector} \qquad \frac{1}{16\pi G}\left[2Y_{\mu\nu}\left(\xi_\rho^c\right)Y^{\mu\nu}\left(C_\sigma^\rho \xi^{c\sigma}\right)\right], \tag{6.4.54}$$

$$\textit{scalar} \qquad \frac{1}{16\pi G}\left[2X_{\mu\nu}\left(\chi\right)X^{\mu\nu}\left(F\chi\right)\right], \tag{6.4.55}$$

$$\textit{vector} \qquad \frac{1}{16\pi G}\left[2Y_{\mu\nu}\left(\eta_\rho^c\right)Y^{\mu\nu}\left(C_\sigma^\rho \eta^{c\sigma}\right)\right], \tag{6.4.56}$$

where the operators in the previous terms are

$$G_{\mu\nu\rho\sigma} = -\bar{g}_{\mu\rho}\bar{g}_{\nu\sigma}\Box - 2R_{\mu\rho\nu\sigma}, \tag{6.4.57}$$

$$X_{\mu\nu}(\psi) = \left(\nabla_\mu\nabla_\nu + \nabla_\nu\nabla_\mu - \frac{1}{2}\bar{g}_{\mu\nu}\Box\right)\psi, \tag{6.4.58}$$

$$F = -\frac{1}{8}\Box, \tag{6.4.59}$$

$$Y_{\mu\nu}(\eta_\rho) = \nabla_\mu\left(\eta_\rho\delta^\rho_\nu\right) + \nabla_\nu\left(\eta_\rho\delta^\rho_\mu\right), \tag{6.4.60}$$

$$C^\rho_\sigma = -\delta^\rho_\sigma\Box. \tag{6.4.61}$$

The metric perturbation $h_{\mu\nu}$ has ten components, of which five correspond to a spin-2 piece $\left(h^{TT}_{\mu\nu}\right)$, three to a spin-1 piece (ξ^c_ν) and two to spin-0 pieces (h,ψ). The Faddeev-Popov terms contain a spin-1 piece (η^c_ν) and a spin-0 piece (χ). If we consider perturbations about Euclidean flat space then the operators G, F and C are all manifestly positive definite. Nevertheless, the term hFh is negative definite, but this is related to some conformal transformations which give the Euclidean action an arbitrarily negative contribution. If we express the partition function in terms of the operators G, F and C, we have

$$Z \sim (\det G)^{-\frac{1}{2}}(\det F)^{-\frac{1}{2}}(\det F)^{-\frac{1}{2}}(\det C)^{-\frac{1}{2}}(\det C)(\det F)$$

$$\sim (\det G)^{-\frac{1}{2}}(\det C)^{\frac{1}{2}}. \tag{6.4.62}$$

Since the degrees of freedom which couple to F are pure gauge, F drops out of this expression. The operator C acting on divergence-free objects is positive semi-definite. The operator G requires a treatment on a case to case basis. If we discover that the operator G admits negative eigenvalues, from Eq.(6.1.2), we discover that this leads to a nonvanishing decay probability Γ related to the imaginary part of the free energy. In a Euclidean Schwarzschild background, we study the solutions of

$$-\Box h^{\mu\nu} - 2R^{\mu\rho\nu\sigma}h_{\rho\sigma} = \lambda h^{\mu\nu}, \tag{6.4.63}$$

where $h^{\mu\nu}$ are transverse, traceless, and normalizable. We decompose the metric perturbation into odd and even perturbations in the sense of Regge and Wheeler[25]. Referred to the (t,r,θ,ϕ) coordinates, the odd perturbations are

$$h^{(odd)}_{\mu\nu} = \begin{pmatrix} 0 & 0 & -h_0(r)\csc\theta\partial_\phi & h_0(r)\sin\theta\partial_\theta \\ . & 0 & -h_1(r)\csc\theta\partial_\phi & h_1(r)\sin\theta\partial_\theta \\ . & . & h_2(r)(\csc\theta\partial_\theta\partial_\phi - \cot\theta\csc\theta\partial_\phi) & \frac{1}{2}h_2(r)\left(\csc\theta\partial^2_\phi + \cos\theta\partial_\theta - \sin\theta\partial^2_\theta\right) \\ . & . & . & -h_2(r)(\sin\theta\partial_\theta\partial_\phi - \cos\theta\partial_\phi) \end{pmatrix}$$

$$\times\exp(i\omega t)\,Y_{lm}(\theta,\phi), \tag{6.4.64}$$

and the even perturbation are

$$h_{\mu\nu}^{(even)}=\begin{pmatrix} H_0(r) & H_1(r) & K_0(r)\partial_\theta & K_0(r)\partial_\phi \\ \cdot & H_2(r) & K_1(r)\partial_\theta & K_1(r)\partial_\phi \\ \cdot & \cdot & r^2\left(G_1(r)+G_2(r)\partial_\theta^2\right) & r^2G_2(r)\left(\partial_\theta\partial_\phi-\cot\theta\partial_\phi\right) \\ \cdot & \cdot & \cdot & \begin{array}{c} r^2\left(G_1(r)\sin^2\theta+\right. \\ \left.G_2(r)\left(\partial_\phi^2+\sin\theta\cos\theta\partial_\theta\right)\right) \end{array} \end{pmatrix}$$

$$\times\exp(i\omega t)\,Y_{lm}(\theta,\phi). \tag{6.4.65}$$

$Y_{lm}(\theta,\phi)$ are the usual spherical harmonics. By applying standard Sturm-Liouville techniques, it can be shown that $G>0$ for $l\geq 2$. For $l=1$, $G\geq 0$ and the zero modes are

$$h_{\mu\nu}^{(i)}=\nabla_\mu\nabla_\nu\Phi^{(i)}, \tag{6.4.66}$$

where

$$\Phi^{(i)}=(r-m)\,Y_{li}(\theta,\phi). \tag{6.4.67}$$

For $l=0$, G is not positive definite. There is only one negative mode which has $\omega=0$ and $l=0$. This mode was discovered as a result of numerical studies. It is most convenient to describe it by referring the mode to an orthonormal tetrad by writing

$$h_{\mu\nu}^{(-)}=\text{diag}\left(\left(1-\frac{2m}{r}\right)\tilde{H}_0(r),\left(1-\frac{2m}{r}\right)^{-1}\tilde{H}_1(r),-\frac{1}{2}r^2\left(\tilde{H}_0(r)+\tilde{H}_1(r)\right),\right.$$

$$\left.-\frac{1}{2}r^2\sin^2\theta\left(\tilde{H}_0(r)+\tilde{H}_1(r)\right)\right). \tag{6.4.68}$$

$\tilde{H}_0(r)$ and $\tilde{H}_1(r)$ obey the following differential equations

$$\left(1-\frac{2m}{r}\right)\frac{d^2}{dr^2}\tilde{H}_1(r)+\frac{4r^2-22mr+24m^2}{r^2(r-3m)}\frac{d}{dr}\tilde{H}_1(r)$$

$$-\frac{8m}{r^2(r-3m)}\tilde{H}_1(r)=-\lambda\tilde{H}_1(r) \tag{6.4.69}$$

$$\tilde{H}_0(r)=\frac{-r(r-2m)}{r-3m}\frac{d}{dr}\tilde{H}_1(r)-\frac{3r-5m}{r-3m}\tilde{H}_1(r). \tag{6.4.70}$$

The eigenvalue of G corresponding to this mode is approximately

$$\lambda^{(-)}=-.19m^{-2}. \tag{6.4.71}$$

At this point, it is possible to compute approximately the partition function, for pure gravity, in a canonical ensemble at a temperature $T=1/\beta$ and zero angular velocity contained in a 3-volume V. This means that functional integrals must be taken over all A.F. metrics. The only known A.F. instantons are periodically-identified flat space and the Schwarzschild instanton. The flat space contribution is

$$Z^{(0)} = \exp\left[\frac{V\pi^2}{45\beta^3}\right]. \tag{6.4.72}$$

This represents the contribution from thermal gravitons in a box of volume V at temperature $T = 1/\beta$. The contribution from the Schwarzschild instanton is

$$Z^{(1)} = \frac{i}{2}\exp\left(-4\pi GM^2\right)\exp\left[\frac{V\pi^2}{45\beta^3}\right](\mu\beta)^{\frac{212}{45}}\left[\frac{M}{2\pi\beta}\right]^{\frac{3}{2}}V. \tag{6.4.73}$$

The factor of $(\mu\beta)^{\frac{212}{45}}$ comes from the anomalous scaling behavior associated with the on shell counterterm, (i.e., when g_{ab} is a solution of Einstein's equation),

$$\Delta I = \frac{1}{16\pi^2}\frac{1}{n-4}\frac{53}{45}\int R_{abcd}R^{abcd}\sqrt{g}d^4x \tag{6.4.74}$$

appearing when one adopts a dimensional regularization scheme to deal with the ultraviolet divergence that occurs in curved space at one-loop. This contains a factor of μ. It is essentially a cutoff parameter due to the non renormalizability of the theory and its value is usually thought of as roughly the Planck Mass. The factors of $\left[\frac{M}{2\pi\beta}\right]^{\frac{3}{2}}V$ emerge from the zero modes and correspond to the one particle partition function of an object of mass m at temperature $1/\beta$ in a box of volume V. The factor of $\frac{i}{2}$ comes from the presence of the negative mode in the spectrum of G. Similarly, the contribution from the N-instanton sector is

$$Z^{(N)} = \left(\frac{i}{2}\right)^N \exp\left(-4\pi NGM^2\right)\exp\left[\frac{V\pi^2}{45\beta^3}\right](\mu\beta)^{\frac{212N}{45}}\frac{1}{N!}\left[\frac{V}{64\pi^3G^{\frac{3}{2}}}\right]^N, \tag{6.4.75}$$

where we have used the relation $\beta = 8\pi GM$. In the dilute-gas approximation, which is valid for widely separated instantons. An estimate for Z is

$$Z = \sum_{N=0}^{\infty} Z^{(N)} \simeq \exp\left[\frac{V\pi^2}{45\beta^3}\right]\exp\left[\left[\frac{i}{2}\right]\left[\frac{V}{64\pi^3G^{\frac{3}{2}}}\right](\mu\beta)^{\frac{212}{45}}\exp\left(-4\pi GM^2\right)\right]. \tag{6.4.76}$$

In the canonical ensemble Z is to be interpreted as the $\exp\left(-F/T\right)$. For systems that contain unstable phases, the imaginary part of F is generically non-zero. The instanton can be interpreted as the thermal nucleation of Schwarzschild metrics-black holes, with a nucleation rate per unit volume given by

$$\Gamma = (\omega_0\beta/\pi)\,\mathrm{Im}\,\mathcal{F}, \tag{6.4.77}$$

where

$$\omega_0^2 = -\lambda = \frac{0.19}{(GM)^2}. \tag{6.4.78}$$

Thus the vacuum decay rate (per unit volume per unit time), at least semiclassically, is

$$\Gamma = A\exp\left(-I_{cl}\right) = \frac{0.87}{\beta}\left(\mu\beta\right)^{\frac{212}{45}}\frac{m_P^3}{64\pi^3}\exp\left(-m_P^2/16\pi T^2\right). \tag{6.4.79}$$

This means that these saddle points are *bounces* instead of being instantons, that is at one loop the approximated functional ceases to be gaussian. Since these backgrounds are periodic in the imaginary time, this periodicity is usually interpreted as finite temperature T. Then the instability of these configurations happens for the hot space. The appearance of a single negative eigenvalue is related to the bounce which shifts the energy of the false ground state[5]. However, as stated in the introduction, we would like to reconsider the stability of flat space without any temperature but by looking at a neutral black hole pair creation, like in a Schwinger process, mediated by a wormhole of the Schwarzschild type.

6.5 Nonconformal Negative Modes Variational Approach to the Negative Mode

Here, we extend the GPY procedure to the Euclidean wormhole described in section 6.3. The Euclidean section of $\mathcal{M} = \mathcal{M}_+ \cup \mathcal{M}_-$ has been denoted by $\bar{\mathcal{M}}$ and the right and left wedges of a Lorentzian eternal black hole (wormhole) are mapped into two complex sectors $\bar{\mathcal{M}}_+$ and $\bar{\mathcal{M}}_-$ to give Eq.(6.1.1), where the integral is performed over all positive definite metrics g. In the case of pure gravity, the Euclidean action comes from Eq.(6.3.29) whose expression is

$$I = -\frac{1}{2\kappa}\int_{\bar{\mathcal{M}}_+} d^4x\,\sqrt{g}\,R + \frac{1}{\kappa}\int_{\partial\bar{\mathcal{M}}_+} d^3x\,\sqrt{\gamma}\,(\Theta - \Theta_0)$$

$$+\frac{1}{2\kappa}\int_{\bar{\mathcal{M}}_-} d^4x\,\sqrt{g}\,R + \frac{1}{\kappa}\int_{\partial\bar{\mathcal{M}}_-} d^3x\,\sqrt{\gamma}\,(\Theta - \Theta_0). \tag{6.5.80}$$

This partition function may be approximated using saddle-point techniques, by Taylor expanding about the known stationary points of the Euclidean action – the solutions to the Einstein field equations

$$R_{\mu\nu} = 0. \tag{6.5.81}$$

The expansions are performed by writing the metric g_{ab} as

$$g_{\mu\nu} = \bar{g}_{\mu\nu} + h_{\mu\nu} \tag{6.5.82}$$

with $h_{\mu\nu}$ treated as a quantum field on the classical fixed background $\bar{g}_{\mu\nu}$ which vanishes on the boundary $\partial\bar{\mathcal{M}}_\pm$. Thus the Euclidean action becomes

$$I[g] = I[\bar{g}]_+ + I_2[h]_+ - I[\bar{g}]_- - I_2[h]_- + \cdots \tag{6.5.83}$$

where the linear term I_1 vanishes precisely because $\bar{g}_{\mu\nu}$ is a classical solution, I_2 is quadratic in the field $h_{\mu\nu}$ in each complex sector, and '$\cdots$' represents terms of higher than quadratic order. The Taylor expansion of the action in $\bar{\mathcal{M}} = \bar{\mathcal{M}}_+ \cup \bar{\mathcal{M}}_-$ leads to the following one loop approximation of the partition function

$$\log Z = -I\left[\bar{g}\right]_+ + \log \int_{\bar{\mathcal{M}}_+} D\left[h\right]_+ \exp\left(-I_2\left[h\right]_+\right) + I\left[\bar{g}\right]_- + \log \int_{\bar{\mathcal{M}}_-} D\left[h\right]_- \exp\left(I_2\left[h\right]_-\right). \tag{6.5.84}$$

It is interesting to note that in this form the gravitational quantum fluctuations on $\bar{\mathcal{M}}_+$ $(\bar{\mathcal{M}}_-)$ act as a sort of source field with respect to $\bar{\mathcal{M}}_-$ $(\bar{\mathcal{M}}_+)$. The quadratic contribution to the action is straightforward to evaluate, and may be written for arbitrary $h_{\mu\nu}$ in each wedge, in the form

$$I_2[h]_\pm = \frac{1}{4\kappa}\int d^4x\sqrt{g}h_\pm^{\mu\nu}G_{\mu\nu\rho\sigma}h_\pm^{\rho\sigma}. \tag{6.5.85}$$

$G_{\mu\nu\rho\sigma}$ is the same operator represented in Eq.(6.4.63); it is a physical gauge invariant spin-2 operator acting on the transverse and trace-free part of $h_{\mu\nu}$. In general, $G_{\mu\nu\rho\sigma}$ will have some finite number – generally zero or one – of negative eigenvalues, which correspond to the nonconformal negative modes of the solution. The eigenvalues of $G_{\mu\nu\rho\sigma}$ are determined by all solutions to the elliptic equation

$$G_{\mu\nu\rho\sigma}h_{(n)}^{\rho\sigma} = \lambda_{(n)}h_{(n)}^{\mu\nu} \tag{6.5.86}$$

where the eigenfunctions $h_{\mu\nu}$ are real, regular, symmetric, transverse, trace-free, and normalizable tensors. Clearly, should one of the eigenvalues of G be negative, then the product of all of the eigenvalues would also be negative. The contribution to $\log Z$ from fluctuations about the classical solution would then contain an imaginary component, leading to an instability in the ensemble similar to the type proposed by Gross, Perry and Yaffe [4]. Naively, then, the one-loop term may be written as $\frac{1}{2}\log Det(\mu^{-2}G)$ where μ is a regularization mass, and the determinant is formally defined as the product of the eigenvalues of G. However, due to the diffeomorphism gauge freedom of the action, A will in general have a large number of zero eigenvalues, and so this procedure as stated is ill-defined. The remedy is to add a gauge fixing term B – such that the operator $A + B$ has no zero eigenvalues – and an associated ghost contribution C, to obtain

$$\log Z = -I[g] - \frac{1}{2}\log Det(\mu^{-2}\{A+B\}) + \log Det(\mu^{-2}C). \tag{6.5.87}$$

Such terms may be dealt with by means of generalized zeta functions, as considered by Gibbons, Hawking, and Perry [24], and extended to include a Λ term by Hawking [23]. In order that this be possible, the terms must be expressed as sums of operators, each with only a finite number of negative eigenvalues. This may be achieved by writing $A + B$ as $-F + G$, where F is a scalar operator acting on the trace of $h_{\mu\nu}$. The ghost term C is a spin-1 operator acting on divergence-free vectors, and An observation of primary significance is that a gauge may be chosen in which the operators F and C have no negative eigenvalues. If the background metric g_{ab} is flat then, in addition, G will be positive-definite, but for a non-flat background this is not the case. Following the treatment given in [4], and subsequently in [6] for the Schwarzschild case, it is clear that only spherically symmetric and τ-independent solutions of Eq.(6.5.86) need be considered as candidate nonconformal negative modes. In static and spherically symmetric backgrounds, modes of higher multipole moment will necessarily have greater eigenvalues. With this assumption, it is then straightforward to write down a construction for such solutions to G valid in a four-dimensional Euclidean wormhole background of the form (6.3.24). Since G acts only on symmetric transverse and trace-free tensors, then clearly the constructed solutions must exhibit all of these properties. If the mode $h^{\pm}_{\mu\nu}$ is written in the manifestly trace-free and symmetric form

$$(h^{\mu}_{\nu})^{\pm} = diag\left(H^{\pm}_{0}(y), H^{\pm}_{1}(y), -\frac{1}{2}\left(H^{\pm}_{0}(y) + H^{\pm}_{1}(y)\right), -\frac{1}{2}\left(H^{\pm}_{0}(y) + H^{\pm}_{1}(y)\right)\right), \tag{6.5.88}$$

then the final property, $\nabla^{a}h^{\pm}_{\mu\nu} = 0$, is guaranteed if it is further assumed that $H^{\pm}_{0}(y)$ and $H^{\pm}_{1}(y)$ are related through the first order equation

$$H^{\pm}_{1,y} + \frac{3}{r}r_{,y}H^{\pm}_{1} + \frac{N_{,y}}{N}H^{\pm}_{1} = \frac{N_{,y}}{N}H^{\pm}_{0} - \frac{3}{r}r_{,y}H^{\pm}_{0}. \tag{6.5.89}$$

In Eqs.(6.5.88) and (6.5.89), we have explicitly written the location of the perturbation on each patch covering one universe. The two patches join at the throat of the wormhole and the transversality condition can be cast into the form

$$\left[\frac{dH^{\pm}_{1}(r)}{dr} + \frac{3}{r}H^{\pm}_{1}(r) + \frac{1}{N}\frac{dN}{dr}H^{\pm}_{1}(r)\right]\frac{dr}{dy} = \left[\frac{1}{N}\frac{dN}{dr}H^{\pm}_{0}(r) - \frac{3}{r}H^{\pm}_{0}(r)\right]\frac{dr}{dy}, \tag{6.5.90}$$

where it is manifest the independence on the wedge location. This is a consequence of the discrete symmetry (6.2.13) between the asymptotically flat wedges. When we apply this symmetry to the the eigenvalue equation (6.5.91) with the help of the following substitution

$$N \to -N$$
$$r_{,y} \to -r_{,y},$$

we find that Eq.(6.5.91) is invariant in form only on the coordinate r. With the ansatz (6.5.88) and (6.5.90), the eigenvalue equation (6.5.86) reduces to a linear second order ordinary differential equation for the component $H^{\pm}_{1}(y)$ and eigenvalue μ^2. To further proceed, we insert Eqs.(6.3.23) and (6.3.25) into Eq.(6.5.86). Since we have chosen y and the Euclidean lapse N_E to be positive in $\mathcal{M}_+$ and negative in

$\bar{\mathcal{M}}_{-}$, then Eq.(6.5.86) is separated in two pieces[3]

$$\frac{d^2 H_1(y)}{dy^2} + b^{\pm}(y)\frac{dH_1(y)}{dy} + a^{\pm}(y) H_1(y) = -\mu^2 H_1(y), \tag{6.5.91}$$

where

$$b^{+}(y) = \sqrt{1 - \frac{2MG}{r(y)}} \frac{\left(4r^2(y) - 23MGr(y) + 27M^2G^2\right)}{r^2(y)\left(r(y) - 2MG\right)\left(r(y) - 3MG\right)}$$

$$a^{+}(y) = \frac{8MG}{\left(r(y) - 3MG\right) r^2(y)} \tag{6.5.92}$$

in $\bar{\mathcal{M}}_{+}$ and

$$b^{-}(y) = -\frac{\left(4r^2(y) - 23MGr(y) + 27M^2G^2\right)}{r^2(y)\left(r(y) - 3MG\right)\sqrt{1 - 2MG/r(y)}}$$

$$a^{-}(y) = \frac{8MG}{\left(r(y) - 3MG\right) r^2(y)} \tag{6.5.93}$$

in $\bar{\mathcal{M}}_{-}$. After having cast Eq.(6.5.91) in the standard Sturm-Liouville form, we apply a variational procedure to establish the existence of an instability by means of trial functions. The standard Sturm-Liouville form of Eq.(6.5.91) is

$$\frac{d}{dy}\left(p^{\pm}(y)\frac{d}{dy}H_1(y)\right) + q^{\pm}(y) H_1(y) + \mu^2 p^{\pm}(y) H_1(y) = 0, \tag{6.5.94}$$

where the integrating factor $p^{\pm}(y)$ is defined by

$$p^{\pm}(y) = \exp\int b^{\pm}(y)\,dy = \exp\pm\int \frac{b^{\pm}(r(y))}{\sqrt{1 - 2MG/r(y)}}dr$$

$$= \frac{\left(r(y)\right)^{9/2}\left(r(y) - 2MG\right)^{3/2}}{\left(r(y) - 3MG\right)^2}. \tag{6.5.95}$$

To compute the integrating factor we used Eq.(6.3.23). The coefficient

$$q^{\pm}(y) = p^{\pm}(y)\, a^{\pm}(y) \tag{6.5.96}$$

and the boundary conditions

$$p^{+}(y)\frac{dH_1(y)}{dy}H_1(y)\Big|_{y=0}^{y=+\infty} \quad \text{and} \quad p^{-}(y)\frac{dH_1(y)}{dy}H_1(y)\Big|_{y=-\infty}^{y=0}, \tag{6.5.97}$$

---

[3]The analogue of Eq.(6.5.91) has been studied in the context of the stability of the Schwarzschild solution for the first time by Regge and Wheeler[25], Vishveshwara[26], Zerilli[27], Press and Teukolsky, [28], Stewart[29] and Chandrasekhar[30]. They concluded that the Schwarzschild solution is *classically* stable.

lead us to write the Sturm-Liouville problem in the following functional form

$$F\left[H_1(y);\lambda\right]=\frac{N\left[H_1(y);\lambda\right]}{D\left[H_1(y);\lambda\right]}=\mu_{\pm}^2(\lambda)$$
$$=\frac{-\int_{I_\pm}dyp^{\pm}(y)\left(\frac{d}{dy}H_1(y)\right)^2+\int_{I_\pm}dyq^{\pm}(y)H_1^2(y)}{\int_{I_\pm}dyp^{\pm}(y)H_1^2(y)}, \tag{6.5.98}$$

where $I_+\equiv[0,+\infty)$ and $I_-\equiv(-\infty,0]$. The choice of a trial function is suggested by the asymptotic behavior of in Eq.(6.5.91). Indeed, when $y\to\pm\infty$, $H_1^{\mp}(y)\simeq\exp(\mp\lambda y)\simeq\exp(-\lambda r)$. The denominator in Eq.(6.5.98) compensates the lack of normalization in the trial function $H_1^{\mp}(y)$. If λ is defined as a variational parameter, then the Rayleigh-Ritz method leads to the following result[4]

$$\mu_{\pm}^2(\lambda_{\min})=-.1374637652 \qquad for \qquad \lambda_{\min}=.4471243737. \tag{6.5.99}$$

However the denominator of Eq.(6.5.98) vanishes for $\bar{\lambda}=1.360771766$ and for $\lambda>\bar{\lambda}$, the "*normalization*" changes sign. The lack of positivity for every λ is related to the appearance of a spurious singularity in $r=3MG$[5]. Therefore, it is necessary to choose a better form for the trial function $H_1^{\mp}(y)$ which partially removes the singularity in $r=3MG$. Such a form is given by $H_1^{\mp}(y)\simeq H_1(r)=\exp(-\lambda r)(r-3MG)$. With this choice, the "*normalization*" is strictly positive for every λ. The minimum of $\mu_{\pm}^2(\lambda)$ is reached for $\lambda_{\min}=.9223711858$ and $\mu_{\pm}^2(\lambda_{\min})=-.3827139389$. It is interesting to note that

$$\mu^2(\lambda_{\min})=\mu_+^2(\lambda_{\min})+\mu_-^2(\lambda_{\min})=-0.765\,44$$
$$\to\mu^2(\lambda_{\min})/(2MG)^2=-0.191\,36/(MG)^2. \tag{6.5.100}$$

This is the same value obtained by GPY for a single Euclidean black hole.

6.6 Hamiltonian formulation of Quantum Gravity

The hamiltonian form of the action can be extracted with the help of Eq.(6.3.29) without the subtraction term of Eq.(6.3.30), which will be considered later in an appropriate form. Following Brown and York[32], we decompose the extrinsic curvature $\Theta^{\mu\nu}$as

$$\Theta^{\mu\nu}=k^{\mu\nu}+u^{\mu}u^{\nu}n_{\alpha}a^{\alpha}+2\sigma^{\alpha(\mu}u^{\nu)}n^{\beta}K_{\alpha\beta}. \tag{6.6.101}$$

The traces of $\Theta^{\mu\nu}$ and $k^{\mu\nu}$ are related by

$$\Theta=k-n_{\alpha}a^{\alpha}, \tag{6.6.102}$$

[4]Details of computation can be found in the Appendix.
[5]See details in Appendix 6.10

where $a^\mu = u^\alpha \nabla_\alpha u^\mu$ is the acceleration of the timelike normal u^μ. In our case, slices corresponding to different time parameter t intersect at the two-dimensional bifurcation surface where the lapse function N vanishes. The vector u^μ is future oriented in M_+ and past oriented in M_-. The complete covariant form of the gravitational action for this foliation with fixed three-geometry at the boundaries of M is

$$S = \frac{1}{2\kappa}\int_{\mathcal{M}_+} d^4x\sqrt{-\,^4g}\,^4R + \frac{1}{\kappa}\int_{(+)t'}^{t''} d^3x\sqrt{g}K - \frac{1}{\kappa}\int_{B_+} d^3x\sqrt{-\gamma}\Theta$$

$$-\frac{1}{2\kappa}\int_{\mathcal{M}_-} d^4x\sqrt{-\,^4g}\,^4R + \frac{1}{\kappa}\int_{(-)t'}^{t''} d^3x\sqrt{g}K - \frac{1}{\kappa}\int_{B_-} d^3x\sqrt{-\gamma}\Theta. \qquad (6.6.103)$$

The notation $\int_{(\pm)t'}^{t''}$ represents an integral over the three-boundary $\Sigma_\pm$ at t'' minus an integral over the three-boundary $\Sigma_\pm$ at t'. The action (6.6.103) has to be complemented with a subtraction term dependent only on the boundary intrinsic geometry[31, 32]. The Hamiltonian can be written following the standard procedure by recognizing the existence of a time vector field t^μ defined at the boundaries B. If we separate time from space, the four dimensional scalar curvature becomes

$$^4R = R + K_{\mu\nu}K^{\mu\nu} - K^2 - 2\nabla_\mu\left(Ku^\mu + a^\mu\right), \qquad (6.6.104)$$

where R denotes the scalar curvature of the three-dimensional spacelike hypersurface Σ. Putting (6.6.104) in (6.6.103), we obtain

$$S = \frac{1}{2\kappa}\int_{\mathcal{M}_+} d^4x\sqrt{-\,^4g}\left[R + K_{\mu\nu}K^{\mu\nu} - K^2 - 2\nabla_\mu\left(Ku^\mu + a^\mu\right)\right]$$

$$-\frac{1}{2\kappa}\int_{\mathcal{M}_-} d^4x\sqrt{-\,^4g}\left[R + K_{\mu\nu}K^{\mu\nu} - K^2 - 2\nabla_\mu\left(Ku^\mu + a^\mu\right)\right] + \frac{1}{\kappa}\int_{(+)t'}^{t''} d^3x\sqrt{g}K$$

$$-\frac{1}{\kappa}\int_{B_+} d^3x\sqrt{-\gamma}\Theta + \frac{1}{\kappa}\int_{(-)t'}^{t''} d^3x\sqrt{g}K - \frac{1}{\kappa}\int_{B_-} d^3x\sqrt{-\gamma}\Theta. \qquad (6.6.105)$$

We now fix our attention on the $\mathcal{M}_+$ surface term of the action

$$S_{\mathcal{M}_+} = \frac{1}{2\kappa}\int_{\mathcal{M}_+} d^4x\sqrt{-\,^4g}\left[-2\nabla_\mu\left(Ku^\mu + a^\mu\right)\right]$$

$$+\frac{1}{\kappa}\int_{(+)t'}^{t''} d^3x\sqrt{g}K - \frac{1}{\kappa}\int_{B_+} d^3x\sqrt{-\gamma}\Theta.$$

With the help of divergence theorem and recalling that $u_\mu u^\mu = -1$, $u \cdot n = 0$, $n_\mu n^\mu = 1$ and $u_\mu a^\mu = 0$, we write

$$S_{\mathcal{M}_+} = -\frac{1}{\kappa}\int_{\Sigma_+} d^3x\sqrt{g}\left[-u_\mu\left(Ku^\mu + a^\mu\right) - K\right]$$

$$-\frac{1}{\kappa}\int_{B_+} d^3x\sqrt{-\gamma}\left[-n_\mu\left(Ku^\mu + a^\mu\right) + k - n_\alpha a^\alpha\right]$$

$$= -\frac{1}{\kappa}\int_{B_+} d^3x\sqrt{-\gamma}\left[n_\mu a^\mu + k - n_\alpha a^\alpha\right] = -\frac{1}{\kappa}\int_{B_+} d^3x\sqrt{-\gamma}k. \qquad (6.6.106)$$

The second negative sign appearing in front of u_μ comes from the fact that the normal (u^μ) is future pointing in the whole manifold, while u_μ is not. On $\mathcal{M}_-$ we consider

$$S_{\mathcal{M}_-} =$$

$$-\frac{1}{2\kappa}\int_{\mathcal{M}_-} d^4x\sqrt{-\,{}^4g}\left[-2\nabla_\mu\left(Ku^\mu + a^\mu\right)\right] + \frac{1}{\kappa}\int_{(-)t'}^{t''} d^3x\sqrt{g}K - \frac{1}{\kappa}\int_{B_-} d^3x\sqrt{-\gamma}\Theta$$

$$= -\frac{1}{\kappa}\int_{\Sigma_-} d^3x\sqrt{g}\left[-u_\mu\left(Ku^\mu + a^\mu\right) - K\right] - \frac{1}{\kappa}\int_{B_-} d^3x\sqrt{-\gamma}\left[-\left(-n_\mu\right)\left(Ku^\mu + a^\mu\right) + \Theta\right]$$

$$= -\frac{1}{\kappa}\int_{B_-} d^3x\sqrt{-\gamma}\left[n_\mu a^\mu + k - n_\alpha a^\alpha\right] = -\frac{1}{\kappa}\int_{B_-} d^3x\sqrt{-\gamma}k. \qquad (6.6.107)$$

This time the extra-minus appearing in the penultimate line in front of n_μ comes from the choice of having the space-like normal inward pointing on B_-. This choice is a direct consequence of the backward Schwarzschild time direction in $\mathcal{M}_-$ and of the coordinate sign of Eq.(6.3.23) related to the space-like normal. Thus the global action evaluated on both boundaries becomes

$$S = \frac{1}{2\kappa}\int_{M_+} d^4x\sqrt{-\,{}^4g}\left[R + K_{\mu\nu}K^{\mu\nu} - (K)^2\right] - \frac{1}{\kappa}\int_{B_+} d^3x\sqrt{-\gamma}k$$

$$-\frac{1}{2\kappa}\int_{M_-} d^4x\sqrt{-\,{}^4g}\left[R + K_{\mu\nu}K^{\mu\nu} - (K)^2\right] - \frac{1}{\kappa}\int_{B_-} d^3x\sqrt{-\gamma}k. \qquad (6.6.108)$$

To build the hamiltonian, we begin to compute the momentum π^{ij} conjugate to the three-metric h_{ij} of Σ, defined as

$$\pi^{ij} = \frac{\sqrt{g}}{2\kappa}\left(Kg^{ij} - K^{ij}\right), \qquad (6.6.109)$$

and the definition of $K_{\mu\nu}$ implies

$$K_{ij} = -\nabla_i u_j = -\frac{1}{2N}\left[\dot{g}_{ij} - 2D_{(i}N_{j)}\right], \qquad (6.6.110)$$

where the dot denotes differentiation with respect to the time t. By performing the usual Legendre transformation, the action becomes

$$S = \int_M d^4x\pi^{ij}\dot{g}_{ij} - \int H dt, \qquad (6.6.111)$$

with the gravitational Hamiltonian H given by

$$H = \int_{\Sigma} d^3x (N\mathcal{H} + N^i \mathcal{H}_i)$$

$$+ \frac{1}{\kappa}\int_{S_+^2} d^2xN\sqrt{\sigma}\left(k - k^0\right) - \frac{1}{\kappa}\int_{S_-^2} d^2xN\sqrt{\sigma}\left(k - k^0\right), \tag{6.6.112}$$

where we have introduced the subtraction terms

$$\frac{1}{\kappa}\int_{S_+^2} d^2xN\sqrt{\sigma}\left(-k^0\right) - \frac{1}{\kappa}\int_{S_-^2} d^2xN\sqrt{\sigma}\left(-k^0\right) \tag{6.6.113}$$

in complete analogy with the terms of Eq.(6.3.30). The reference space (in this case flat space) is a fixed hypersurface of this spacetime and k^0 the trace of extrinsic curvature in this space of a two-dimensional surface whose induced metric is σ_{ab}. The Hamiltonian has both volume and boundary contributions. The volume part involves the Hamiltonian

$$\mathcal{H} = (2\kappa)\, G_{ijkl}\pi^{ij}\pi^{kl}/\sqrt{g} - \sqrt{g}R/\left(2\kappa\right) = 0 \tag{6.6.114}$$

and momentum constraints

$$\mathcal{H}_i = -2\pi^j_{i|j} = 0, \tag{6.6.115}$$

where

$$G_{ijkl} = \frac{1}{2}\left(g_{ik}g_{jl} + g_{il}g_{jk} - g_{ij}g_{kl}\right) \tag{6.6.116}$$

is the supermetric and R denotes the scalar curvature of the surface Σ. The volume part of the Hamiltonian(6.6.112) is zero for the flat and the Schwarzschild space on shell, then in this context the total Hamiltonian reduces to

$$H = \frac{1}{\kappa}\int_{S_+^2} d^2xN\sqrt{\sigma}\left(k - k^0\right) - \frac{1}{\kappa}\int_{S_-^2} d^2xN\sqrt{\sigma}\left(k - k^0\right). \tag{6.6.117}$$

6.6.1 Quasilocal Energy

Quasilocal energy is defined as the value of the Hamiltonian that generates unit time translations orthogonal to the two-dimensional boundaries, i.e.

$$E_{\text{tot}} = H_{\text{quasilocal}} = H_+ - H_- = E_+ - E_-,$$

$$E_+ = \frac{1}{\kappa}\int_{S_+^2} d^2x\sqrt{\sigma}\left(k - k^0\right)$$

$$E_- = -\frac{1}{\kappa}\int_{S_-^2} d^2x\sqrt{\sigma}\left(k - k^0\right). \tag{6.6.118}$$

where $|N| = 1$ at both S_+ and S_-. E_{tot} is the quasilocal energy of a spacelike hypersurface $\Sigma = \Sigma_+ \cup \Sigma_-$ bounded by two boundaries S^3_+ and S^3_- located in the two disconnected regions $\mathcal{M}_+$ and $\mathcal{M}_-$ respectively. Let us consider the case of the static Einstein-Rosen bridge whose metric is defined as:

$$ds^2 = -N^2 dt^2 + g_{yy} dy^2 + r^2(y)\, d\Omega^2, \tag{6.6.119}$$

where N, g_{yy}, and r are functions of the radial coordinate y continuously defined on $\mathcal{M}$ and expressed by Eq.(6.3.23). If we make the identification $N^2 = 1 - \frac{2MG}{r}$, the line element (6.6.119) reduces to the Schwarzschild metric written in another form. The boundaries S^2_+ and S^2_- are located at coordinate values $y = y_+$ and $y = y_-$ respectively. The normal to the boundaries is $n^\mu = (h^{yy})^{\frac{1}{2}} \delta^\mu_y$. Since this normal is defined continuously along Σ, the value of k depends on the function $r_{,y}$, which is positive for B^2_+ and negative for B^2_-. The application of the quasilocal energy definition gives

$$\begin{aligned} E &= E_+ - E_- \\ &= \frac{1}{8\pi G}\left\{4\pi R^2\left[\frac{2}{R}\left(\left(1 - r_{,y}\right)_{y=y_+} - \left(1 - r_{,y}\right)_{y=y_-}\right)\right]\right\}. \\ &= \frac{R}{G}\left[\left(1 - \sqrt{1 - \frac{2MG}{r(y_+)}}\right) - \left(1 - \sqrt{1 - \frac{2MG}{r(y_-)}}\right)\right]. \end{aligned} \tag{6.6.120}$$

As an illustration, consider the case when the boundary B_+ is located at right-hand infinity ($y_+ = +\infty$) and the boundary B_- is located at y_-, then the quasilocal energy gives

$$M - \frac{R}{G}\left[1 - \left(1 - \frac{2MG}{R}\right)^{\frac{1}{2}}\right]. \tag{6.6.121}$$

It is easy to see that E_+ and E_- tend individually to the $\mathcal{ADM}$ mass $\mathcal{M}$ when the boundaries B^3_+ and B^3_- tend respectively to right and left spatial infinity. It should be noted that the total energy is zero for boundary conditions symmetric with respect to the bifurcation surface. Indeed, if we set $y_- = y_+ = \bar{y}$ in Eq. (6.6.120), we obtain

$$\frac{R}{G}\left[\left(1 - \sqrt{1 - \frac{2MG}{r(\bar{y})}}\right) - \left(1 - \sqrt{1 - \frac{2MG}{r(\bar{y})}}\right)\right] = 0. \tag{6.6.122}$$

In particular, when $\bar{y} \to \pm\infty$

$$E = E_+ - E_- = M + (-M) = 0, \tag{6.6.123}$$

where the asymptotic contribution has been considered. We can also see Eq.(6.6.123) from a different point of view. In fact this Equation say that flat space can be thought of as a composition of two pieces: the former, with positive energy, in the region Σ_+ and the latter, with negative energy, in the region Σ_-, where the positive and negative concern the bifurcation surface (hole) which is formed due to a topology change of the manifold. The most appropriate mechanism to explain this splitting seems to be a black hole pair creation. Note that in analogy with Eq.(6.3.44), we have found a particular situation where the classical energy vanishes. Thus at the tree level, we cannot obtain any information about a possible transition from a false vacuum to a true one. It is therefore appropriate to investigate the perturbed WDW equation in its Sturm-Liouville form.

6.7 The Wheeler-De Witt equation and the cosmological constant

To better understand the role of the cosmological constant in our context, we recall that the modified Einstein field equations are

$$R_{\mu\nu} - \frac{1}{2}g_{\mu\nu}R + \Lambda_c g_{\mu\nu} = 8\pi G T_{\mu\nu}, \tag{6.7.124}$$

where Λ_c is the cosmological constant, G is the gravitational constant and $T_{\mu\nu}$ is the energy-momentum tensor. By redefining

$$T_{\mu\nu}^{tot} \equiv T_{\mu\nu} - \frac{\Lambda_c}{8\pi G}g_{\mu\nu}, \tag{6.7.125}$$

one can regain the original form of the field equations

$$R_{\mu\nu} - \frac{1}{2}g_{\mu\nu}R = 8\pi G T_{\mu\nu}^{tot} = 8\pi G\left(T_{\mu\nu} + T_{\mu\nu}^{\Lambda}\right), \tag{6.7.126}$$

at the prize of introducing a vacuum energy density and vacuum stress-energy tensor

$$\rho_\Lambda = \frac{\Lambda_c}{8\pi G}; \qquad T_{\mu\nu}^{\Lambda} = -\rho_\Lambda g_{\mu\nu}. \tag{6.7.127}$$

Alternatively, Eq.(6.7.124) can be cast into the form,

$$R_{\mu\nu} - \frac{1}{2}g_{\mu\nu}R + \Lambda_{eff} g_{\mu\nu} = 0, \tag{6.7.128}$$

where we have included the contribution of the vacuum energy density in the form $T_{\mu\nu} = -\langle\rho\rangle g_{\mu\nu}$. In this case Λ_c can be considered as the bare cosmological constant

$$\Lambda_{eff} = 8\pi G\rho_{eff} = \Lambda_c + 8\pi G\langle\rho\rangle. \tag{6.7.129}$$

Experimentally, we know that the effective energy density of the universe ρ_{eff} is of the order $10^{-47}GeV^4$ and the estimate of Eq.(6.1.4) approximately gives $\approx 10^{71}GeV^4$. This gives a difference of about 118 orders[33]. In General Relativity, the approach to quantization is based on Eqs.(6.6.114, 6.6.115). In particular, when we include a cosmological constant, Eq.(6.6.114), considered as an operatorial equation, modifies into

$$\left[\frac{2\kappa}{\sqrt{g}}G_{ijkl}\pi^{ij}\pi^{kl}-\frac{\sqrt{g}}{2\kappa}\left(R-2\Lambda_c\right)\right]\Psi\left[g_{ij}\right]=0 \tag{6.7.130}$$

and

$$-2\nabla_i\pi^{ij}\Psi\left[g_{ij}\right]=0, \tag{6.7.131}$$

where R is the three-scalar curvature, Λ_c is the bare cosmological constant and $\kappa=8\pi G$, is known as Wheeler-De Witt equation (WDW)[34]. Eqs.(6.7.130) and (6.7.131) describe the *wave function of the universe*. The WDW equation represents invariance under *time* reparametrization in an operatorial form, while Eq.(6.7.131) represents invariance under diffeomorphism. Note that the WDW equation can be cast into the form

$$\left[\frac{2\kappa}{\sqrt{g}}G_{ijkl}\pi^{ij}\pi^{kl}-\frac{\sqrt{g}}{2\kappa}R\right]\Psi\left[g_{ij}\right]=-\frac{\sqrt{g}}{\kappa}\Lambda_c\Psi\left[g_{ij}\right], \tag{6.7.132}$$

which formally looks like an eigenvalue equation. Therefore, the WDW equation can be used to estimate $\langle\rho\rangle$. The WDW equation (6.7.130), written as an eigenvalue equation, can be cast into the form

$$\hat{\Lambda}_\Sigma\Psi\left[g_{ij}\right]=\Lambda'\Psi\left[g_{ij}\right], \tag{6.7.133}$$

where

$$\left\{\begin{array}{l}\hat{\Lambda}_\Sigma=\frac{2\kappa}{\sqrt{g}}G_{ijkl}\pi^{ij}\pi^{kl}-\frac{\sqrt{g}}{2\kappa}R\\ \\ \Lambda'=-\frac{\Lambda}{\kappa}\end{array}\right. . \tag{6.7.134}$$

We, now multiply Eq.(6.7.133) by $\Psi^*\left[g_{ij}\right]$ and we functionally integrate over the three spatial metric g_{ij}, then after an integration over the hypersurface Σ, one can formally re-write the WDW equation as

$$\frac{1}{V}\frac{\int\mathcal{D}\left[g_{ij}\right]\Psi^*\left[g_{ij}\right]\int_\Sigma d^3x\hat{\Lambda}_\Sigma\Psi\left[g_{ij}\right]}{\int\mathcal{D}\left[g_{ij}\right]\Psi^*\left[g_{ij}\right]\Psi\left[g_{ij}\right]}=\frac{1}{V}\frac{\left\langle\Psi\left|\int_\Sigma d^3x\hat{\Lambda}_\Sigma\right|\Psi\right\rangle}{\langle\Psi|\Psi\rangle}=\Lambda'. \tag{6.7.135}$$

The formal eigenvalue equation is a simple manipulation of Eq.(6.7.130). However, we gain more information if we consider a separation of the spatial part of the metric into a background term, $\bar{g}_{ij}$, and a perturbation, h_{ij},

$$g_{ij}=\bar{g}_{ij}+h_{ij}. \tag{6.7.136}$$

Thus eq.(6.7.135) becomes

$$\frac{\left\langle\Psi\left|\int_\Sigma d^3x\left[\hat{\Lambda}_\Sigma^{(0)}+\hat{\Lambda}_\Sigma^{(1)}+\hat{\Lambda}_\Sigma^{(2)}+\ldots\right]\right|\Psi\right\rangle}{\langle\Psi|\Psi\rangle}=\Lambda'\Psi\left[g_{ij}\right], \tag{6.7.137}$$

where $\hat{\Lambda}_{\Sigma}^{(i)}$ represents the ith order of perturbation in h_{ij}. By observing that the kinetic part of $\hat{\Lambda}_{\Sigma}$ is quadratic in the momenta, we only need to expand the three-scalar curvature $\int d^3x\sqrt{g}R^{(3)}$ up to quadratic order and we get

$$\int_{\Sigma} d^3x\sqrt{\bar{g}}\left[-\frac{1}{4}h\triangle h + \frac{1}{4}h^{li}\triangle h_{li} - \frac{1}{2}h^{ij}\nabla_l\nabla_i h^l_j\right.$$

$$\left.+\frac{1}{2}h\nabla_l\nabla_i h^{li} - \frac{1}{2}h^{ij}R_{ia}h^a_j + \frac{1}{2}hR_{ij}h^{ij} + \frac{1}{4}h\left(R^{(0)}\right)h\right] \tag{6.7.138}$$

where h is the trace of h_{ij} and $R^{(0)}$ is the three dimensional scalar curvature. To explicitly make calculations, we need an orthogonal decomposition for both π_{ij} and h_{ij} to disentangle gauge modes from physical deformations. We define the inner product

$$\langle h, k\rangle := \int_{\Sigma} \sqrt{g}G^{ijkl}h_{ij}(x)\,k_{kl}(x)\,d^3x, \tag{6.7.139}$$

by means of the inverse WDW metric G_{ijkl}, to have a metric on the space of deformations, i.e. a quadratic form on the tangent space at h_{ij}, with

$$G^{ijkl} = (g^{ik}g^{jl} + g^{il}g^{jk} - 2g^{ij}g^{kl}). \tag{6.7.140}$$

The inverse metric is defined on cotangent space and it assumes the form

$$\langle p, q\rangle := \int_{\Sigma} \sqrt{g}G_{ijkl}p^{ij}(x)\,q^{kl}(x)\,d^3x, \tag{6.7.141}$$

so that

$$G^{ijnm}G_{nmkl} = \frac{1}{2}\left(\delta^i_k\delta^j_l + \delta^i_l\delta^j_k\right). \tag{6.7.142}$$

Note that in this scheme the "inverse metric" is actually the WDW metric defined on phase space. The desired decomposition on the tangent space of 3-metric deformations[35, 36, 37, 38] is:

$$h_{ij} = \frac{1}{3}hg_{ij} + (L\xi)_{ij} + h^{\perp}_{ij} \tag{6.7.143}$$

where the operator L maps ξ_i into symmetric tracefree tensors

$$(L\xi)_{ij} = \nabla_i\xi_j + \nabla_j\xi_i - \frac{2}{3}g_{ij}(\nabla\cdot\xi). \tag{6.7.144}$$

Thus the inner product between three-geometries becomes

$$\langle h, h\rangle := \int_{\Sigma} \sqrt{g}G^{ijkl}h_{ij}(x)\,h_{kl}(x)\,d^3x =$$

$$\int_{\Sigma} \sqrt{g}\left[-\frac{2}{3}h^2 + (L\xi)^{ij}(L\xi)_{ij} + h^{ij\perp}h^{\perp}_{ij}\right]. \tag{6.7.145}$$

With the orthogonal decomposition in hand we can define the trial wave functional as

$$\Psi\left[h_{ij}\left(\overrightarrow{x}\right)\right] = \mathcal{N}\Psi\left[h^{\perp}_{ij}\left(\overrightarrow{x}\right)\right]\Psi\left[h^{\|}_{ij}\left(\overrightarrow{x}\right)\right]\Psi\left[h^{trace}_{ij}\left(\overrightarrow{x}\right)\right], \tag{6.7.146}$$

where

$$\begin{aligned}\Psi\left[h^{\perp}_{ij}\left(\overrightarrow{x}\right)\right] &= \exp\left\{-\tfrac{1}{4}\left\langle hK^{-1}h\right\rangle^{\perp}_{x,y}\right\}\\ \Psi\left[h^{\|}_{ij}\left(\overrightarrow{x}\right)\right] &= \exp\left\{-\tfrac{1}{4}\left\langle (L\xi)K^{-1}(L\xi)\right\rangle^{\|}_{x,y}\right\}.\\ \Psi\left[h^{trace}_{ij}\left(\overrightarrow{x}\right)\right] &= \exp\left\{-\tfrac{1}{4}\left\langle hK^{-1}h\right\rangle^{Trace}_{x,y}\right\}\end{aligned} \tag{6.7.147}$$

The symbol "$\perp$" denotes the transverse-traceless tensor (TT) (spin 2) of the perturbation, while the symbol "$\|$" denotes the longitudinal part (spin 1) of the perturbation. Finally, the symbol "*trace*" denotes the scalar part of the perturbation. $\mathcal{N}$ is a normalization factor, $\langle\cdot,\cdot\rangle_{x,y}$ denotes space integration and K^{-1} is the inverse "*propagator*". We will fix our attention to the TT tensor sector of the perturbation representing the graviton. Therefore, representation (6.7.146) reduces to

$$\begin{aligned}&\Psi\left[h_{ij}\left(\overrightarrow{x}\right)\right]\\ &= \mathcal{N}\exp\left\{-\frac{1}{4}\left\langle hK^{-1}h\right\rangle^{\perp}_{x,y}\right\}.\end{aligned} \tag{6.7.148}$$

Actually there is no reason to neglect longitudinal perturbations. However, following the analysis of Mazur and Mottola of Ref.[37] on the perturbation decomposition, we can discover that the relevant components can be restricted to the TT modes and to the trace modes. Moreover, for certain backgrounds TT tensors can be a source of instability as shown in Refs.[39]. Even the trace part can be regarded as a source of instability. Indeed this is usually termed *conformal* instability. The appearance of an instability on the TT modes is known as non conformal instability. This means that does not exist a gauge choice that can eliminate negative modes. To proceed with Eq.(6.7.137), we need to know the action of some basic operators on $\Psi[h_{ij}]$. The action of the operator h_{ij} on $|\Psi\rangle = \Psi[h_{ij}]$ is realized by[40]

$$h_{ij}(x)|\Psi\rangle = h_{ij}\left(\overrightarrow{x}\right)\Psi[h_{ij}]. \tag{6.7.149}$$

The action of the operator π_{ij} on $|\Psi\rangle$, in general, is

$$\pi_{ij}(x)|\Psi\rangle = -i\frac{\delta}{\delta h_{ij}\left(\overrightarrow{x}\right)}\Psi[h_{ij}], \tag{6.7.150}$$

while the inner product is defined by the functional integration:

$$\langle\Psi_1 \mid \Psi_2\rangle = \int[\mathcal{D}h_{ij}]\Psi^*_1[h_{ij}]\Psi_2[h_{kl}]. \tag{6.7.151}$$

We demand that

$$\frac{1}{V}\frac{\left\langle \Psi \left| \int_{\Sigma} d^3x \hat{\Lambda}_{\Sigma} \right| \Psi \right\rangle}{\langle \Psi | \Psi \rangle} = \frac{1}{V}\frac{\int \mathcal{D}\left[g_{ij}\right] \Psi^*\left[h_{ij}\right] \int_{\Sigma} d^3x \hat{\Lambda}_{\Sigma} \Psi\left[h_{ij}\right]}{\int \mathcal{D}\left[g_{ij}\right] \Psi^*\left[h_{ij}\right] \Psi\left[h_{ij}\right]} \tag{6.7.152}$$

be stationary against arbitrary variations of $\Psi\left[h_{ij}\right]$. Note that Eq.(6.7.152) can be considered as the variational analog of a Sturm-Liouville problem with the cosmological constant regarded as the associated eigenvalue. Therefore the solution of Eq.(6.7.135) corresponds to the minimum of Eq.(6.7.152). The form of $\left\langle \Psi \left| \hat{\Lambda}_{\Sigma} \right| \Psi \right\rangle$ can be computed with the help of the wave functional (6.7.148) and with the help of

$$\left\{ \begin{array}{c} \frac{\langle \Psi | h_{ij}(\overrightarrow{x}) | \Psi \rangle}{\langle \Psi | \Psi \rangle} = 0 \\ \\ \frac{\langle \Psi | h_{ij}(\overrightarrow{x}) h_{kl}(\overrightarrow{y}) | \Psi \rangle}{\langle \Psi | \Psi \rangle} = K_{ijkl}\left(\overrightarrow{x}, \overrightarrow{y}\right) \end{array} \right. . \tag{6.7.153}$$

Since the wave functional (6.7.148) separates the degrees of freedom, we assume that

$$\Lambda' = \Lambda'^{\perp}, \tag{6.7.154}$$

then Eq.(6.7.152) becomes

$$\frac{1}{V}\frac{\left\langle \Psi \left| \hat{\Lambda}_{\Sigma}^{\perp} \right| \Psi \right\rangle}{\langle \Psi | \Psi \rangle} = \Lambda'^{\perp}. \tag{6.7.155}$$

Extracting the TT tensor contribution, we get

$$\hat{\Lambda}_{\Sigma}^{\perp} = \frac{1}{4V}\int_{\Sigma} d^3x \sqrt{g} G^{ijkl}\left[(2\kappa) K^{-1\perp}(x,x)_{ijkl} \right.$$
$$\left. + \frac{1}{(2\kappa)}\left(\triangle_2\right)_j^a K^{\perp}(x,x)_{iakl}\right]. \tag{6.7.156}$$

The propagator $K^{\perp}(x,x)_{iakl}$ can be represented as

$$K^{\perp}\left(\overrightarrow{x}, \overrightarrow{y}\right)_{iakl} := \sum_{\tau} \frac{h_{ia}^{(\tau)\perp}\left(\overrightarrow{x}\right) h_{kl}^{(\tau)\perp}\left(\overrightarrow{y}\right)}{2\lambda(\tau)}, \tag{6.7.157}$$

where $h_{ia}^{(\tau)\perp}\left(\overrightarrow{x}\right)$ are the eigenfunctions of $\triangle_2$. τ denotes a complete set of indices and $\lambda(\tau)$ are a set of variational parameters to be determined by the minimization of Eq.(6.7.156). The expectation value of $\hat{\Lambda}_{\Sigma}^{\perp}$ is easily obtained by inserting the form of the propagator into Eq.(6.7.156)

$$\Lambda'^{\perp}(\lambda_i)$$

$$= \frac{1}{4}\sum_{\tau}\sum_{i=1}^{2}\left[(2\kappa)\,\lambda_i(\tau) + \frac{\omega_i^2(\tau)}{(2\kappa)\,\lambda_i(\tau)}\right]. \tag{6.7.158}$$

By minimizing with respect to the variational function $\lambda_i(\tau)$, we obtain the total one loop energy density for TT tensors

$$\Lambda^{TT}(\lambda_i) = \frac{1}{4}\sum_{\tau}\left[\sqrt{\omega_1^2(\tau)} + \sqrt{\omega_2^2(\tau)}\right], \tag{6.7.159}$$

where

$$\Lambda^{TT}(\lambda_i) = \Lambda'^{\perp}(\lambda_i) = -\Lambda^{\perp}(\lambda_i)/\kappa. \tag{6.7.160}$$

The above expression makes sense only for $\omega_i^2(\tau) > 0$.

6.8 The transverse traceless (TT) spin 2 operator for the Schwarzschild metric and the W.K.B. approximation

The Spin-two operator for the Schwarzschild metric is defined by

$$\left(\triangle_2 h^{TT}\right)_i^j := -\left(\triangle_T h^{TT}\right)_i^j + 2\left(Rh^{TT}\right)_i^j, \tag{6.8.161}$$

where the transverse-traceless (TT) tensor for the quantum fluctuation is obtained by the following decomposition

$$h_i^j = h_i^j - \frac{1}{3}\delta_i^j h + \frac{1}{3}\delta_i^j h = \left(h^T\right)_i^j + \frac{1}{3}\delta_i^j h. \tag{6.8.162}$$

This implies that $\left(h^T\right)_i^j \delta_j^i = 0$. The transversality condition is applied on $\left(h^T\right)_i^j$ and becomes $\nabla_j \left(h^T\right)_i^j = 0$. Thus

$$-\left(\triangle_T h^{TT}\right)_i^j = -\triangle_S \left(h^{TT}\right)_i^j + \frac{6}{r^2}\left(1 - \frac{2MG}{r}\right), \tag{6.8.163}$$

where $\triangle_S$ is the scalar curved Laplacian, whose form is

$$\triangle_S = \left(1 - \frac{2MG}{r}\right)\frac{d^2}{dr^2} + \left(\frac{2r - 3MG}{r^2}\right)\frac{d}{dr} - \frac{L^2}{r^2} \tag{6.8.164}$$

and R_j^a is the mixed Ricci tensor whose components are:

$$R_i^a = \left\{-\frac{2MG}{r^3}, \frac{MG}{r^3}, \frac{MG}{r^3}\right\}, \tag{6.8.165}$$

This implies that the scalar curvature is traceless. We are therefore led to study the following eigenvalue equation

$$\left(\triangle_2 h^{TT}\right)_i^j = \omega^2 h_j^i \tag{6.8.166}$$

where ω^2 is the eigenvalue of the corresponding equation. In doing so, we follow Regge and Wheeler in analyzing the equation as modes of definite frequency, angular momentum and parity[25]. In particular, our choice for the three-dimensional gravitational perturbation is represented by its even-parity form

$$\begin{aligned} &\left(h^{even}\right)_j^i (r,\vartheta,\phi) \\ &= diag\left[H(r), K(r), L(r)\right] Y_{lm}(\vartheta,\phi), \end{aligned} \tag{6.8.167}$$

with

$$\left\{ \begin{array}{l} H(r) = h_1^1(r) - \frac{1}{3}h(r) \\ K(r) = h_2^2(r) - \frac{1}{3}h(r) \\ L(r) = h_3^3(r) - \frac{1}{3}h(r) \end{array} \right. . \tag{6.8.168}$$

From the transversality condition we obtain $h_2^2(r) = h_3^3(r)$. Then $K(r) = L(r)$. For a generic value of the angular momentum L, representation (6.8.167) joined to Eq.(6.8.163) lead to the following system of PDE's

$$\left\{ \begin{array}{l} \left(-\triangle_S + \frac{6}{r^2}\left(1 - \frac{2MG}{r}\right) - \frac{4MG}{r^3}\right) H(r) = \omega_{1,l}^2 H(r) \\ \\ \left(-\triangle_S + \frac{6}{r^2}\left(1 - \frac{2MG}{r}\right) + \frac{2MG}{r^3}\right) K(r) = \omega_{2,l}^2 K(r) \end{array} \right. . \tag{6.8.169}$$

Defining reduced fields

$$H(r) = \frac{f_1(r)}{r}; \qquad K(r) = \frac{f_2(r)}{r}, \tag{6.8.170}$$

and passing to the proper geodesic distance from the *throat* of the bridge

$$dx = \pm \frac{dr}{\sqrt{1 - \frac{2MG}{r}}}, \tag{6.8.171}$$

the system (6.8.169) becomes

$$\left\{ \begin{array}{l} \left[-\frac{d^2}{dx^2} + V_1(r)\right] f_1(x) = \omega_{1,l}^2 f_1(x) \\ \\ \left[-\frac{d^2}{dx^2} + V_2(r)\right] f_2(x) = \omega_{2,l}^2 f_2(x) \end{array} \right. \tag{6.8.172}$$

with

$$\left\{ \begin{array}{l} V_1(r) = \frac{l(l+1)}{r^2} + U_1(r) \\ \\ V_2(r) = \frac{l(l+1)}{r^2} + U_2(r) \end{array} \right. ; \tag{6.8.173}$$

where we have defined $r \equiv r(x)$ and

$$\left\{\begin{array}{l} U_1(r) = \left[\frac{6}{r^2}\left(1 - \frac{2MG}{r}\right) - \frac{3MG}{r^3}\right] \\ \\ U_2(r) = \left[\frac{6}{r^2}\left(1 - \frac{2MG}{r}\right) + \frac{3MG}{r^3}\right] \end{array}\right. . \tag{6.8.174}$$

Note that

$$\left\{\begin{array}{c} U_1(r) \geq 0 \qquad \text{when } r \geq \frac{5MG}{2} \\ U_1(r) < 0 \qquad \text{when } 2MG \leq r < \frac{5MG}{2} \\ \\ U_2(r) > 0 \ \forall r \in [2MG, +\infty) \end{array}\right. . \tag{6.8.175}$$

The functions $U_1(r)$ and $U_2(r)$ play the rôle of two r-dependent effective masses $m_1^2(r)$ and $m_2^2(r)$, respectively. In order to use the WKB approximation, we define two r-dependent radial wave numbers $k_1(r, l, \omega_{1,nl})$ and $k_2(r, l, \omega_{2,nl})$

$$\left\{\begin{array}{l} k_1^2(r, l, \omega_{1,nl}) = \omega_{1,nl}^2 - \frac{l(l+1)}{r^2} - m_1^2(r) \\ \\ k_2^2(r, l, \omega_{2,nl}) = \omega_{2,nl}^2 - \frac{l(l+1)}{r^2} - m_2^2(r) \end{array}\right. \tag{6.8.176}$$

for $r \geq \frac{5MG}{2}$. When $2MG \leq r < \frac{5MG}{2}$, $k_1^2(r, l, \omega_{1,nl})$ becomes

$$k_1^2(r, l, \omega_{1,nl}) = \omega_{1,nl}^2 - \frac{l(l+1)}{r^2} + m_1^2(r). \tag{6.8.177}$$

6.9 One loop energy Regularization and Renormalization

In this section, we proceed to evaluate Eq.(6.7.159). The method is equivalent to the scattering phase shift method and to the same method used to compute the entropy in the brick wall model. We begin by counting the number of modes with frequency less than ω_i, $i = 1, 2$. This is given approximately by

$$\tilde{g}(\omega_i) = \int \nu_i(l, \omega_i)(2l + 1), \tag{6.9.178}$$

where $\nu_i(l, \omega_i)$, $i = 1, 2$ is the number of nodes in the mode with (l, ω_i), such that $(r \equiv r(x))$

$$\nu_i(l, \omega_i) = \frac{1}{2\pi}\int_{-\infty}^{+\infty} dx \sqrt{k_i^2(r, l, \omega_i)}. \tag{6.9.179}$$

Here it is understood that the integration with respect to x and l is taken over those values which satisfy $k_i^2(r, l, \omega_i) \geq 0$, $i = 1, 2$. With the help of Eqs.(6.9.178, 6.9.179), we obtain the one loop total energy for TT tensors

$$\frac{1}{8\pi}\sum_{i=1}^{2}\int_{-\infty}^{+\infty} dx \left[\int_0^{+\infty} \omega_i \frac{d\tilde{g}(\omega_i)}{d\omega_i} d\omega_i\right]. \tag{6.9.180}$$

By extracting the energy density contributing to the cosmological constant, we get

$$\Lambda^{TT} = \frac{1}{16\pi^2} \int_0^{+\infty} \omega_i^2 \sqrt{\omega_i^2 + m_i^2(r)} d\omega_i, \tag{6.9.181}$$

where we have included an additional 4π coming from the angular integration. We use the zeta function regularization method to compute the energy densities ρ_1 and ρ_2. Note that this procedure is completely equivalent to the subtraction procedure of the Casimir energy computation where the zero point energy (ZPE) in different backgrounds with the same asymptotic properties is involved. To this purpose, we introduce the additional mass parameter μ in order to restore the correct dimension for the regularized quantities. Such an arbitrary mass scale emerges unavoidably in any regularization scheme. Then we have

$$\rho_i(\varepsilon) = \frac{1}{16\pi^2} \mu^{2\varepsilon} \int_0^{+\infty} d\omega_i \frac{\omega_i^2}{\left(\omega_i^2 + m_i^2(r)\right)^{\varepsilon - \frac{1}{2}}}, \tag{6.9.182}$$

where

$$\begin{cases} \rho_1 = \frac{1}{16\pi^2} \int_0^{+\infty} \omega_1^2 \sqrt{\omega_1^2 + m_1^2(r)} d\omega_1 \\ \rho_2 = \frac{1}{16\pi^2} \int_0^{+\infty} \omega_2^2 \sqrt{\omega_2^2 + m_2^2(r)} d\omega_2 \end{cases}. \tag{6.9.183}$$

The integration has to be meant in the range where $\omega_i^2 + m_i^2(r) \geq 0$[6]. One gets

$$\rho_i(\varepsilon) = -\frac{m_i^2(r)}{256\pi^2} \left[\frac{1}{\varepsilon} + \ln\left(\frac{\mu^2}{m_i^2(r)}\right) + 2\ln 2 - \frac{1}{2}\right], \tag{6.9.184}$$

$i = 1, 2$. In order to renormalize the divergent ZPE, we observe that from Eq.(6.9.181), after reinserting the gravitational constant, we write

$$\Lambda^{TT} = -8\pi G\left(\rho_1(\varepsilon) + \rho_2(\varepsilon)\right). \tag{6.9.185}$$

To handle with the divergent energy density we extract the divergent part of Λ^{TT}, in the limit $\varepsilon \to 0$ and we set

$$\Lambda^{TT,div} = \frac{G}{32\pi\varepsilon}\left(m_1^4(r) + m_2^4(r)\right). \tag{6.9.186}$$

Thus, the renormalization is performed via the absorption of the divergent part into the re-definition of the bare classical constant Λ^{TT}

$$\Lambda^{TT} \to \Lambda_0^{TT} - \Lambda^{TT,div}. \tag{6.9.187}$$

The remaining finite value for the cosmological constant reads

$$\frac{\Lambda_0^{TT}}{8\pi G} = \frac{1}{256\pi^2}\left\{m_1^4(r)\left[\ln\left(\frac{\mu^2}{m_1^2(r)}\right) + 2\ln 2 - \frac{1}{2}\right]\right.$$

[6] Details of the calculation can be found in the Appendix.

$$+m_2^4(r)\left[\ln\left(\frac{\mu^2}{m_2^2(r)}\right)+2\ln 2-\frac{1}{2}\right]\right\} = \rho_{eff}^{TT}(\mu,r). \tag{6.9.188}$$

The quantity in Eq.(6.9.188) depends on the arbitrary mass scale μ. It is appropriate to use the renormalization group equation to eliminate such a dependence. To this aim, we impose that[41]

$$\frac{1}{8\pi G}\mu\frac{\partial\Lambda_0^{TT}(\mu)}{\partial\mu} = \mu\frac{d}{d\mu}\rho_{eff}^{TT}(\mu,r). \tag{6.9.189}$$

Solving it we find that the renormalized constant Λ_0^{TT} should be treated as a running one in the sense that it varies provided that the scale μ is changing

$$\Lambda_0^{TT}(\mu,r) = \Lambda_0^{TT}(\mu_0,r)+\frac{G}{16\pi}\left(m_1^4(r)+m_2^4(r)\right)\ln\frac{\mu}{\mu_0}. \tag{6.9.190}$$

Substituting Eq.(6.9.190) into Eq.(6.9.188) we find

$$\frac{\Lambda_0^{TT}(\mu_0,r)}{8\pi G} = \frac{1}{256\pi^2}\left\{m_1^4(r)\left[\ln\left(\frac{m_1^2(r)}{\mu_0^2}\right)-2\ln 2+\frac{1}{2}\right]\right. \tag{6.9.191}$$

$$\left.+m_2^4(r)\left[\ln\left(\frac{m_2^2(r)}{\mu_0^2}\right)-2\ln 2+\frac{1}{2}\right]\right\}. \tag{6.9.192}$$

In order to fix the dependence of Λ on r and M, we find the minimum of $\Lambda_0^{TT}(\mu_0,r)$. To this aim, last equation can be cast into the form[7]

$$\frac{\Lambda_0^{TT}(\mu_0,r)}{8\pi G} = \frac{\mu_0^4}{256\pi^2}\left\{x^2(r)\left[\ln\left(\frac{|x(r)|}{4}\right)+\frac{1}{2}\right]\right. \tag{6.9.193}$$

$$\left.+y^2(r)\left[\ln\left(\frac{y(r)}{4}\right)+\frac{1}{2}\right]\right\}, \tag{6.9.194}$$

where $x(r) = \pm m_1^2(r)/\mu_0^2$ and $y(r) = m_2^2(r)/\mu_0^2$. Now we find the extrema of $\Lambda_0(\mu_0; x(r), y(r))$ in the range $\frac{5MG}{2} \leq r$ and we get

$$\begin{cases} x(r) = 0 \\ y(r) = 0 \end{cases}, \tag{6.9.195}$$

which is never satisfied and

$$\begin{cases} x(r) = 4/e \\ y(r) = 4/e \end{cases} \quad\Longrightarrow\quad \begin{cases} m_1^2(r) = 4\mu_0^2/e \\ m_2^2(r) = 4\mu_0^2/e \end{cases}. \tag{6.9.196}$$

---

[7]Recall Eqs.(6.8.176, 6.8.177), showing a change of sign in $m_1^2(r)$. Even if this is not the most appropriate notation to indicate a change of sign in a quantity looking like a "*square effective mass*", this reveals useful in the zeta function regularization and in the serch for extrema.

which implies $M = 0$ and $\bar{r} = \sqrt{3e}/2\mu_0$. On the other hand, in the range $2MG \leq r < \frac{5MG}{2}$, we get again

$$\begin{cases} x(r) = 0 \\ y(r) = 0 \end{cases}, \tag{6.9.197}$$

which has no solution and

$$\begin{cases} -m_1^2(r) = 4\mu_0^2/e \\ m_2^2(r) = 4\mu_0^2/e \end{cases}, \tag{6.9.198}$$

which implies

$$\begin{cases} \bar{M} = 4\mu_0^2\bar{r}^3/3eG \\ \bar{r} = \sqrt{6e}/4\mu_0 \end{cases}. \tag{6.9.199}$$

Eq.(6.9.194) evaluated on the minimum, now becomes

$$\Lambda_0^{TT}\left(\bar{M}, \bar{r}\right) = -\frac{\mu_0^4 G}{2e^2\pi}. \tag{6.9.200}$$

It is interesting to note that thanks to the renormalization group equation (6.9.189), we can directly compute Λ_0^{TT} at the scale μ_0 and only with the help of Eq.(6.9.190), we have access at the scale μ.

6.10 Summary and Conclusions

In this paper we have analyzed, the one loop contribution coming from quantum fluctuations around a background metric describing a Schwarzschild wormhole which has boundary terms in both wedges of the Penrose diagram. Usually, this kind of analysis is performed only in the R_+ wedge of diagram (6.3.2). However, due to the symmetry property of transformation (6.2.13), a contribution from region R_- is not unexpected. Moreover, the possibility of a fine tuning between boundaries in the opposite regions gives the opportunity to vanish the classical gravitational action. The vanishing of the classical action leads to the interpretation that we are dealing with flat space. On the other hand the presence in each wedge of a term proportional to the ADM mass is a signal that a non flat configuration has been taken under examination. An interpretation of this puzzling situation can be given in terms of black hole pair creation mediated by a wormhole, with the pair elements residing in the different universes. This proposal is not completely new and it has been investigated in Refs.[42] in a Hamiltonian formulation, where a foliation of the manifold $\mathcal{M}$ is crucial. On the other hand, the action formalism has the advantage of being fully covariant. However, the vanishing of the classical action is not sufficient to establish if spacetime will produce a pair or will persist in its flat configuration. It is, therefore necessary to discover if the one loop approximation will produce an instability. The variational approach we have used for the Sturm-Liouville problem in the TT sector has shown an unstable mode in the s-wave approximation in both wedges. This is the signal that flat space is not only unstable when the space is furnished with a temperature $T = 1/(8\pi MG)$, but is even unstable with respect to Schwinger pair creation. One possible conclusion is that flat space can be no more considered as the general accepted vacuum of quantum gravity. The other aspect is that, instead of rejecting this result, we can further investigate the meaning of such an instability in a restricted bounding box. Indeed, we know that a partial success in this direction has been obtained by Allen[6] in terms of hot quantum gravity, showing that the instability can disappear if the bounding box is restricted enough. To further proceed, we have considered how to extract information on the cosmological constant using the Wheeler-De Witt equation. In particular, by means of a variational approach and a orthogonal decomposition of the modes, we have studied the contribution of the transverse-traceless tensors and the trace in a Schwarzschild background. The use of the zeta function and a renormalization group equation have led to Eq.(6.9.200) and recalling Eq.(6.7.160), we have obtained

$$\Lambda_0^{\perp}\left(\bar{M}, \bar{r}\right) = \frac{\mu_0^4 G}{2e^2\pi}. \tag{6.10.201}$$

If we choose to fix the renormalization point $\mu_0 = m_p$, we obtain approximately $\Lambda_0^{\perp}\left(\bar{M}, \bar{r}\right) \simeq 10^{37} GeV^2$ which, in terms of energy density is in agreement with the estimate of Eq.(6.1.4). Once fixed the scale μ_0, we can see what happens at the cosmological constant at the scale μ, by means of Eq.(6.9.190). What we see is that the cosmological constant is vanishing at the sub-planckian scale $\mu = m_p \exp\left(-\frac{1}{4}\right)$, but unfortunately is a scale which is very far from the nowadays observations. However, our analysis is not complete. Indeed, the appearance of a nonconformal unstable mode, like the one computed in Section 6.5 does not mean that we have to reject the solution. In fact in Ref.[42], we have shown how to cure such a problem. In that context, a model of "space-time foam" has been introduced in a large N wormhole approach reproducing a correct decreasing of the cosmological constant and simultaneously a stabilization of the system under examination. Unfortunately in that approach a renormalization scheme was missing and a W.K.B. approximation on the wave function has been used to recover a Schrödinger-like equation. The possible next step is to repeat the scheme we have adopted here in a large N context, to recover the correct vanishing behavior of the cosmological constant.

A. Searching for Eigenvalues with the Rayleigh-Ritz method

In this Appendix, we will explicitly solve the eigenvalue problem of Eq.(6.5.98) with the Rayleigh-Ritz method. We here report two principal choices of trial wave functions:

1. $H_1^{\pm}(y) = \exp(\mp\lambda y) \rightarrow \exp(-\lambda r)$. For this choice Eq.(6.5.98) becomes,

$$\mu^2(\lambda) = \lambda^2 - 8MG\frac{\int_{2MG}^{+\infty} dr e^{-2\lambda r}\frac{r^3(r-2MG)}{(r-3MG)^3}}{\int_{2MG}^{+\infty} dr e^{-2\lambda r}\frac{r^5(r-2MG)}{(r-3MG)^2}} = \lambda^2 - \frac{A(\lambda)}{B(\lambda)}, \tag{6.10.202}$$

where

$$\begin{aligned} A(\lambda) &= 8MG\int_{2MG}^{+\infty} dr e^{-2\lambda r}\frac{r^3(r-2MG)}{(r-3MG)^3} \\ &= 256(MG)^3\int_1^{+\infty} d\rho \exp\left(-2\tilde{\lambda}\rho\right)\frac{\rho^3(\rho-1)}{(2\rho-3)^3} \\ &= 256(MG)^3\left[I_3^4\left(\tilde{\lambda}\right) - I_3^3\left(\tilde{\lambda}\right)\right] \end{aligned} \tag{6.10.203}$$

and

$$\begin{aligned} B(\lambda) &= \int_{2MG}^{+\infty} dr e^{-2\lambda r}\frac{r^5(r-2MG)}{(r-3MG)^2} \\ &= 128(MG)^5\int_1^{+\infty} d\rho \exp\left(-2\tilde{\lambda}\rho\right)\frac{\rho^5(\rho-1)}{(2\rho-3)^2} \\ &= 128(MG)^5\left[I_2^6\left(\tilde{\lambda}\right) - I_2^5\left(\tilde{\lambda}\right)\right]. \end{aligned} \tag{6.10.204a}$$

We have defined $\rho = r/(2MG)$, $\tilde{\lambda} = \lambda 2MG$ and introduced the following function

$$I_m^n\left(\tilde{\lambda}\right) = \int_1^{+\infty} d\rho \exp\left(-2\tilde{\lambda}\rho\right)\frac{\rho^n}{(2\rho-3)^m} = (-1)^n\frac{d^n I_m^0\left(\tilde{\lambda}\right)}{d\tilde{\lambda}^n}. \tag{6.10.205}$$

$I_m^0\left(\tilde{\lambda}\right)$ can be easily integrated with the formula[43]

$$\int_u^{+\infty} dx \frac{\exp(-\nu x)}{(x+\beta)^m} = \tag{6.10.206}$$

$$\exp(-\nu u)\sum_{k=1}^{m-1}\frac{(k-1)!}{(m-1)!}\frac{(-\nu)^{m-k-1}}{(u+\beta)^k} - \frac{(-\nu)^{m-1}}{(m-1)!}\exp(\beta\nu)\,\mathrm{Ei}\left[-(u+\beta)\nu\right]$$

with $m \geq 2$, $|\arg(u+\beta)| < \pi$ and $\mathrm{Re}\,\nu > 0$. $\mathrm{Ei}(x)$ is the exponential function. We get $A(\lambda) =$

$$\frac{\left(-18\lambda e^{\lambda} - 108\lambda^3\,\mathrm{Ei}(\lambda) + 72\lambda^2\,\mathrm{Ei}(\lambda) + 81\lambda^2 e^{\lambda} - 27\lambda^3 e^{\lambda} + 27\lambda^4\,\mathrm{Ei}(\lambda) - 2e^{\lambda}\right)e^{-3\lambda}}{8\lambda^2} \tag{6.10.207}$$

and $B(\lambda) =$

$$\frac{\left(24e^{\lambda} - 243\lambda^5 e^{\lambda} + 243\lambda^6\,\mathrm{Ei}(\lambda) - 648\lambda^5\,\mathrm{Ei}(\lambda) + 194\lambda^3 e^{\lambda} + 72\lambda e^{\lambda} + 126\lambda^2 e^{\lambda} + 405\lambda^4 e^{\lambda}\right)e^{-3\lambda}}{128\lambda^5} \tag{6.10.208}$$

$B(\lambda)$ vanishes for $\lambda = 1.360771766$ and $\mu^2(\lambda)$ gets its minimum for $\lambda_{\min} = .4471243737$ leading to $\mu^2(\lambda_{\min}) = -.1374637652$

2. With the help of Eq.(6.10.207) and choosing $H_1^{\pm}(y) = \exp(\mp\lambda y)(r(y) - 3MG) \rightarrow \exp(-\lambda r)(r - 3MG)$ we get a more complicated form in $N[H_1(y);\lambda]$ of Eq.(6.5.98), while $D[H_1(y);\lambda]$ is simpler. Indeed

$$N[H_1(y);\lambda] = -\int_{2MG}^{+\infty} dr \frac{r^5(r-2\,MG)}{(r-3\,MG)^2}\left(\frac{d}{dr}\left[e^{-2\lambda r}(r-3MG)\right]\right)^2$$
$$+8MG\int_{2MG}^{+\infty} dr e^{-2\lambda r}\frac{r^3(r-2MG)}{(r-3MG)}$$
$$= -\frac{(2MG)^7}{(2MG)^2}\int_1^{+\infty} d\rho\frac{\rho^5(\rho-1)}{(2\rho-3)^2}\left(\frac{d}{d\rho}\left[e^{-2\tilde{\lambda}\rho}(2\rho-3)\right]\right)^2$$
$$+8MG\frac{(2MG)^5}{MG}\int_1^{+\infty} d\rho e^{-2\tilde{\lambda}\rho}\frac{\rho^3(\rho-1)}{(2\rho-3)}$$
$$= -\,(2MG)^5$$
$$\times\left\{\frac{\left(243\,\lambda^6 Ei(\lambda) - 239\,\lambda^5 e^{\lambda} + 42\,e^{\lambda} + 225\,\lambda^4 e^{\lambda} - 432\,\lambda^5 Ei(\lambda)\right)e^{-3\,\lambda}}{16\lambda^5}\right.$$
$$\left. + \frac{\left(86\,\lambda^2 e^{\lambda} + 102\,\lambda^3 e^{\lambda} + 78\,\lambda\, e^{\lambda}\right)e^{-3\,\lambda}}{16\lambda^5}\right\} \qquad (6.10.209)$$

and

$$D[H_1(y);\lambda] = \int_{2MG}^{+\infty} dr\left[e^{-2\lambda r} r^5(r-2\,MG)\right]$$
$$= (2MG)^7\int_1^{+\infty} d\rho \exp\left(-2\tilde{\lambda}\rho\right)\rho^5(\rho-1)$$
$$= (2MG)^7\,\frac{e^{-2\lambda}\left(45 + 2\,\lambda^5 + 10\,\lambda^4 + 30\,\lambda^3 + 60\,\lambda^2 + 75\,\lambda\right)}{8\lambda^7}. \qquad (6.10.210)$$

Note the positiveness of $D[H_1(y);\lambda]$ for every λ, ensuring a correct normalization. The minimum of $\mu^2(\lambda)$ is reached in this case for $\lambda_{\min} = .9223711858$ and $\mu^2(\lambda_{\min}) = -.3827139389$.

B. The zeta function regularization

In this appendix, we report details on computation leading to expression (6.9.182). We begin with the following integral

$$\rho(\varepsilon) = \begin{cases} I_+ = \mu^{2\varepsilon}\int_0^{+\infty} d\omega\frac{\omega^2}{(\omega^2+m^2(r))^{\varepsilon-\frac{1}{2}}} \\ \\ I_- = \mu^{2\varepsilon}\int_0^{+\infty} d\omega\frac{\omega^2}{(\omega^2-m^2(r))^{\varepsilon-\frac{1}{2}}} \end{cases}, \qquad (6.10.211)$$

with $m^2(r) > 0$.

I_+ computation

If we define $t = \omega/\sqrt{m^2(r)}$, the integral I_+ in Eq.(6.10.211) becomes

$$\rho(\varepsilon) = \mu^{2\varepsilon} m^{4-2\varepsilon}(r) \int_0^{+\infty} dt \frac{t^2}{(t^2+1)^{\varepsilon-\frac{1}{2}}}$$

$$= \frac{1}{2}\mu^{2\varepsilon} m^{4-2\varepsilon}(r) B\left(\frac{3}{2}, \varepsilon - 2\right)$$

$$\frac{1}{2}\mu^{2\varepsilon} m^{4-2\varepsilon}(r) \frac{\Gamma\left(\frac{3}{2}\right)\Gamma(\varepsilon-2)}{\Gamma\left(\varepsilon-\frac{1}{2}\right)}$$

$$= \frac{\sqrt{\pi}}{4} m^4(r) \left(\frac{\mu^2}{m^2(r)}\right)^{\varepsilon} \frac{\Gamma(\varepsilon-2)}{\Gamma\left(\varepsilon-\frac{1}{2}\right)}, \qquad (6.10.212)$$

where we have used the following identities involving the beta function

$$B(x,y) = 2\int_0^{+\infty} dt \frac{t^{2x-1}}{(t^2+1)^{x+y}} \qquad \mathrm{Re}\, x > 0, \mathrm{Re}\, y > 0 \qquad (6.10.213)$$

related to the gamma function by means of

$$B(x,y) = \frac{\Gamma(x)\Gamma(y)}{\Gamma(x+y)}. \qquad (6.10.214)$$

Taking into account the following relations for the Γ-function

$$\begin{array}{c} \Gamma(\varepsilon-2) = \frac{\Gamma(1+\varepsilon)}{\varepsilon(\varepsilon-1)(\varepsilon-2)} \\ \Gamma\left(\varepsilon-\frac{1}{2}\right) = \frac{\Gamma\left(\varepsilon+\frac{1}{2}\right)}{\varepsilon-\frac{1}{2}} \end{array}, \qquad (6.10.215)$$

and the expansion for small ε

$$\begin{array}{rl} \Gamma(1+\varepsilon) = & 1-\gamma\varepsilon+O\left(\varepsilon^2\right) \\ \Gamma\left(\varepsilon+\frac{1}{2}\right) = & \Gamma\left(\frac{1}{2}\right)-\varepsilon\Gamma\left(\frac{1}{2}\right)(\gamma+2\ln 2)+O\left(\varepsilon^2\right) \\ x^{\varepsilon} = & 1+\varepsilon\ln x+O\left(\varepsilon^2\right) \end{array}, \qquad (6.10.216)$$

where γ is the Euler's constant, we find

$$\rho(\varepsilon) = -\frac{m^4(r)}{16}\left[\frac{1}{\varepsilon}+\ln\left(\frac{\mu^2}{m^2(r)}\right)+2\ln 2-\frac{1}{2}\right]. \qquad (6.10.217)$$

I_- computation

If we define $t = \omega/\sqrt{m^2(r)}$, the integral I_- in Eq.(6.10.211) becomes

$$\rho(\varepsilon) = \mu^{2\varepsilon} m^{4-2\varepsilon}(r) \int_0^{+\infty} dt \frac{t^2}{(t^2-1)^{\varepsilon-\frac{1}{2}}}$$

$$= \frac{1}{2}\mu^{2\varepsilon} m^{4-2\varepsilon}(r)\, B\left(\varepsilon-2, \frac{3}{2}-\varepsilon\right)$$

$$\frac{1}{2}\mu^{2\varepsilon} m^{4-2\varepsilon}(r) \frac{\Gamma\left(\frac{3}{2}-\varepsilon\right)\Gamma(\varepsilon-2)}{\Gamma\left(-\frac{1}{2}\right)}$$

$$= -\frac{1}{4\sqrt{\pi}} m^4(r) \left(\frac{\mu^2}{m^2(r)}\right)^{\varepsilon} \Gamma\left(\frac{3}{2}-\varepsilon\right)\Gamma(\varepsilon-2), \tag{6.10.218}$$

where we have used the following identity involving the beta function

$$\frac{1}{p}B\left(1-\nu-\frac{\mu}{p},\nu\right) = \int_1^{+\infty} dt t^{\mu-1}\left(t^p-1\right)^{\nu-1}$$

$$p>0, \mathrm{Re}\,\nu>0, \mathrm{Re}\,\mu<p-p\,\mathrm{Re}\,\nu \tag{6.10.219}$$

and the reflection formula

$$\Gamma(z)\Gamma(1-z) = -z\Gamma(-z)\Gamma(z) \tag{6.10.220}$$

From the first of Eqs.(6.10.215) and from the expansion for small ε

$$\Gamma\left(\frac{3}{2}-\varepsilon\right) = \Gamma\left(\frac{3}{2}\right)\left(1-\varepsilon\left(-\gamma-2\ln 2+2\right)\right)+O\left(\varepsilon^2\right)$$

$$x^{\varepsilon} = 1+\varepsilon\ln x+O\left(\varepsilon^2\right), \tag{6.10.221}$$

we find

$$\rho(\varepsilon) = -\frac{m^4(r)}{16}\left[\frac{1}{\varepsilon}+\ln\left(\frac{\mu^2}{m^2(r)}\right)+2\ln 2-\frac{1}{2}\right]. \tag{6.10.222}$$

References

[1] R. Schoen, S.T. Yau, *Commun. Math. Phys.* **65**, 45 (1979); *Commun. Math. Phys.* **79**, 231 (1981).

[2] R. Schoen and S.T. Yau, *Phys. Rev. Lett.* **42**, 547 (1979).

[3] E. Witten, *Commun. Math. Phys.* **80**, 381 (1981).

[4] D.J. Gross, M.J. Perry, & L.G. Yaffe, *Phys. Rev.* **D 25** 330 (1982).

[5] S. Coleman, *Nucl. Phys.* **B 298** (1988), 178; S. Coleman, *Aspects of Symmetry* (Cambridge University Press, Cambridge, 1985).

[6] B. Allen, *Phys. Rev.* **D 30** 1153 (1984).

[7] R. Bousso and S.W. Hawking, *Phys. Rev.* **D 52**, 5659 (1995), gr-qc/9506047; R. Bousso and S.W. Hawking, *Phys. Rev.* **D 54**, 5659 (1995), gr-qc/9606052.

[8] P. Ginsparg and M.J. Perry, *Nucl. Phys.* **B 222** (1983) 245.

[9] R.E. Young, *Phys. Rev.* **D 28, (**1983) 2436; R.E. Young, *Phys. Rev.* **D 28** (1983) 2420.

[10] M.S. Volkov and A. Wipf, *Nucl. Phys.* **B 582** (2000), 313; hep-th/0003081.

[11] S.W. Hawking and D.N. Page, *Commun. Math. Phys.* **87**, 577 (1983).

[12] T. Prestidge, *Phys. Rev.* **D 61** (2000) 084002, hep-th/9907163.

[13] E. Witten, *Nucl Phys* **B195** 481 (1982).

[14] R. Gregory and R. Laflamme, *Phys. Rev.* **D 37** 305 (1988).

[15] S.S. Gubser and I. Mitra, *Instability of charged black holes in Anti-de Sitter space*, hep-th/0009126; S.S. Gubser and I. Mitra, *JHEP* **8** (2001) 18. J.P. Gregory and S.F. Ross, *Phys. Rev.* **D 64** (2001) 124006, hep-th/0106220. H.S. Reall, *Phys. Rev.* **D 64** (2001) 044005, hep-th/0104071. G. Gibbons and S.A. Hartnoll, *Phys. Rev.* **D 66** (2001) 064024, hep-th/0206202. I.P. Neupane, *Thermodynamic and Gravitational Instability on Hyperbolic Spaces*, hep-th/0302132.

[16] O.J.C. Dias and J.P. Lemos, *Phys. Rev.* **D 69**, 084006 (2004); hep-th/0310068.

[17] R. Arnowitt, S. Deser, and C. W. Misner, in *Gravitation: An Introduction to Current Research,* edited by L. Witten (John Wiley & Sons, Inc., New York, 1962); B. S. DeWitt, *Phys. Rev.* **160**, 1113 (1967).

[18] C.W. Misner, K.S. Thorne and J.A. Wheeler, *Gravitation* (Freeman, San Francisco, 1973) 842.

[19] S.W. Hawking and G.F.R. Ellis, *The Large Scale Structure of Spacetime* (Cambridge Univ. Press, 1973).

[20] M.D. Kruskal, *Phys. Rev.* **119**, (1960) 1743; G. Szekeres, Publ. Math. Debreceni **7** (1960) 285.

[21] V.P. Frolov and E.A. Martinez, *Class. Quant. Grav.* 13 (1996) 481, gr-qc/9411001.

[22] E.A. Martinez, *Phys. Rev.* **D 51**, 5732 (1995).

[23] S.W Hawking, The Path-Integral Approach to Quantum Gravity, in *General Relativity: An Einstein Centenary Survey*, Ed. S.W Hawking & W. Israel (Cambridge University Press, Cambridge, 1979).

[24] G.W Gibbons, S.W Hawking, & M.J. Perry, *Nucl Phys* **B138** 141 (1978).

[25] T. Regge and J. A. Wheeler, *Phys. Rev.* **108**, 1063 (1957).

[26] C.V. Vishveshwara, *Phys. Rev.* **D 1**, 2870 (1970).

[27] F. Zerilli, *Phys. Rev. Lett.* **24**, 737 (1970).

[28] W. Press and S.A. Teukolsky, *Astrophys. J.* **185**, 649 (1973); **193**, 443 (1974).

[29] J.M. Stewart, *Proc. R. Soc. London* **A 334**, 51 (1975).

[30] S. Chandrasekhar, *The Mathematical Theory of Black Holes,* (Oxford Science Publications, 1983).

[31] G. W. Gibbons and S.W. Hawking, *Phys. Rev.* **D 15** 2752, (1977).

[32] J.D. Brown and J.W. York, *Phys. Rev.* **D 47**, 1407 (1993).

[33] For a pioneering review on this problem see S. Weinberg, *Rev. Mod. Phys.* **61**, 1 (1989). For more recent and detailed reviews see V. Sahni and A. Starobinsky, *Int. J. Mod. Phys.* **D 9**, 373 (2000), astro-ph/9904398; N. Straumann, *The history of the cosmological constant problem* gr-qc/0208027; T.Padmanabhan, *Phys.Rept.* **380**, 235 (2003), hep-th/0212290.

[34] B. S. DeWitt, *Phys. Rev.* **160**, 1113 (1967).

[35] M. Berger and D. Ebin, *J. Diff. Geom.* **3**, 379 (1969).

[36] J. W. York Jr., *J. Math. Phys.*, **14**, 4 (1973); *Ann. Inst. Henri Poincaré* **A 21**, 319 (1974).

[37] P. O. Mazur and E. Mottola, *Nucl. Phys.* **B 341**, 187 (1990).

[38] D.V. Vassilevich, *Int. J. Mod. Phys.* **A 8**, 1637 (1993). D. V. Vassilevich, *Phys. Rev.* **D 52** 999 (1995); gr-qc/9411036.

[39] D.J. Gross, M.J. Perry and L.G. Yaffe, *Phys. Rev.* **D 25**, (1982) 330. B. Allen, *Phys. Rev.* **D 30** 1153 (1984). E. Witten, *Nucl Phys* **B 195** 481 (1982). P. Ginsparg and M.J. Perry, *Nucl. Phys.* **B 222** (1983) 245. R.E. Young, *Phys. Rev.* **D 28,** (1983) 2436. R.E. Young, *Phys. Rev.* **D 28** (1983) 2420. S.W. Hawking and D.N. Page, *Commun. Math. Phys.* **87**, 577 (1983). R. Gregory and R. Laflamme, *Phys. Rev.* **D 37** 305 (1988). R. Garattini, *Int. J. Mod. Phys.* **A 14**, 2905 (1999); gr-qc/9805096. E Elizalde, S Nojiri and S D Odintsov, *Phys. Rev.* **D 59** (1999) 061501; hep-th 9901026. M.S. Volkov and A. Wipf, *Nucl. Phys.* **B 582** (2000), 313; hep-th/0003081. R. Garattini, *Class. Quant. Grav.* **17**, 3335 (2000); gr-qc/0006076. T. Prestidge, *Phys. Rev.* **D 61**, (2000) 084002; hep-th/9907163. S.S. Gubser and I. Mitra, *Instability of charged black holes in Anti-de Sitter space*, hep-th/0009126; R. Garattini, *Class. Quant. Grav.* **18**, 571 (2001); gr-qc/0012078. S.S. Gubser and I. Mitra, *JHEP* **8** (2001) 18. J.P. Gregory and S.F. Ross, *Phys. Rev.* **D 64** (2001) 124006, hep-th/0106220. H.S. Reall, *Phys. Rev.* **D 64** (2001) 044005, hep-th/0104071. G. Gibbons and S.A. Hartnoll, *Phys. Rev.* **D 66** (2001) 064024, hep-th/0206202.

[40] A. K. Kerman and D. Vautherin, *Ann. Phys.*, **192**, 408 (1989); J. M. Cornwall, R. Jackiw and E. Tomboulis, *Phys. Rev.* **D 8**, (1974) 2428; R. Jackiw in *Séminaire de Mathématiques Supérieures, Montréal, Québec, Canada- June 1988 - Notes by P. de Sousa Gerbert*; M. Consoli and G. Preparata, *Phys. Lett.* **B**, **154**, 411 (1985).

[41] J.Perez-Mercader and S.D. Odintsov, *Int. J. Mod. Phys.* **D 1**, 401 (1992). I.O. Cherednikov, *Acta Physica Slovaca*, **52**, (2002), 221. I.O. Cherednikov, *Acta Phys. Polon.* **B 35**, 1607 (2004). M. Bordag, U. Mohideen and V.M. Mostepanenko, *Phys. Rep.* **353**, 1 (2001). Inclusion of non-perturbative effects, namely beyond one-loop, in de Sitter Quantum Gravity have been discussed in S. Falkenberg and S. D. Odintsov, *Int. J. Mod. Phys.* **A 13**, 607 (1998); hep-th 9612019.

[42] R. Garattini, *Nuovo Cimento* **B 113** (1998) 963, gr-qc/9609006. R. Garattini, *Mod. Phys. Lett.* **A 13** (1998) 159, gr-qc/9801045. R. Garattini, *Int.*

J. Mod. Phys. **A 14** (1999) 2905, gr-qc/9805096. R. Garattini, *Class. Quant. Grav.* **18** (2001) 571, gr-qc/0012078. R. Garattini, *Nucl. Phys. Proc. Suppl.* **88**, 297 (2000), gr-qc/9910037. R. Garattini, *Int. J. Mod. Phys.* **D 4**, 635 (2002); Archive:gr-qc/0003090. R. Garattini, *Int. J. Mod. Phys.* **A 17**, 1965 (2002), Archive: gr-qc/0111068.

[43] I.S. Gradshteyn and I.M. Ryzhik, *Table of Integrals, Series, and Products* (corrected and enlarged edition), edited by A. Jeffrey (Academic Press, Inc.).

In: Focus on Quantum Gravity Research
Editor: David C. Moore, pp. 261-310

ISBN 1-59454-660-6

Chapter 7

Manifestly Covariant Canonical Quantization of Gravity and Diffeomorphism Anomalies in Four Dimensions

T.A. Larsson
Vanadisvägen 29, S-113 23 Stockholm, Sweden
email: thomas.larsson@hdd.se

Canonical quantization of gravity requires knowledge about the representation theory of its constraint algebra, which is physically equivalent to the algebra of arbitrary 4-diffeomorphisms. All interesting lowest-energy representations are projective, making the relevant algebra into a four-dimensional generalization of the Virasoro algebra. Such diffeomorphism anomalies are invisible in field theory, because the relevant cocycles are functionals of the observer's trajectory in spacetime. The multi-dimensional Virasoro algebra acts naturally in the phase space of arbitrary histories, with dynamics playing the role of first-class constraints. General relativity is regularized by expanding all fields in Taylor series around the observer's trajectory, and truncating at some fixed order. This regularized but manifestly general-covariant theory is quantized in the history phase space, and dynamics is imposed afterwards, in analogy with BRST quantization. Infinities arise when the regularization is removed; it is presently unclear how these should be dealt with.

7.1 Introduction

After its invention by Dirac in 1927, local relativistic quantum field theory (QFT) has undergone three major crises.

The first crisis occurred in the 1930s, when it became clear that quantum mechanics (QM) was naïvely incompatible with special relativity, and in particular that quantization of electromagnetism gave infinite answers. This led many people to believe that radical new physics was needed. Alas, the problems were solved by renormalization, which involves no new physics, but merely a reinterpretation (albeit a radical one) of old physics.

The second crisis occurred in the 1960s, when QFT appeared to be incapable of describing the strong and weak interactions, in particular asymptotic freedom. Therefore many people turned to other, allegedly more fundamental ideas, such as analyticity of the S-matrix or string theory. However, it again turned out that straightforward QFT was the correct answer once it was applied to non-abelian Yang-Mills theory.

The third crisis is of course the apparent incompatibility between QM and gravity. This problem was realized already around 1960, when it was found that perturbative quantum gravity is not renormalizable and hence not a predictive theory, but it only became acute when the standard model was completed 25 years ago, and gravity remained the only interaction not described by a consistent QFT. As during the previous crises, the difficulties with gravity has led many people to abandon QFT in favor of more fashionable ideas.

The philosophy of this author is that quantum gravity should also be described by straightforward QFT, but that a small but profound fix is needed. The fix should be small, in order to preserve the experimental successes of the standard model and general relativity, but it must also be profound, because otherwise it would already have been found. This philosophy is in many ways similar to renormalization, which can be regarded as a small but profound fix which made QM compatible with locality.

The key idea in the present work is that we need to quantize not only the fields, but also the observer's trajectory in spacetime. In non-relativistic QM, observation is a complicated, non-local process which assigns numbers to experiments. However, in a relativistic theory, a process must be localized; it happens *somewhere*. In order to maintain locality, we must assign the process of observation to some definite event in spacetime. As time proceeds, the observer (or detector or test particle) traces out a curve $q(t)$ in spacetime. Like the quantum fields, the observer's trajectory should be treated as a material, quantized object; it has a conjugate momentum, it is represented on the Hilbert space, etc.

To consider the observer's trajectory as a material object is certainly a very small fix, which does not introduce any new physical ideas. One may wonder why such a small, almost trivial modification should be important. The reason is that it makes it possible for new types of anomalies to arise. If the field theory has a gauge symmetry of Yang-Mills type, there is a gauge anomaly proportional to the quadratic Casimir, and the gauge (or current) algebra becomes a higher-dimensional generalization of affine Kac-Moody algebras. Similarly, a general-covariant theory, in any number of dimensions, acquires a diffeomorphism anomaly, which is described by a higher-dimensional generalization of the Virasoro algebra. The reason why these anomalies can not be seen in conventional QFT, without explicit reference to the observer, is that the relevant cocycles are functionals of the observer's trajectory. If this trajectory has not been introduced, it is of course impossible to write down the relevant anomalies.

At this point it is necessary to discuss the issue of gauge anomalies and consistency, in particular unitarity. Gauge and diffeomorphism anomalies are usually considered as a sign of inconsistency and should therefore be cancelled [1, 21]. There is ample evidence that this is the correct prescription for conventional gauge anomalies, arising from chiral fermions coupled to gauge fields. However, the anomalies discussed in this paper are of a completely different type, depending on the observer's trajectory, and intuition derived from conventional anomalies needs not apply. Instead, the situation is similar to conformal field theory applied to two-dimensional statistical physics, where it is well known that infinite spacetime symmetry (gauge or not) is compatible with locality (in the sense of correlation functions depending on separation) only in the presence of a conformal anomaly. Hence gauge anomalies are viewed as a means to gauge symmetry breaking.

Be that as it is. Even if we would want to mod out the new gauge and diff anomalies (which this author believes is wrong), we still need to know about them. A good analogy is bosonic string theory, whose conformal anomaly cancels precisely in 26 dimensions. However, if we did not know about conformal anomalies, there would be no condition that singled out the number 26. Similarly, we need to know about the new anomalies that arise when we quantize gauge-invariant or diff-invariant theories in the presence of the observer's trajectory, even if all we wanted to do with these anomalies were to cancel them.

Anomalies manifest themselves as extensions of the constraint algebras. Extensions of the diffeomorphism algebra were classified by Dzhumadildaev [7], and their representation theory was developed in [2, 3, 4, 5, 14, 15, 16, 23]; see [17] for a recent review. Unfortunately, the canonical formalism is not very well suited for quantization of relativistic theories, because the foliation of spacetime into fixed time slices breaks manifest covariance. This problem becomes especially serious in general-covariant theories, where the very constraint algebra is modified, from the algebra of arbitrary 4-diffeomorphisms into the Dirac algebra of constraints. However, this modification is a consequence of the chosen formalism rather than a true physical effect, and in covariant approaches the constraint algebra of general relativity is indeed the 4-diffeomorphism algebra.

It is well known that phase space is a covariant concept; it is the space of histories which solve the dynamics. Each phase space point (q, p) generates a unique history $(q(t), p(t))$ under Hamiltonian evolution, and thus we may view $(q, p) = (q(0), p(0))$ as a particular coordinatization of phase space. We can now understand how the Dirac algebra arises. A canonical transformation in the history phase space maps the history $(q(t), p(t)) \to (q'(t'), p'(t'))$. If in particular

$$(q, p) = (q(0), p(0)) \stackrel{\text{symm}}{\longrightarrow} (q'(0), p'(0)) = (q', p'),$$

the transformation preserves the standard coordinatization; this is the situation with spatial diffeomorphisms in general relativitiy. However, temporal diffeomorphisms break the standard coordinatization, and if we insist on keeping it, we must add a compensating transformation to move back to the $t = 0$ surface. Schematically,

$$(q, p) = (q(0), p(0)) \stackrel{\text{symm}}{\longrightarrow} (q'(t), p'(t)) \stackrel{\text{comp}}{\longrightarrow} (q'(0), p'(0)) = (q', p').$$

The combination of symmetry and compensating transformations generate the Dirac algebra.

It is clear that we can avoid this complication if we work in the history phase space directly, because then we do not need to worry about compensating transformations. To this end, a novel quantization scheme has recently been proposed, called manifest covariant canonical quantization (MCCQ) [18, 19]. The idea is to make the space of arbitrary histories $(q(t), p(t))$ into a phase space $\mathcal{P}$ by defining the Poisson brackets

$$[p(t), q(t')] = \delta(t - t'), \qquad [p(t), p(t')] = [q(t), q(t')] = 0.$$

The Euler-Lagrange equations now define a constraint $\mathcal{E}(t) \approx 0$ in $\mathcal{P}$; since $\mathcal{E}(t)$ only depends on $q(t)$ this constraint is first class. This observation allows us to apply powerful cohomological methods from BRST quantization of theories with first class constraints. In other words, the idea in MCCQ is to quantize in the history phase space first and to impose dynamics afterwards, by passing to cohomology. Since dynamics is regarded as a constraint, the cohomology is nontrivial even for systems without gauge symmetries, like the harmonic oscillator and the free scalar field.

It is important to understand which ideas are crucial and which are merely convenient. The formulation of QFT using MCCQ is "just formalism". It is very convenient to have a canonical quantization scheme which respects general covariance, but it is probably possible to reexpress the results in this paper using non-covariant canonical quantization, at the price of great complications. In contrast, adding the observer's trajectory to the quantum fields is absolutely essential. Without it, the new observer-dependent anomalies can not be formulated, and this is a hard obstruction to quantization.

History methods have recently been advocated by Savvidou and Isham [11, 24, 25]; in particular, the last reference contains a very good summary of the conceptual problems involved in non-covariant canonical quantization. Their formalism differs in details from MCCQ, e.g. because they do not use cohomological methods. There is also a substantial difference, namely that the observer's trajectory is not introduced, and hence no diffeomorphism anomalies are seen.

Finally, let us emphasize the philosophical motivation for introducing MCCQ. The representations considered here are of lowest-energy type, i.e. there is a natural Hamiltonian whose eigenvalues are bounded from below. This is the kind of representations expected to be relevant to quantum theory. However, they do not look like standard formulations of QFT, because their natural habitat is in history space. To apply them to physics, we must first recast physics in a suitable, history-oriented form. It is the same argument that leads us to use tensor calculus in general relativity; the classical irreps of the diffeomorphism group act on modules of tensor fields, so we should formulate physics in terms of those.

This article is organized as follows.

In the next section, the relation between anomalies, consistency, locality and unitarity is further discussed, using the infinite conformal symmetry in two-dimensional spacetime as a paradigm. The new anomalies, i.e. the higher-dimensional generalizations of the affine and Virasoro algebras, are reviewed in Section 7.3. For easy comparison to the one-dimensional case, we describe these extensions in a Fourier basis on the N-dimensional torus. However, the geometrical content is clearer in a real-space basis, which is introduced in Section 7.4. It turns out that in addition to the diffeomorphism algebra $\mathfrak{vect}(N)$ (algebra of vector fields in N dimensions) and the gauge algebra $\mathfrak{map}(N, \mathfrak{g})$ (algebra of maps from N-dimensional spacetime to the finite-dimensional Lie algebra $\mathfrak{g}$), we must also introduce the observer's trajectory and the algebra of reparametrizations; the full algebra is called the DGRO (Diffeomorphism, Gauge, Reparametrization, Observer) algebra $DGRO(N, \mathfrak{g})$.

Its representation theory is developed in Section 7.5. Rather than starting from the fields themselves, as is done in one dimension, the right approach is to first expand all fields in a Taylor series around the observer's trajectory and truncate at some finite order p before quantization. This gives us a non-linear realization of the diffeomorphism algebra on finitely many functions of a *single* variable, which is precisely the situation where normal ordering works. We also get an action of a $\mathfrak{vect}(1)$ algebra describing reparametrizations for free, i.e. without enlarging the realization.

In Section 7.6 we introduce the manifestly covariant canonical quantization scheme mentioned above, and apply it to the free scalar field in Section 7.7. However, the Hamiltonian still singles out a preferred time direction. To remedy this, we introduce in Section 7.8 the observer and define the Hamiltonian covariantly as the operator which translates the fields relative to the observer. The scalar field is again used as an example in Section 7.9. The formalism is then extended to theories with gauge symmetries in Section 7.10. As examples we treat the free Maxwell field in Section 7.11 and pure Einstein gravity in Section 7.12. The $\mathfrak{u}(1)$ gauge symmetry of the Maxwell field is anomaly free in the absense of matter, but there are diffeomorphism anomalies already in pure gravity, because it is an interacting theory.

Truncating the Taylor expansion at order p is a regularization, and at the end we want to remove the regulator by taking the limit $p \to \infty$. This limit is problematic and poorly understood, as is discussed in Section 7.13. We conclude with a conceptual discussion in Sections 7.14 and 7.15.

7.2 Anomalies, consistency, locality, and unitarity

A quantum theory is defined by a Hilbert space and a Hamiltonian which generates time evolution. The main conditions for consistency are unitarity and lack of infinities. If the theory has some symmetries, these must be realized as unitary operators acting on the Hilbert space as well. In particular, if time translation is included among the symmetries, which is the case for the Poincaré and diffeomorphism algebras, a unitary representation of the symmetry algebra is usually enough for consistency. From this viewpoint, there is a 1-1 correspondence between general-covariant QFTs and unitary representations of the diffeomorphism group on a conventional Hilbert space. Namely, given the QFT, its Hilbert space carries a unitary representation of the diffeomorphism group. Conversely, if we have a unitary representation of the diffeomorphism group, the Hilbert space on which it acts can be interpreted as the Hilbert space of some general-covariant QFT.

Unfortunately, this observation is not yet so powerful, because no non-trivial, unitary, lowest-energy irreps of the diffeomorphism algebra are known except in one dimension. Nevertheless, we are able to make some very general observations. Assume that some algebra (or group) $\mathfrak{g}$ has a unitary representation R and a subalgebra $\mathfrak{h}$. Then the restriction of R to $\mathfrak{h}$ is still unitary, and this must hold for every subalgebra $\mathfrak{h}$ of $\mathfrak{g}$. In particular, let $\mathfrak{g} = \mathfrak{vect}(N)$ be the diffeomorphism algebra in N dimensions and $\mathfrak{h} = \mathfrak{vect}(1)$ the diffeomorphism algebra in one dimension. There are infinitely many such subalgebras, and the restriction of R to each and every one of them must be unitary. Fortunately, the unitary irreps of the diffeomorphism algebra in one dimension are known. The result is that the only proper unitary irrep is the trivial one, but there are many unitary irreps with a diffeomorphism anomaly. From this it follows that the trivial representation is the only unitary representation also in N dimensions."There are no local observables in quantum gravity".

However, there is one well-known case where we know how to combine locality and infinite spacetime symmetry with quantum theory: conformal field theory (CFT). This is usually thought of as a theory of conformally invariant quantum fields in two dimensions, but since the local conformal group is the same as (twice) the diffeomorphism group in one dimension, it is also about diffeomorphism invariant QFT in one dimension. Locality means that the correlation functions depend on separation. For two points z and w in $\mathbb{R}$ or $\mathbb{C}$, the correlator is

$$\langle \phi(z)\phi(w) \rangle \sim \frac{1}{(z-w)^{2h}} + more, \qquad (7.2.1)$$

where *more* stands for less singular terms when $z \to w$. That the correlation function has this form is a diffeomorphism-invariant statement. The *more* terms will change under an arbitary diffeomorphism, but the leading singularity will always have the same form, and in particular the anomalous dimension h is well defined.

We can phrase this slightly differently. The short-distance singularity only depends on two points being infinitesimally close. This is good, because we cannot determine the finite distance between two points without knowing about the metric. General relativity does not have a background metric structure, but it does have a background differentiable structure (locally at least), and that is enough for defining anomalous dimensions. Diffemorphisms move points around, but they do not separate two points which are infinitesimally close.

The relevant algebra in CFT is not really the one-dimensional diffeomorphism algebra (or the two-dimensional conformal algebra), but rather its central extension known as the Virasoro algebra:

$$[L_m, L_n] = (n-m)L_{m+n} - \frac{c}{12}(m^3 - m)\delta_{m+n}. \tag{7.2.2}$$

A lowest-energy representation is characterized by a vacuum satisfying

$$L_0|0\rangle = h|0\rangle, \qquad L_m|0\rangle = 0 \text{ for all } m < 0. \tag{7.2.3}$$

In particular, the lowest L_0 eigenvalue can be identified with the anomalous dimension h in the correlation function (7.2.1). This means that locality, in the sense of correlation functions depending on separation, requires that $h > 0$. It is well known [8] that unitarity either implies that

$$c = 1 - \frac{6}{m(m+1)}, \qquad h = h_{rs}(m) = \frac{[(m+1)r - ms]^2 - 1}{4m(m+1)}, \tag{7.2.4}$$

where $m \geqslant 2$ and $1 \leqslant r < m, 1 \leqslant s < r$ are positive integers, or that $c \geqslant 1$, $h \geqslant 0$. In particular, the central extension c is non-zero for any non-trivial, unitary irrep with $h \neq 0$. This leads to the important observation:

Locality and unitarity are compatible with diffeomorphism (and local conformal) symmetry only in the presence of an anomaly.

This is true in higher dimensions as well. Consider the correlator $\langle\phi(x)\phi(y)\rangle$, where x and y are points in $\mathbb{R}^N$. We could take some one-dimensional curve $q(t)$ passing through x and y, such that $x = q(t)$ and $y = q(t')$. Then the short-distance behaviour is of the form

$$\langle\phi(x)\phi(y)\rangle \sim \frac{1}{(t-t')^{2h}} + \textit{more}, \tag{7.2.5}$$

and h is independent of the choice of curve, provided that it is sufficiently regular. The subalgebra of $\mathfrak{vect}(N)$ which preserves $q(t)$ is a Virasoro algebra, so $h > 0$ implies that $c > 0$.

This observation is completely standard in the application of CFT to statistical physics in two dimensions. The simplest example of a unitary model is the Ising model, which consists of three irreps, with $c = 1/2$ and $h = 0$, $h = 1/16$, and $h = 1/2$. The Ising model is perfectly consistent despite the anomaly, both mathematically (unitarity), and more importantly physically (it is realized in nature, in soft condensed matter systems). The standard counter-argument is that infinite conformal symmetry in condensed matter is not a gauge symmetry, but rather an anomalous global symmetry. However, if we could take the classical limit of such a system, the conformal symmetry would seem to be a gauge symmetry. Namely, the anomaly vanishes in the classical limit, and we can write down a classical BRST operator which is nilpotent, and the symmetry is gauge on the classical level. There is no classical way to distinguish between such a "fake" gauge symmetry and a genuine gauge symmetry which extends to the quantum level.

More generally, let us assume that we have some phase space, and a Lie algebra $\mathfrak{g}$ with generators J_a, satisfying

$$[J_a, J_b] = f_{ab}{}^c J_c, \tag{7.2.6}$$

acts on this phase space. The Einstein convention is used; repeated indices, one up and one down, are implicitly summed over. If the bracket with the Hamiltonian gives us a new element in $\mathfrak{g}$,

$$[J_a, H] = C_a^b J_b, \tag{7.2.7}$$

we say that $\mathfrak{g}$ is a *symmetry* of the Hamiltonian system. If $\mathfrak{g}$ in addition contains arbitary functions of time, the symmetry is a *gauge symmetry*. In this case, a solution to Hamilton's equations depends on arbitrary functions of time and is thus not fully specified by the positions and momenta at time $t = 0$. The standard example is electromagnetism, where the zeroth component A_0 of the vector potential is arbitrary, because its canonical momentum $F^{00} = 0$. An arbitary time evolution is of course not acceptable. The reason why this seems to happen is that a gauge symmetry is a redundancy of the description; the true dynamical degrees of freedom are fewer than what one naïvely expects. In electromagnetism, the gauge potential has four components but the photon has only two polarizations.

There are various ways to handle quantization of gauge systems. One is to eliminate the gauge degrees of freedom first and then quantize. This is cumbersome and it is usually preferable to quantize first and eliminate the gauge symmetries afterwards. The simplest way is to require that the gauge generators annihilate physical states,

$$J_a \big| phys \big\rangle = 0, \tag{7.2.8}$$

and also that two physical states are equivalent if the differ by some gauge state, $J_a| >$. This procedure produces a Hilbert space of physical states.

However, one thing may go wrong. Upon quantization, a symmetry may acquire some quantum corrections, so that $\mathfrak{g}$ is replaced by

$$[J_a, J_b] = f_{ab}{}^c J_c + \hbar D_{ab} + O(\hbar^2). \tag{7.2.9}$$

The operator D_{ab} is called an *anomaly*. We can also have anomalies of the type

$$[J_a, H] = C_a^b J_b + \hbar E_a + O(\hbar^2). \tag{7.2.10}$$

If we now try to keep the definition of a physical state, we see that we must also demand that

$$D_{ab}|phys\rangle = 0. \tag{7.2.11}$$

This implies further reduction of the Hilbert space. In the case that D_{ab} is invertible, there are no physical states at all, so the Hilbert space is empty. However, this does not necessarily mean that the anomaly by itself is inconsistent, only that our definition of physical states is. In the presence of an anomaly, additional states become physical. So our Hilbert space becomes larger, containing some, or even all, of the previous gauge degrees of freedom. A gauge anomaly implies that the gauge symmetry is broken on the quantum level.

Such a "fake" gauge symmetry may well be consistent. The Virasoro algebra is obviously anomalous, with the central charge playing the role of the D_{ab}, and still it has unitary representations with non-zero c. Of course, a gauge anomaly *may* be inconsistent, if the anomalous algebra does not possess any unitary representations. This is apparently what happens for the chiral-fermion type anomaly which is relevant e.g. in the standard model.

7.3 Multi-dimensional Virasoro algebra

All non-trivial, unitary, lowest-energy irreps of the diffeomorphism algebra are anomalous, in any number of dimensions. This is well-known in one dimension, where the diffeomorphism algebra acquires an extension known as the Virasoro algebra. It is also true in several dimensions, which one proves by considering the restriction to the many Virasoro subalgebras living on lines in spacetime. This is perhaps rather surprising, in view of the following two no-go theorems:

- The diffeomorphism algebra has no central extension except in one dimension.
- In field theory, there are no pure gravitational anomalies in four dimension.

However, the assumptions in these no-go theorems are too strong; the Virasoro extension is not central except in one dimension, and one needs to go slightly beyond field theory by explicitly specifying where observation takes place.

To make contact with the Virasoro algebra in its most familiar form, we describe its multi-dimensional sibling in a Fourier basis on the N-dimensional torus. Recall first that the algebra of diffeomorphisms on the circle, $\mathfrak{vect}(1)$, has generators

$$L_m = -i\exp(imx)\frac{d}{dx}, \tag{7.3.1}$$

where $x \in S^1$. $\mathfrak{vect}(1)$ has a central extension, known as the Virasoro algebra:

$$[L_m, L_n] = (n-m)L_{m+n} - \frac{c}{12}(m^3 - m)\delta_{m+n}, \tag{7.3.2}$$

where c is a c-number known as the central charge or conformal anomaly. This means that the Virasoro algebra is a Lie algebra; anti-symmetry and the Jacobi identities still hold. The term linear in m is unimportant, because it can be removed by a redefinition of L_0. The cubic term m^3 is a non-trivial extension which cannot be removed by any redefinition.

The generators (7.3.1) immediately generalize to vector fields on the N-dimensional torus:

$$L_\mu(m) = -i\exp(im_\rho x^\rho)\partial_\mu, \tag{7.3.3}$$

where $x = (x^\mu)$, $\mu = 1, 2, ..., N$ is a point in N-dimensional space and $m = (m_\mu) \in \mathbb{Z}^N$. These operators generate the algebra $\mathfrak{vect}(N)$:

$$[L_\mu(m), L_\nu(n)] = n_\mu L_\nu(m+n) - m_\nu L_\mu(m+n). \tag{7.3.4}$$

The question is now whether the Virasoro extension, i.e. the m^3 term in (7.3.2), also generalizes to higher dimensions.

Rewrite the ordinary Virasoro algebra (7.3.2) as

$$\begin{aligned}
[L_m, L_n] &= (n-m)L_{m+n} + cm^2 n S_{m+n}, \\
[L_m, S_n] &= (n+m)S_{m+n}, \\
[S_m, S_n] &= 0, \\
mS_m &\equiv 0.
\end{aligned} \tag{7.3.5}$$

It is easy to see that the two formulations of the Virasoro algebra are equivalent (the linear cocycle has been absorbed into a redefinition of L_0). The second formulation immediately generalizes to N dimensions. The defining relations are

$$\begin{aligned}
[L_\mu(m), L_\nu(n)] &= n_\mu L_\nu(m+n) - m_\nu L_\mu(m+n) \\
&\quad + (c_1 m_\nu n_\mu + c_2 m_\mu n_\nu) m_\rho S^\rho(m+n), \\
[L_\mu(m), S^\nu(n)] &= n_\mu S^\nu(m+n) + \delta^\nu_\mu m_\rho S^\rho(m+n), \\
[S^\mu(m), S^\nu(n)] &= 0, \\
m_\mu S^\mu(m) &\equiv 0.
\end{aligned} \tag{7.3.6}$$

This is an extension of $\mathfrak{vect}(N)$ by the abelian ideal with basis $S^\mu(m)$. Geometrically, we can think of $L_\mu(m)$ as a vector field and $S^\mu(m) = \epsilon^{\mu\nu_2..\nu_N} S_{\nu_2..\nu_N}(m)$ as a dual one-form (and $S_{\nu_2..\nu_N}(m)$ as an $(N-1)$-form); the last condition expresses closedness. The cocycle proportional to c_1 was discovered by Rao and Moody [23], and the one proportional to c_2 by this author [13].

There is also a similar multi-dimensional generalization of affine Kac-Moody algebras, presumably first written down by Kassel [12]. It is sometimes called the central extension, but this term is somewhat misleading because the extension does not commute with diffeomorphisms, although it does commute with all gauge transformations.

Let $\mathfrak{g}$ be a finite-dimensional Lie algebra with structure constants $f_{ab}{}^c$ and Killing metric δ_{ab}. The Kassel extension of the current algebra $\mathfrak{map}(N, \mathfrak{g})$ is defined by the brackets

$$\begin{aligned} [J_a(m), J_b(n)] &= f_{ab}{}^c J_c(m+n) + k\delta_{ab} m_\rho S^\rho(m+n), \\ [J_a(m), S^\mu(n)] &= [S^\mu(m), S^\nu(n)] = 0, \\ m_\mu S^\mu(m) &\equiv 0. \end{aligned} \tag{7.3.7}$$

This algebra admits an intertwining action of the N-dimensional Virasoro algebra (7.3.6):

$$[L_\mu(m), J_a(n)] = n_\mu J_a(m+n). \tag{7.3.8}$$

The current algebra $\mathfrak{map}(N, \mathfrak{g})$ also admits another type of extension in some dimensions. The best known example is the Mickelsson-Faddeev algebra, relevant for the conventional anomalies in field theory, which arise when chiral fermions are coupled to gauge fields in three spatial dimensions. Let $d_{abc} = \mathrm{tr}\{T_a, T_b\}T_c$ be the totally symmetric third Casimir operator, and let $\epsilon^{\mu\nu\rho}$ be the totally anti-symmetric epsilon tensor in three dimensions. The Mickelsson-Faddeev algebra [20] reads in a Fourier basis:

$$\begin{aligned} [J_a(m), J_b(n)] &= f_{ab}{}^c J_c(m+n) + d_{abc}\epsilon^{\mu\nu\rho} m_\mu n_\nu A^c_\rho(m+n), \\ [J_a(m), A^b_\nu(n)] &= -f_{ac}{}^b A^c_\nu(m+n) + \delta^b_a m_\nu \delta(m+n), \\ [A^a_\mu(m), A^b_\nu(n)] &= 0. \end{aligned} \tag{7.3.9}$$

$A^a_\mu(m)$ are the Fourier components of the gauge connection.

Note that $Q_a \equiv J_a(0)$ generates a Lie algebra isomorphic to $\mathfrak{g}$, whose Cartan subalgebra is identified with the charges. Moreover, the subalgebra of (7.3.7) spanned by $J_a(m_0) \equiv J_a(m)$, where $m = (m_0, 0, ..., 0) \in \mathbb{Z}$, reads

$$[J_a(m_0), J_b(n_0)] = f_{ab}{}^c J_c(m_0+n_0) + k\delta_{ab} m_0 \delta(m_0+n_0), \tag{7.3.10}$$

which we recognize as the affine algebra $\widehat{\mathfrak{g}}$. Since all non-trivial unitary irreps of $\widehat{\mathfrak{g}}$ has $k > 0$ [9], it is impossible to combine unitary and non-zero $\mathfrak{g}$ charges also for the higher-dimensional algebra (7.3.7). This follows immediately from the fact that the restriction of a unitary irrep to a subalgebra is also unitary (albeit in general reducible).

In contrast, the Mickelsson-Faddeev algebra (7.3.9) has apparently no faithful unitary representations on a separable Hilbert space [22]. A simple way to understand this is to note that the restriction to every loop subalgebra is proper and hence lacks unitary representations of lowest-weight type. This presumably means that this kind of extension should be avoided. Indeed, Nature appears to abhor this kind of anomaly, which is proportional to the third Casimir.

7.4 DGRO algebra

The Fourier formalism in the previous section makes the analogy with the usual Virasoro algebra manifest, but it is neither illuminating nor a useful starting point for representation theory. To bring out the geometrical content, we introduce the DGRO *(Diffeomorphism, Gauge, Reparametrization, Observer)* algebra $DGRO(N, \mathfrak{g})$, whose ingredients are spacetime diffeomorphisms which generate $\mathfrak{vect}(N)$, reparametrizations of the observer's trajectory which form an additional $\mathfrak{vect}(1)$ algebra, and gauge transformations which generate a current algebra. Classically, the algebra is $\mathfrak{vect}(N) \ltimes \mathfrak{map}(N, \mathfrak{g}) \oplus \mathfrak{vect}(1)$.

Let $\xi = \xi^\mu(x)\partial_\mu$, $x \in \mathbb{R}^N$, $\partial_\mu = \partial/\partial x^\mu$, be a vector field, with commutator $[\xi, \eta] \equiv \xi^\mu\partial_\mu\eta^\nu\partial_\nu - \eta^\nu\partial_\nu\xi^\mu\partial_\mu$, and greek indices $\mu, \nu = 1, 2, .., N$ label the spacetime coordinates. The Lie derivatives $\mathcal{L}_\xi$ are the generators of $\mathfrak{vect}(N)$.

Let $f = f(t)d/dt$, $t \in S^1$, be a vector field in one dimension. The commutator reads $[f, g] = (f\dot{g} - g\dot{f})d/dt$, where the dot denotes the t derivative: $\dot{f} \equiv df/dt$. We will also use $\partial_t = \partial/\partial t$ for the partial t derivative. The choice that t lies on the circle is physically unnatural and is made for technical simplicity only (quantities can be expanded in Fourier series). However, this seems to be a minor problem at the present level of understanding. Denote the reparametrization generators L_f.

Let $\mathfrak{map}(N, \mathfrak{g})$ be the current algebra corresponding to the finite-dimensional semisimple Lie algebra $\mathfrak{g}$ with basis J_a, structure constants $f_{ab}{}^c$, and Killing metric δ_{ab}. The brackets in $\mathfrak{g}$ are given by (7.2.6). A basis for $\mathfrak{map}(N, \mathfrak{g})$ is given by $\mathfrak{g}$-valued functions $X = X^a(x)J_a$ with commutator $[X, Y] = f_{ab}{}^c X^a Y^b J_c$. The intertwining $\mathfrak{vect}(N)$ action is given by $\xi X = \xi^\mu\partial_\mu X^a J_a$. Denote the $\mathfrak{map}(N, \mathfrak{g})$ generators by $\mathcal{J}_X$.

Finally, let $Obs(N)$ be the space of local functionals of the observer's trajectory $q^\mu(t)$, i.e. polynomial functions of $q^\mu(t)$, $\dot{q}^\mu(t)$, ... $d^k q^\mu(t)/dt^k$, k finite, regarded as a commutative algebra. $Obs(N)$ is a $\mathfrak{vect}(N)$ module in a natural manner.

$DGRO(N, \mathfrak{g})$ is an abelian but non-central Lie algebra extension of $\mathfrak{vect}(N) \ltimes \mathfrak{map}(N, \mathfrak{g}) \oplus \mathfrak{vect}(1)$ by $Obs(N)$:

$$0 \longrightarrow Obs(N) \longrightarrow DGRO(N, \mathfrak{g}) \longrightarrow \mathfrak{vect}(N) \ltimes \mathfrak{map}(N, \mathfrak{g}) \oplus \mathfrak{vect}(1) \longrightarrow 0.$$

The brackets are given by

$$\begin{aligned}
[\mathcal{L}_\xi, \mathcal{L}_\eta] &= \mathcal{L}_{[\xi,\eta]} + \frac{1}{2\pi i}\int dt\ \dot{q}^\rho(t)\Big\{c_1\partial_\rho\partial_\nu\xi^\mu(q(t))\partial_\mu\eta^\nu(q(t)) + \\
&\quad + c_2\partial_\rho\partial_\mu\xi^\mu(q(t))\partial_\nu\eta^\nu(q(t))\Big\}, \\
[\mathcal{L}_\xi, \mathcal{J}_X] &= \mathcal{J}_{\xi X}, \\
[\mathcal{J}_X, \mathcal{J}_Y] &= \mathcal{J}_{[X,Y]} - \frac{c_5}{2\pi i}\delta_{ab}\int dt\ \dot{q}^\rho(t)\partial_\rho X^a(q(t))Y^b(q(t)), \\
[L_f, \mathcal{L}_\xi] &= \frac{c_3}{4\pi i}\int dt\ (\ddot{f}(t) - i\dot{f}(t))\partial_\mu\xi^\mu(q(t)), \\
[L_f, \mathcal{J}_X] &= 0, \\
[L_f, L_g] &= L_{[f,g]} + \frac{c_4}{24\pi i}\int dt(\dddot{f}(t)\dot{g}(t) - \dot{f}(t)g(t)), \\
[\mathcal{L}_\xi, q^\mu(t)] &= \xi^\mu(q(t)), \\
[L_f, q^\mu(t)] &= -f(t)\dot{q}^\mu(t), \\
[\mathcal{J}_X, q^\mu(t)] &= [q^\mu(s), q^\nu(t)] = 0,
\end{aligned} \tag{7.4.1}$$

extended to all of $Obs(N)$ by Leibniz' rule and linearity. The numbers $c_1 - c_5$ are called *abelian charges*, in analogy with the central charge of the Virasoro algebra. In [14, 15] slightly more complicated extensions were considered, which depend on three additional abelian charges $c_6 - c_8$. However, these vanish automatically when $\mathfrak{g}$ is semisimple.

7.5 Representations of the DGRO algebra

To construct Fock representations of the ordinary Virasoro algebra is straightforward:

- Start from classical modules, i.e. primary fields = scalar densities.
- Introduce canonical momenta.
- Normal order.

The first two steps of this procedure generalize nicely to higher dimensions. The classical representations of the DGRO algebra are tensor fields over $\mathbb{R}^N \times S^1$ valued in $\mathfrak{g}$ modules. The basis of a classical DGRO module $\mathcal{Q}$ is thus a field $\phi^\alpha(x,t)$, $x \in \mathbb{R}^N$, $t \in S^1$, where α is a collection of all kinds of indices. The $DGRO(N, \mathfrak{g})$ action on $\mathcal{Q}$ can be succinctly summarized as

$$\begin{aligned}
[\mathcal{L}_\xi, \phi^\alpha(x,t)] &= -\xi^\mu(x)\partial_\mu\phi^\alpha(x,t) - \partial_\nu\xi^\mu(x)T^{\alpha\nu}_{\beta\mu}\phi^\beta(x,t), \\
[\mathcal{J}_X, \phi^\alpha(x,t)] &= -X^a(x)J^\alpha_{\beta a}\phi^\beta(x,t), \\
[L_f, \phi^\alpha(x,t)] &= -f(t)\partial_t\phi^\alpha(x,t) - \lambda(\dot{f}(t) - if(t))\phi^\alpha(x,t).
\end{aligned} \tag{7.5.1}$$

Here $J_a = (J^\alpha_{\beta a})$ and $T^\mu_\nu = (T^{\alpha\mu}_{\beta\nu})$ are matrices satisfying $\mathfrak{g}$ (7.2.6) and $gl(N)$, respectively:

$$[T^\mu_\nu, T^\sigma_\tau] = \delta^\sigma_\nu T^\mu_\tau - \delta^\mu_\tau T^\sigma_\mu. \tag{7.5.2}$$

The tensor field representations of the DGRO algebra can thus be expressed in matrix form as

$$\begin{aligned} \mathcal{L}_\xi &= -\int d^N x \int dt\ (\xi^\mu(x)\partial_\mu \phi^\alpha(x,t) + \partial_\nu \xi^\mu(x) T^{\alpha\nu}_{\beta\mu}\phi^\beta(x,t))\pi_\alpha(x,t), \\ \mathcal{J}_X &= -\int d^N x \int dt\ X^a(x) J^\alpha_{\beta a}\phi^\beta(x,t)\pi_\alpha(x,t), \\ L_f &= -\int d^N x \int dt\ (f(t)\partial_t \phi^\alpha(x,t) - \lambda(\dot{f}(t) - i f(t))\phi^\alpha(x,t))\pi_\alpha(x,t), \end{aligned} \tag{7.5.3}$$

where the conjugate momentum $\pi_\alpha(x,t) = \delta/\delta\phi^\alpha(x,t)$ satisfies

$$[\pi_\alpha(x,t), \phi^\beta(x',t')] = \delta^\beta_\alpha \delta(x-x')\delta(t-t'). \tag{7.5.4}$$

However, the normal-ordering step simply does not work in several dimensions, because

- It requires that a foliation of spacetime into space and time has been introduced, which runs against the idea of diffeomorphism invariance.
- Normal ordering of bilinear expressions always results in a *central* extension, but the Virasoro cocycle is non-central when $N \geqslant 2$.
- It is ill defined. Formally, attempts to normal order result in an *infinite* central extension, which of course makes no sense.

To avoid this problem, the crucial idea in [14] was to expand all fields in a Taylor series around the observer's trajectory and truncate at order p, before introducing canonical momenta. Hence we expand e.g.,

$$\phi^\alpha(x,t) = \sum_{|\mathbf{m}|\leqslant p} \frac{1}{\mathbf{m}!}\phi^\alpha_{,\mathbf{m}}(t)(x-q(t))^{\mathbf{m}}, \tag{7.5.5}$$

where $\mathbf{m} = (m_1, m_2, ..., m_N)$, all $m_\mu \geqslant 0$, is a multi-index of length $|\mathbf{m}| = \sum_{\mu=1}^N m_\mu$, $\mathbf{m}! = m_1! m_2! ... m_N!$, and

$$(x-q(t))^{\mathbf{m}} = (x^1 - q^1(t))^{m_1}(x^2 - q^2(t))^{m_2} ... (x^N - q^N(t))^{m_N}. \tag{7.5.6}$$

Denote by μ a unit vector in the μ:th direction, so that $\mathbf{m} + \mu = (m_1, ..., m_\mu + 1, ..., m_N)$, and let

$$\phi^\alpha_{,\mathbf{m}}(t) = \partial_{\mathbf{m}}\phi^\alpha(q(t),t) = \underbrace{\partial_1..\partial_1}_{m_1}..\underbrace{\partial_N..\partial_N}_{m_N}\phi^\alpha(q(t),t) \tag{7.5.7}$$

be the $|\mathbf{m}|$:th order derivative of $\phi^\alpha(x,t)$ evaluated on the observer's trajectory $q^\mu(t)$.

Given two jets $\phi_{,\mathbf{m}}(t)$ and $\psi_{,\mathbf{m}}(t')$, we define their product

$$(\phi(t)\psi(t'))_{,\mathbf{m}} = \sum_{\mathbf{n}} \binom{\mathbf{m}}{\mathbf{n}} \phi_{,\mathbf{n}}(t)\psi_{,\mathbf{m}-\mathbf{n}}(t'). \tag{7.5.8}$$

It is clear that $(\phi(t)\psi(t'))_{,\mathbf{m}}$ is the jet corresponding to the field $\phi(x,t)\psi(x,t')$. For brevity, we also denote $(\phi\psi)_{,\mathbf{m}}(t) = (\phi(t)\psi(t))_{,\mathbf{m}}$.

p-jets transform under $DGRO(N,\mathfrak{g})$ as

$$\begin{aligned}
[\mathcal{L}_\xi, \phi^\alpha_{,\mathbf{m}}(t)] &= \partial_{\mathbf{m}}([\mathcal{L}_\xi, \phi^\alpha(q(t),t)]) + [\mathcal{L}_\xi, q^\mu(t)]\partial_\mu\partial_{\mathbf{m}}\phi^\alpha(q(t),t) \\
&\equiv -\sum_{|\mathbf{n}|\leqslant|\mathbf{m}|\leqslant p} T^{\alpha\mathbf{n}}_{\beta\mathbf{m}}(\xi(q(t)))\phi^\beta_{,\mathbf{n}}(t), \\
[\mathcal{J}_X, \phi^\alpha_{,\mathbf{m}}(t)] &= \partial_{\mathbf{m}}([\mathcal{J}_X, \phi^\alpha(q(t),t)]) \\
&\equiv -\sum_{|\mathbf{n}|\leqslant|\mathbf{m}|\leqslant p} J^{\alpha\mathbf{n}}_{\beta\mathbf{m}}(X(q(t)))\phi^\beta_{,\mathbf{n}}(t), \\
[L_f, \phi^\alpha_{,\mathbf{m}}(t)] &= -f(t)\dot\phi^\alpha_{,\mathbf{m}}(t) - \lambda(\dot f(t) - if(t))\phi^\alpha_{,\mathbf{m}}(t),
\end{aligned} \tag{7.5.9}$$

where

$$\begin{aligned}
T^{\mathbf{m}}_{\mathbf{n}}(\xi) &\equiv (T^{\alpha\mathbf{m}}_{\beta\mathbf{n}}(\xi)) = \sum_{\mu\nu}\binom{\mathbf{n}}{\mathbf{m}}\partial_{\mathbf{n}-\mathbf{m}+\nu}\xi^\mu T^\nu_\mu \\
&\quad + \sum_\mu \binom{\mathbf{n}}{\mathbf{m}-\mu}\partial_{\mathbf{n}-\mathbf{m}+\mu}\xi^\mu - \sum_\mu \delta^{\mathbf{m}-\mu}_{\mathbf{n}}\xi^\mu,
\end{aligned} \tag{7.5.10}$$

$$J^{\mathbf{m}}_{\mathbf{n}}(X) \equiv (J^{\alpha\mathbf{m}}_{\beta\mathbf{n}}(X)) = \binom{\mathbf{n}}{\mathbf{m}}\partial_{\mathbf{n}-\mathbf{m}}X^a J_a,$$

and

$$\binom{\mathbf{m}}{\mathbf{n}} = \frac{\mathbf{m}!}{\mathbf{n}!(\mathbf{m}-\mathbf{n})!} = \binom{m_1}{n_1}\binom{m_2}{n_2}\dots\binom{m_N}{n_N}. \tag{7.5.11}$$

We thus obtain a non-linear realization of $\mathfrak{vect}(N)$ on the space of trajectories in the space of tensor-valued p-jets[1]; denote this space by J^pQ. Note that J^pQ is spanned by $q^\mu(t)$ and $\{\phi^\alpha_{,\mathbf{m}}(t)\}_{|\mathbf{m}|\leqslant p}$ and thus not a $DGRO(N,\mathfrak{g})$ module by itself, because diffeomorphisms act non-linearly on $q^\mu(t)$, as can be seen in (7.4.1). However, the space $C(J^pQ)$ of functionals on J^pQ (local in t) *is* a module, because

[1] p-jets are usually defined as an equivalence class of functions: two functions are equivalent if all derivatives up to order p, evaluated at q^μ, agree. However, each class has a unique representative which is a polynomial of order at most p, namely the Taylor expansion around q^μ, so we may canonically identify jets with truncated Taylor series. Since $q^\mu(t)$ depends on a parameter t, we deal in fact with trajectories in jet space, but these will also be called jets for brevity.

the action on a p-jet can never produce a jet of order higher than p. The space $C(q) \otimes_q J^p\mathcal{Q}$, where only the trajectory itself appears non-linearly, is a submodule.

The crucial observation is that the jet space $J^p\mathcal{Q}$ consists of finitely many functions of a single variable t, which is precisely the situation where the normal ordering prescription works. After normal ordering, denoted by double dots : :, we obtain a Fock representation of the DGRO algebra:

$$\begin{aligned}
\mathcal{L}_\xi &= \int dt\,\Big\{ :\xi^\mu(q(t))p_\mu(t): - \sum_{|\mathbf{n}|\leqslant|\mathbf{m}|\leqslant p} T^{\alpha\mathbf{n}}_{\beta\mathbf{m}}(\xi(q(t)))\,:\phi^\beta_{,\mathbf{n}}(t)\pi^{;\mathbf{m}}_\alpha(t): \Big\}, \\
\mathcal{J}_X &= -\int dt\,\Big\{ \sum_{|\mathbf{n}|\leqslant|\mathbf{m}|\leqslant p} J^{\alpha\mathbf{n}}_{\beta\mathbf{m}}(\xi(q(t)))\,:\phi^\beta_{,\mathbf{n}}(t)\pi^{;\mathbf{m}}_\alpha(t): \Big\}, \qquad (7.5.12)\\
L_f &= \int dt\,\Big\{ -f(t):\dot\phi^\alpha_{,\mathbf{m}}(t)\pi^{;\mathbf{m}}_\alpha(t): - \lambda(\dot f(t) - if(t))\,:\phi^\alpha_{,\mathbf{m}}(t)\pi^{;\mathbf{m}}_\alpha(t): \Big\},
\end{aligned}$$

where we have introduced canonical momenta $p_\mu(t) = \delta/\delta q^\mu(t)$ and $\pi^{;\mathbf{m}}_\alpha(t) = \delta/\delta\phi^\alpha_{,\mathbf{m}}(t)$. The field $\phi^\alpha(x,t)$ can be either bosonic or fermionic but the trajectory $q^\mu(t)$ is of course always bosonic.

Normal ordering is defined with respect to frequency; any function of $t \in S^1$ can be expanded in a Fourier series, e.g.

$$p_\mu(t) = \sum_{m=-\infty}^{\infty} \hat p_\mu(m)\mathrm{e}^{-imt} \equiv p^<_\mu(t) + \hat p_\mu(0) + p^>_\mu(t), \qquad (7.5.13)$$

where $p^<_\mu(t)$ ($p^>_\mu(t)$) is the sum over negative (positive) frequency modes only. Then

$$:\xi^\mu(q(t))p_\mu(t): \equiv \xi^\mu(q(t))p^<_\mu(t) + p^>_\mu(t)\xi^\mu(q(t)), \qquad (7.5.14)$$

where the zero mode has been included in $p^<_\mu(t)$.

It is clear that (7.5.12) defines a Fock representation for every $gl(N)$ irrep ϱ and every $\mathfrak{g}$ irrep M; denote this Fock space by $J^p\mathcal{F}$, which indicates that it also depends on the truncation order p. Namely, introduce a Fock vacuum $\big|0\big\rangle$ which is annihilated by half of the oscillators, i.e.

$$\phi^{\alpha<}_{,\mathbf{m}}(t)\big|0\big\rangle = \pi^{;\mathbf{m}}_{\alpha<}(t)\big|0\big\rangle = q^\mu_<(t)\big|0\big\rangle = p^<_\mu(t)\big|0\big\rangle = 0. \qquad (7.5.15)$$

Then $DGRO(N,\mathfrak{g})$ acts on the space of functionals $C(q^\mu_>, p^>_\mu, \phi^{\alpha>}_{,\mathbf{m}}, \pi^{;\mathbf{m}}_{\alpha>})$ of the remaining oscillators; this is the Fock module. Define numbers $k_0(\varrho)$, $k_1(\varrho)$, $k_2(\varrho)$ and y_M by

$$\begin{aligned}
\mathrm{tr}_\varrho\, T^\mu_\nu &= k_0(\varrho)\delta^\mu_\nu, \\
\mathrm{tr}_\varrho\, T^\mu_\nu T^\sigma_\tau &= k_1(\varrho)\delta^\mu_\tau\delta^\sigma_\nu + k_2(\varrho)\delta^\mu_\nu\delta^\sigma_\tau, \qquad (7.5.16)\\
\mathrm{tr}_M\, J_a J_b &= y_M\delta_{ab}.
\end{aligned}$$

For an unconstrained tensor with p upper and q lower indices and weight κ, we have

$$\begin{aligned} \dim(\varrho) &= N^{p+q}, & k_0(\varrho) &= -(p-q-\kappa N)N^{p+q-1}, \\ k_1(\varrho) &= (p+q)N^{p+q-1}, & k_2(\varrho) &= ((p-q-\kappa N)^2 - p - q)N^{p+q-2}. \end{aligned} \tag{7.5.17}$$

Note that if $\kappa = (p-q)/N$, ϱ is an $sl(N)$ representation. For the symmetric representations on ℓ lower indices, S_ℓ, and on ℓ upper indices, S^ℓ, we have

$$\begin{aligned} \dim(S_\ell) = \dim(S^\ell) &= \binom{N-1+\ell}{\ell}, \\ k_0(S_\ell) = -k_0(S^\ell) &= \binom{N-1+\ell}{\ell-1}, \\ k_1(S_\ell) = k_1(S^\ell) &= \binom{N+\ell}{\ell-1}, \\ k_2(S_\ell) = k_2(S^\ell) &= \binom{N-1+\ell}{\ell-2}. \end{aligned} \tag{7.5.18}$$

The values of the abelian charges $c_1 - c_5$ (7.4.1) were calculated in [14], Theorems 1 and 3, and in [15], Theorem 1:

$$\begin{aligned} c_1 &= 1 - u\binom{N+p}{N} - x\binom{N+p+1}{N+2}, \\ c_2 &= -v\binom{N+p}{N} - 2w\binom{N+p}{N+1} - x\binom{N+p}{N+2}, \\ c_3 &= 1 + (1-2\lambda)(w\binom{N+p}{N} + x\binom{N+p}{N+1}), \\ c_4 &= 2N - x(1 - 6\lambda + 6\lambda^2)\binom{N+p}{N}, \\ c_5 &= y\binom{N+p}{N}. \end{aligned} \tag{7.5.19}$$

where

$$\begin{aligned} u &= \mp k_1(\varrho) \dim M, & x &= \mp \dim \varrho \dim M, \\ v &= \mp k_2(\varrho) \dim M, & y &= \mp \dim \varrho \, y_M; \\ w &= \mp k_0(\varrho) \dim M, \end{aligned} \tag{7.5.20}$$

and the sign factor depends on the Grassmann parity of ϕ^α; the upper sign holds for bosons and the lower for fermions, respectively. The p-independent contributions to c_1, c_3 and c_4 come from the trajectory $q^\mu(t)$ itself.

7.6 MCCQ: Manifestly Covariant Canonical Quantization

In the previous sections we constructed representations of gauge and diffeomorphism algebras. These representations are of lowest-energy type, i.e. there is a Hamiltonian whose eigenvalues are bounded from below; this is the kind of representations relevant to quantum theory. Now it is time to apply them to physics. To this end, we first reformulate quantum physics in the history phase space, which is the representation theory's natural habitat.

Consider a classical dynamical system with action S and degrees of freedom ϕ^α. As is customary in the antifield literature, we use an abbreviated notation where the index α stands for both discrete indices and spacetime coordinates. Dynamics is governed by the Euler-Lagrange (EL) equations,

$$\mathcal{E}_\alpha = \partial_\alpha S \equiv \frac{\delta S}{\delta \phi^\alpha} = 0. \tag{7.6.1}$$

An important role is also played by the Hessian, i.e. the symmetric second functional-derivative matrix

$$K_{\alpha\beta} = K_{\beta\alpha} = \partial_\beta \mathcal{E}_\alpha = \frac{\delta \mathcal{E}_\alpha}{\delta \phi^\beta} = \frac{\delta^2 S}{\delta \phi^\alpha \delta \phi^\beta}. \tag{7.6.2}$$

The Hessian is assumed non-singular, so it has an inverse $M^{\alpha\beta}$ satisfying

$$K_{\beta\gamma} M^{\gamma\alpha} = M^{\alpha\gamma} K_{\gamma\beta} = \delta^\alpha_\beta. \tag{7.6.3}$$

Introduce an antifield ϕ^*_α for each EL equation (7.6.1), and replace the space of ϕ-histories $\mathcal{Q}$ by the extended history space $\mathcal{Q}^*$, spanned by both ϕ and ϕ^*. In $\mathcal{Q}^*$ we define the Koszul-Tate (KT) differential δ by

$$\delta \phi^\alpha = 0, \qquad \delta \phi^*_\alpha = \mathcal{E}_\alpha. \tag{7.6.4}$$

One checks that δ is nilpotent, $\delta^2 = 0$. Define the antifield number $\operatorname{afn} \phi^\alpha = 0$, $\operatorname{afn} \phi^*_\alpha = 1$. The KT differential clearly has antifield number $\operatorname{afn} \delta = -1$.

The space $C(\mathcal{Q}^*)$ decomposes into subspaces $C^k(\mathcal{Q}^*)$ of fixed antifield number

$$C(\mathcal{Q}^*) = \sum_{k=0}^{\infty} C^k(\mathcal{Q}^*) \tag{7.6.5}$$

The KT complex is

$$0 \xleftarrow{\delta} C^0 \xleftarrow{\delta} C^1 \xleftarrow{\delta} C^2 \xleftarrow{\delta} \dots \tag{7.6.6}$$

The cohomology spaces are defined as usual by $H^\bullet_{cl}(\delta) = \ker\delta/\mathrm{im}\ \delta$, i.e. $H^k_{cl}(\delta) = (\ker\delta)_k/(\mathrm{im}\ \delta)_k$, where the subscript *cl* indicates that we deal with a classical phase space. It is easy to see that

$$\begin{aligned} (\ker\delta)_0 &= C(\mathcal{Q}), \\ (\mathrm{im}\ \delta)_0 &= C(\mathcal{Q})\mathcal{E}_\alpha \equiv \mathcal{N}. \end{aligned} \tag{7.6.7}$$

Thus $H^0_{cl}(\delta) = C(\mathcal{Q})/\mathcal{N} = C(\Sigma)$. Since we assume that there are no non-trivial relations among the $\mathcal{E}_\alpha$, the higher cohomology groups vanish. This is a standard result [10]. The complex (7.6.6) thus gives us a resolution of the covariant phase space $C(\Sigma)$, which by definition means that $H^0_{cl}(\delta) = C(\Sigma)$, $H^k_{cl}(\delta) = 0$, for all $k > 0$.

Alas, the antifield formalism is not suited for canonical quantization. We can define an antibracket in $\mathcal{Q}^*$, but in order to do canonical quantization we need an honest Poisson bracket. To this end, we introduce canonical momenta conjugate to the history and its antifield, and obtain an even larger space $\mathcal{P}^*$, which may be thought of as the phase space corresponding to the extended history space $\mathcal{Q}^*$.

Introduce canonical momenta $\pi_\alpha = \delta/\delta\phi^\alpha$ and $\pi^\alpha_* = \delta/\delta\phi^*_\alpha$ for both the fields and antifields. The momenta satisfy by definition the graded canonical commutation relations (ϕ^α is assumed bosonic),

$$\begin{aligned} &[\pi_\beta, \phi^\alpha] = \delta^\alpha_\beta, && [\phi^\alpha, \phi^\beta] = [\pi_\alpha, \pi_\beta] = 0, \\ &\{\pi^\beta_*, \phi^*_\alpha\} = \delta^\beta_\alpha, && \{\phi^*_\alpha, \phi^*_\beta\} = \{\pi^\alpha_*, \pi^\beta_*\} = 0, \end{aligned} \tag{7.6.8}$$

where $\{\cdot,\cdot\}$ is the symmetric bracket. Let $\mathcal{P}$ be the phase space of histories with basis (ϕ^α, π_β), and let $\mathcal{P}^*$ be the extended phase space with basis $(\phi^\alpha, \pi_\beta, \phi^*_\alpha, \pi^\beta_*)$.

The definition of the KT differential extends to $\mathcal{P}^*$ by requiring that $\delta F = [Q_{KT}, F]$ for every $F \in C(\mathcal{P}^*)$, where the KT operator is

$$Q_{KT} = \mathcal{E}_\alpha \pi^\alpha_*. \tag{7.6.9}$$

It acts on the various fields as

$$\begin{aligned} \delta\phi^\alpha &= 0, \\ \delta\phi^*_\alpha &= \mathcal{E}_\alpha, \\ \delta\pi_\alpha &= -\frac{\delta\mathcal{E}_\beta}{\delta\phi^\alpha}\pi^\beta_* = -K_{\alpha\beta}\pi^\beta_*, \\ \delta\pi^\alpha_* &= 0, \end{aligned} \tag{7.6.10}$$

where $K_{\alpha\beta}$ is the Hessian (7.6.2). We check that δ is still nilpotent: $\delta^2 = \{Q_{KT}, Q_{KT}\} = 0$.

Like $C(\mathcal{Q})$, the function space $C(\mathcal{P}^*)$ decomposes into subspaces of fixed antifield number, $C(\mathcal{P}^*) = \sum_{k=-\infty}^{\infty} C^k(\mathcal{P}^*)$. We can therefore define a KT complex in $C(\mathcal{P}^*)$

$$\ldots \xleftarrow{\delta} C^{-2} \xleftarrow{\delta} C^{-1} \xleftarrow{\delta} C^0 \xleftarrow{\delta} C^1 \xleftarrow{\delta} C^2 \xleftarrow{\delta} \ldots \tag{7.6.11}$$

Because the Hessian (7.6.2) is non-singular by assumption with inverse $M^{\alpha\beta}$, we can invert the relation $\delta\pi_\alpha = -K_{\alpha\beta}\pi_*^\beta$ and get

$$\pi_*^\alpha = -M^{\alpha\beta}\delta\pi_\beta = \delta(-M^{\alpha\beta}\pi_\beta), \tag{7.6.12}$$

since $M^{\alpha\beta}$ depends on ϕ alone.

Let us now compute the cohomology. Any function which contains π_α is not closed, so $\ker\delta = C(\phi, \phi^*, \pi_*)$. Moreover, $\mathrm{im}\,\delta$ is generated by the two ideals $\mathcal{E}_\alpha$ and π_*^α. The momenta π_α and π_*^α thus vanish in cohomology, and the part with zero antifield number is thus still $H_{cl}^0(\delta) = C(\mathcal{Q})/\mathcal{N} = C(\Sigma)$. The higher cohomology groups $H_{cl}^k(\delta) = 0$ by the same argument as above. Hence the complex (7.6.11) yields a different resolution of the function space $C(\Sigma)$.

It is important that the spaces C^k in (7.6.11) are phase spaces, equipped with the Poisson bracket (7.6.8). Unlike the resolution (7.6.6), the new resolution (7.6.11) therefore allows us to do canonical quantization: replace Poisson brackets by commutators and represent the graded Heisenberg algebra (7.6.8) on a Hilbert space. However, the Heisenberg algebra can be represented on different Hilbert spaces; there is no Stone-von Neumann theorem in infinite dimension. To pick the correct one, we must impose the physical condition that there is an energy which is bounded on below.

To define the Hamiltonian, we must single out a privileged variable t among the α's, and declare it to be time. Thus replace $\alpha = (i, t)$, so e.g. $\phi^\alpha = \phi^i(t)$, $\mathcal{E}_\alpha = \mathcal{E}_i(t)$, etc. This step means of course that we sacrifice covariance. The Hamiltonian reads

$$H = -i \int dt\; \dot{\phi}^i(t)\pi_i(t) + \dot{\phi}_i^*(t)\pi_*^i(t). \tag{7.6.13}$$

It satisfies

$$\begin{aligned} [H, \phi^i(t)] &= -i\dot{\phi}^i(t), & [H, \pi_i(t)] &= -i\dot{\pi}_i(t), \\ [H, \phi_i^*(t)] &= -i\dot{\phi}_i^*(t), & [H, \pi_*^i(t)] &= -i\dot{\pi}_*{}^i(t). \end{aligned} \tag{7.6.14}$$

Expand all fields in a Fourier series with respect to time, e.g,

$$\phi^i(t) = \sum_{m=-\infty}^{\infty} \phi^i(m)e^{imt}. \tag{7.6.15}$$

The Fourier modes $\pi_i(m)$, $\phi_i^*(m)$ and $\pi_*^i(m)$ are defined analogously. The Hamiltonian acts on the Fourier modes as

$$\begin{aligned} [H, \phi^i(m)] &= m\phi^i(m), & [H, \pi_i(m)] &= m\pi_i(m), \\ [H, \phi_i^*(m)] &= m\phi_i^*(m), & [H, \pi_*^i(m)] &= m\pi_*^i(m). \end{aligned} \tag{7.6.16}$$

Now quantize. In the spirit of BRST quantization, our strategy is to quantize first and impose dynamics afterwards. In the extended history phase space $\mathcal{P}^*$, we define a Fock vacuum $|0\rangle$ which is annihilated by all negative frequency modes, i.e.

$$\phi^i(-m)|0\rangle = \pi_i(-m)|0\rangle = \phi_i^*(-m)|0\rangle = \pi_*^i(-m)|0\rangle = 0, \tag{7.6.17}$$

for all $-m < 0$. We must also decide which of the zero modes that annihilate the vacuum, but the decision is not important unless zero-momentum modes will survive in cohomology, and even then it will not affect the eigenvalues of the Hamiltonian.

The Hamiltonian (7.6.13) does not act in a well-defined manner, because it assigns an infinite energy to the Fock vacuum. To correct for that, we replace the Hamiltonian by

$$H = -i \int dt\ :\dot{\phi}^i(t)\pi_i(t): + :\dot{\phi}_i^*(t)\pi_*^i(t):, \tag{7.6.18}$$

where normal ordering $:\cdot:$ moves negative frequency modes to the right and positive frequency modes to the left. The vacuum has zero energy as measured by the normal-ordered Hamiltonian, $H|0\rangle = 0$. The Hilbert space can be identified with

$$\mathcal{H}(\mathcal{P}^*) = C(\phi^i(m>0), \pi_i(m>0), \phi_i^*(m>0), \pi_*^i(m>0)). \tag{7.6.19}$$

The energy of a state in $\mathcal{H}(\mathcal{P}^*)$ follows from

$$H\phi^{i_1}(m_1)...\pi_*^{i_n}(m_n)|0\rangle = (m_1 + ... + m_n)\phi^{i_1}(m_1)...\pi_*^{i_n}(m_n)|0\rangle. \tag{7.6.20}$$

It is important that the KT operator

$$Q_{KT} = \mathcal{E}_\alpha \pi_*^\alpha = \int dt\ \mathcal{E}_i(t)\pi_*^i(t) = \sum_{m=-\infty}^{\infty} \mathcal{E}_i(m)\pi_*^i(-m) \tag{7.6.21}$$

is already normal ordered, because $\mathcal{E}_\alpha$ and π_*^α commute. This means that $Q_{KT}^2 = 0$ also quantum mechanically; there are no anomalies. Moreover, Q_{KT} still commutes with the Hamiltonian, $[Q_{KT}, H] = 0$, and this property is not destroyed by normal ordering. Hence the Hilbert space $\mathcal{H}(\mathcal{P}^*)$ has also a well-defined decomposition into subspaces of definite antifield number,

$$\mathcal{H}(\mathcal{P}^*) = ... + \mathcal{H}^{-2} + \mathcal{H}^{-1} + \mathcal{H}^0 + \mathcal{H}^1 + \mathcal{H}^2 + ... \tag{7.6.22}$$

There is a KT complex in $\mathcal{H}(\mathcal{P}^*)$

$$\dots \xleftarrow{\delta} \mathcal{H}^{-2} \xleftarrow{\delta} \mathcal{H}^{-1} \xleftarrow{\delta} \mathcal{H}^0 \xleftarrow{\delta} \mathcal{H}^1 \xleftarrow{\delta} \mathcal{H}^2 \xleftarrow{\delta} \dots \tag{7.6.23}$$

The physical Hilbert space is identified with $\mathcal{H}(\Sigma) = \mathcal{H}^0_{qm}(Q_{KT}) = (\ker Q_{KT})_0/(\mathrm{im}\ Q_{KT})_0$. The action of the Hamiltonian on the physical Hilbert space is still given by (7.6.20), restricted to $\mathcal{H}(\Sigma) \subset \mathcal{H}(\mathcal{P}^*)$, and that coincides with the conventional action of the Hamiltonian.

Hence we have quantized the theory given by the EL equation (7.6.1) by first quantizing the space of phase space histories $\mathcal{P}^*$, and then imposing dynamics through KT cohomology.

7.7 Scalar field I: non-covariant quantization

The action, Euler-Lagrange equations, and Hessian read

$$\begin{aligned} S &= \frac{1}{2}\int d^N x\ (\partial_\mu\phi(x)\partial^\mu\phi(x) - \omega^2\phi^2(x)), \\ \mathcal{E}(x) &\equiv -\frac{\delta S}{\delta\phi(x)} = \Box\phi(x) + \omega^2\phi(x) = 0, \\ K(x,x') &\equiv -\frac{\delta^2 S}{\delta\phi(x)\delta\phi(x')} = \Box\delta(x-x') + \omega^2\delta(x-x'), \end{aligned} \tag{7.7.1}$$

where $\Box = \partial_\mu\partial^\mu$.

Introduce antifields $\phi^*(x)$ and canonical momenta $\pi(x) = \delta/\delta\phi(x)$ and $\pi_*(x) = \delta/\delta\phi^*(x)$. The non-zero brackets are

$$[\pi(x), \phi(x')] = \{\pi_*(x), \phi^*(x')\} = \delta(x-x'). \tag{7.7.2}$$

The KT differential reads

$$Q_{KT} = \int d^N x\ (\Box\phi(x) + \omega^2\phi(x))\pi_*(x). \tag{7.7.3}$$

Q_{KT} acts as $\delta F = [Q_{KT}, F]$, where

$$\begin{aligned} \delta\phi(x) &= 0, \\ \delta\phi^*(x) &= \Box\phi(x) + \omega^2\phi(x), \\ \delta\pi(x) &= -(\Box\pi_*(x) + \omega^2\pi_*(x)), \\ \delta\pi_*(x) &= 0. \end{aligned} \tag{7.7.4}$$

Now we do a Fourier transformation. The extended phase space $\mathcal{P}^*$ is spanned by modes $\phi(k)$, $\phi^*(k)$, $\pi(k)$ and $\pi_*(k)$, and the EL equation becomes

$$\mathcal{E}(k) = -(k^2 - \omega^2)\phi(k) = 0. \tag{7.7.5}$$

The non-zero brackets are

$$[\pi(k), \phi(k')] = \{\pi_*(k), \phi^*(k')\} = \delta(k+k'). \tag{7.7.6}$$

The KT differential is

$$Q_{KT} = \int d^N k\ (k^2 - \omega^2)\phi(k)\pi_*(-k). \tag{7.7.7}$$

Q_{KT} acts as $\delta F = [Q_{KT}, F]$, where

$$\begin{aligned} \delta\phi(k) &= 0, \\ \delta\phi^*(k) &= (k^2 - \omega^2)\phi(k), \\ \delta\pi(k) &= -(k^2 - \omega^2)\pi_*(k), \\ \delta\pi_*(k) &= 0. \end{aligned} \tag{7.7.8}$$

The cohomology is computed as follows. Since the equations (7.7.8) decouple, we can consider each value of k separately. First assume that $k^2 \neq \omega^2$. $\phi(k)$ and $\pi_*(k)$ are closed for all k, but $\phi^*(k)$ and $\pi(k)$ are not closed since $\delta\phi^*(k) \neq 0$, etc. We can invert the second and third equations to read

$$\begin{aligned} \phi(k) &= \frac{1}{k^2 - \omega^2}\delta\phi^*(k), \\ \pi_*(k) &= -\frac{1}{k^2 - \omega^2}\delta\pi(k). \end{aligned} \tag{7.7.9}$$

Hence $\phi(k)$ and $\pi_*(k)$ lie in the image of δ, and the cohomology vanishes completely: only $\phi(k)$ and $\pi_*(k)$ lie in the kernel, but they also lie in the image.

Now turn to the case $k^2 = \omega^2$, say $k = (\omega, 0, 0, 0)$. Clearly, $\delta\phi(k) = \delta\pi(k) = \delta\phi^*(k) = \delta\pi_*(k) = 0$, so all four variables lie in the kernel but not in the image. Thus the cohomology spaces are too big; the classical cohomology spaces can be identified with $H^\bullet_{cl}(\delta) = C(\phi(k), \pi(k), \phi^*(k), \pi_*(k))$. The zeroth cohomology space consists of such functions with total antifield number zero, i.e. $H^0_{cl}(\delta) = C(\phi(k), \pi(k), (\phi^*(k)\pi_*(k')))$. In [18] it was proposed that this problem could be handled by adding a small perturbation to make the Hessian non-singular, so the momenta can be killed in cohomology. This unwanted cohomology is an embarassment, especially since it reappears in Maxwell theory, but we have nothing more to say about it.

To quantize the theory we must specify a Hamiltonian. Let it be

$$\begin{aligned} H &= -i\int d^N x\ (\partial_0\phi(x)\pi(x) + \partial_0\phi^*(x)\pi_*(x)) \\ &= \int d^N k\ k_0(\phi(k)\pi(-k) + \phi^*(k)\pi_*(-k)). \end{aligned} \tag{7.7.10}$$

Note that at this stage we break Poincaré invariance, since the Hamiltonian treats the x^0 coordinate differently from the other x^μ. Quantize by introducing a Fock vacuum $\big|0\big\rangle$ satisfying

$$\phi(k)\big|0\big\rangle = \pi(k)\big|0\big\rangle = \phi^*(k)\big|0\big\rangle = \pi_*(k)\big|0\big\rangle = 0, \tag{7.7.11}$$

for all k such that $k_0 < 0$. After adding a small perturbation to make the Hessian invertible, $\pi(k)$ and $\pi_*(k)$ vanish in cohomology, as do the off-shell components of $\phi(k)$ and $\phi^*(k)$. The classical cohomology $H^\bullet_{cl}(Q_{KT}) = C(\phi(k; k^2 = \omega^2), \phi^*(k; k^2 = \omega^2))$ consists of functions of the on-shell components of ϕ and ϕ^*, and $H^0_{cl}(Q_{KT}) = C(\phi(k; k^2 = \omega^2))$ is the classical phase space. The quantization step eliminates the components $\phi(k)$ with $k_0 < 0$, which leaves us with the physical Hilbert space $\mathcal{H} = H^0_{qm}(Q_{KT}) = C(\phi(k; k^2 = \omega^2 \text{ and } k_0 > 0))$. A basis for $\mathcal{H}$ consists of multi-quanta states

$$\big|k, k', ..., k^{(n)}\big\rangle = \phi(k)\phi(k')...\phi(k^{(n)})\big|0\big\rangle \tag{7.7.12}$$

with energy $H = k + k' + ... + k^{(n)}$.

7.8 MCCQ: Jets and covariant quantization

A covariant definition of the phase space was given in the Section 7.6, but the Hamiltonian and thus the quantum Hilbert space broke covariance, due to the selection of a privileged time coordinate. In this section we correct this defect.

The compact notation is not very useful here, because the notion of covariance does not make sense unless some indices are identified with spacetime coordinates. So we assume that we have some fields $\phi^\alpha(x)$, where $x = (x^\mu) \in \mathbb{R}^N$ is the spacetime coordinate. The EL equations read

$$\mathcal{E}_\alpha(x) \equiv \frac{\delta S}{\delta\phi^\alpha(x)} = 0. \tag{7.8.1}$$

We also need the Hessian

$$K_{\alpha\beta}(x, x') = K_{\beta\alpha}(x', x) = \frac{\delta\mathcal{E}_\alpha(x)}{\delta\phi^\beta(x')} = \frac{\delta^2 S}{\delta\phi^\alpha(x)\delta\phi^\beta(x')}. \tag{7.8.2}$$

which we assume is non-singular.

Now let all fields depend on an additional parameter t. It will eventually be identified with time, but so far it is completely unrelated to the x^μ. Upon the substitution $\phi^\alpha(x) \to \phi^\alpha(x, t)$, the EL equations are replaced by

$$\mathcal{E}_\alpha(x, t) = 0. \tag{7.8.3}$$

The Hessian (7.8.2) becomes

$$K_{\alpha\beta}(x, t, x', t') = K_{\beta\alpha}(x', t', x, t) = \frac{\delta\mathcal{E}_\alpha(x, t)}{\delta\phi^\beta(x', t')}, \tag{7.8.4}$$

which has the inverse $M^{\alpha\beta}(x, t, x', t')$ satisfying

$$\int d^N x'' \int dt''\ K_{\beta\gamma}(x, t, x'', t'')M^{\gamma\alpha}(x'', t'', x', t') = \delta^\alpha_\beta\delta(x - x')\delta(t - t').$$

To remove the condition (7.8.3) in cohomology we introduce antifields $\phi^*_\alpha(x, t)$. But the fields in the physical phase space do not depend on the parameter t, which gives rise to the extra condition

$$\partial_t\phi^\alpha(x, t) \equiv \frac{\partial\phi^\alpha(x, t)}{\partial t} = 0. \tag{7.8.5}$$

We can implement this condition by introducing new antifields $\bar\phi^\alpha(x, t)$. However, the identities $\partial_t\mathcal{E}_\alpha(x, t) \equiv 0$ give rise to unwanted cohomology. To kill this condition, we must introduce yet another antifield $\bar\phi^*_\alpha(x, t)$. The KT differential δ is defined by

$$\begin{aligned} \delta\phi^\alpha(x, t) &= 0, \\ \delta\phi^*_\alpha(x, t) &= \mathcal{E}_\alpha(x, t), \\ \delta\bar\phi^\alpha(x, t) &= \partial_t\phi^\alpha(x, t), \\ \delta\bar\phi^*_\alpha(x, t) &= \partial_t\phi^*_\alpha(x, t) - \int d^N x' \int dt'\ K_{\alpha\beta}(x, t, x', t')\bar\phi^\beta(x', t'). \end{aligned} \tag{7.8.6}$$

The zeroth cohomology group $H^0_{cl}(\delta)$ equals $C(\phi)$, modulo the ideals generated by $\mathcal{E}_\alpha(x,t)$ and $\partial_t\phi^\alpha(x,t)$. Moreover, the wouldbe cohomology related to the identity

$$\delta\Big\{\partial_t\phi^*_\alpha(x,t) - \int d^Nx' \int dt' \,\frac{\delta\mathcal{E}_\alpha(x,t)}{\delta\phi^\beta(x',t')}\bar\phi^\beta(x',t')\Big\} \equiv 0 \tag{7.8.7}$$

is killed because the expression equals $\delta\bar\phi^*_\alpha(x,t)$.

Introduce canonical momenta for all fields and antifields: $\pi_\alpha(x,t) = \delta/\delta\phi^\alpha(x,t)$, $\pi^\alpha_*(x,t) = \delta/\delta\phi^*_\alpha(x,t)$, $\bar\pi_\alpha(x,t) = \delta/\delta\bar\phi^\alpha(x,t)$, and $\bar\pi^\alpha_*(x,t) = \delta/\delta\bar\phi^*_\alpha(x,t)$. The KT operator takes the explicit form

$$\begin{aligned} Q_{KT} &= \int d^Nx \int dt\, \Big\{\mathcal{E}_\alpha(x,t))\pi^\alpha_*(x,t) + \partial_t\phi^\alpha(x,t)\bar\pi_\alpha(x,t) \\ &\quad + (\partial_t\phi^*_\alpha(x,t) - \int d^Nx' \int dt'\; K_{\alpha\beta}(x,t,x',t')\bar\phi^\beta(t'))\bar\pi^\alpha_*(x,t)\Big\}. \end{aligned} \tag{7.8.8}$$

From this we can read off the action of δ on the momenta. As in the previous section, the zeroth cohomology group consists of functions $\phi^\alpha(x,t)$ which satisfy $\mathcal{E}_\alpha(x,t) = 0$ and $\partial_t\phi^\alpha(x,t) = 0$. Hence $H^0_{cl}(\delta) = C(\Sigma)$, as desired.

At this point, we must define a Hamiltonian. The candidate

$$\begin{aligned} H_0 &= -i\int dt\, \Big\{\partial_t\phi^\alpha(x,t)\pi_\alpha(x,t) + \partial_t\phi^*_\alpha(x,t)\pi^\alpha_*(x,t) \\ &\quad + \partial_t\bar\phi^\alpha(x,t)\bar\pi_\alpha(x,t) + \partial_t\bar\phi^*_\alpha(x,t)\bar\pi^\alpha_*(x,t)\Big\} \end{aligned} \tag{7.8.9}$$

might seem natural, but it is not acceptable. The action of the Hamiltonian is KT exact, e.g.

$$[H_0, \phi^\alpha(x,t)] = \partial_t\phi^\alpha(x,t) = \delta\bar\phi^\alpha(x,t), \tag{7.8.10}$$

and thus $H_0 \approx 0$. H_0 is not a genuine Hamiltonian, but rather a Hamiltonian constraint $H_0 \approx 0$, familiar from canonical quantization of general relativity.

However, we can construct a well-defined and physical Hamiltonian with some extra work. The crucial idea is to introduce the observer's trajectory $q^\mu(t) \in \mathbb{R}^N$, and then expand all fields in a Taylor series around this trajectory as in (7.5.5). Expand also the Euler-Lagrange equations and the antifields in a similar Taylor series, e.g. $\mathcal{E}_\alpha(x,t) = \sum_{\mathbf{m}} \frac{1}{\mathbf{m}!}\mathcal{E}_{\alpha,\mathbf{m}}(t)(x-q(t))^{\mathbf{m}}$. Such relations define the jets $\mathcal{E}_{\alpha,\mathbf{m}}(t)$, $\phi^*_{\alpha,\mathbf{m}}(t)$, $\bar\phi^\alpha_{,\mathbf{m}}(t)$ and $\bar\phi^*_{\alpha,\mathbf{m}}(t)$. Jets of antifields will sometimes be called antijets.

The equation of motion and the time-independence condition translate into

$$\begin{aligned} \mathcal{E}_{\alpha,\mathbf{m}}(t) &= 0, \\ D_t\phi^\alpha_{,\mathbf{m}}(t) &\equiv \dot\phi^\alpha_{,\mathbf{m}}(t) - \sum_\mu \dot q^\mu(t)\phi^\alpha_{,\mathbf{m}+\mu}(t) = 0. \end{aligned} \tag{7.8.11}$$

The KT differential δ which implements these conditions is

$$\begin{aligned}
\delta\phi^\alpha_{,\mathbf{m}}(t) &= 0, \\
\delta\phi^*_{\alpha,\mathbf{m}}(t) &= \mathcal{E}_{\alpha,\mathbf{m}}(t), \\
\delta\bar\phi^\alpha_{,\mathbf{m}}(t) &= D_t\phi^\alpha_{,\mathbf{m}}(t), \\
\delta\bar\phi^*_{\alpha,\mathbf{m}}(t) &= D_t\phi^*_{\alpha,\mathbf{m}}(t) - \sum_{\mathbf{n}} \int dt'\ K^{\mathbf{n}}_{\mathbf{m};\alpha\beta}(t,t')\bar\phi^\beta_{,\mathbf{n}}(t').
\end{aligned} \tag{7.8.12}$$

The cohomology group $H^0_{cl}(\delta)$ consists of linear combinations of jets $\phi^\alpha_{,\mathbf{m}}(t)$ satisfying $\mathcal{E}_{\alpha,\mathbf{m}}(t) = 0$ and $D_t\phi^\alpha_{,\mathbf{m}}(t) = 0$.

The Taylor expansion requires that we introduce the observer's trajectory as a physical field, but what equation of motion does it obey? The obvious answer is the geodesic equation, which we compactly write as $\mathcal{G}_\mu(t) = 0$. The geodesic operator $\mathcal{G}_\mu(t)$ is a function of the metric $g_{\mu\nu}(q(t), t)$ and its derivatives on the curve $q^\mu(t)$. To eliminate this ideal in cohomology we introduce the trajectory antifield $q^*_\mu(t)$, and extend the KT differential to it:

$$\begin{aligned}
\delta q^\mu(t) &= 0, \\
\delta q^*_\mu(t) &= \mathcal{G}_\mu(t).
\end{aligned} \tag{7.8.13}$$

For models defined over Minkowski spacetime, the geodesic equation simply becomes $\ddot q^\mu(t) = 0$, and the KT differential reads

$$\delta q^*_\mu(t) = \eta_{\mu\nu}\ddot q^\nu(t). \tag{7.8.14}$$

$H^0_{cl}(\delta)$ only contains trajectories which are straight lines,

$$q^\mu(t) = u^\mu t + a^\mu, \tag{7.8.15}$$

where u^μ and a^μ are constant vectors. We may also require that u^μ has unit length, $u_\mu u^\mu = 1$. This condition fixes the scale of the parameter t in terms of the Minkowski metric, so we may regard it as proper time rather than as an arbitrary parameter.

Now introduce the canonical momenta $\pi^{;\mathbf{m}}_\alpha(t) = \delta/\delta\phi^\alpha_{,\mathbf{m}}(t)$, $\pi^{\alpha,\mathbf{m}}_*(t) = \delta/\delta\phi^*_{\alpha,\mathbf{m}}(t)$, $\bar\pi^{;\mathbf{m}}_\alpha(t) = \delta/\delta\bar\phi^\alpha_{,\mathbf{m}}(t)$, $\bar\pi^{\alpha,\mathbf{m}}_*(t) = \delta/\delta\bar\phi^*_{\alpha,\mathbf{m}}(t)$ for the jets and antijets (jet and antijet momenta), and momenta $p_\mu(t) = \delta/\delta q^\mu(t)$ and $p^\mu_*(t) = \delta/\delta q^*_\mu(t)$ for the observer's trajectory and its antifield. We can now define a genuine Hamiltonian H, which translates the fields relative to the observer or vice versa. Since the formulas are shortest when H acts on the trajectory but not on the jets, we make that choice, and define

$$H = i\int dt\ (\dot q^\mu(t)p_\mu(t) + \dot q^*_\mu(t)p^\mu_*(t)). \tag{7.8.16}$$

Note the sign; moving the fields forward in t is equivalent to moving the observer backwards. From (7.5.5) we get the energy of the fields:

$$[H, \phi^\alpha(x,t)] = -i\dot{q}^\mu(t)\partial_\mu\phi^\alpha(x,t). \tag{7.8.17}$$

This a crucial result, because it allows us to define a genuine energy operator in a covariant way. In Minkowski space, the trajectory is a straight line (7.8.15), and $\dot{q}^\mu(t) = u^\mu$. If we take u^μ to be the constant four-vector $u^\mu = (1,0,0,0)$, then (7.8.17) reduces to

$$[H, \phi^\alpha(x,t)] = -i\frac{\partial}{\partial x^0}\phi^\alpha(x,t). \tag{7.8.18}$$

Equation (7.8.16) is thus a genuine covariant generalization of the energy operator.

Now we quantize the theory. Since all operators depend on the parameter t, we can define the Fourier components as in (7.5.13). The the Fock vacuum $\big|0\big\rangle$ is defined to be annihilated by all negative frequency modes, $\phi^\alpha_{,\mathbf{m}}(-m)$, $q^\mu(-m)$, etc. with $m < 0$. The normal-ordered form of the Hamiltonian (7.8.16) reads, in Fourier space,

$$H = -\sum_{m=-\infty}^{\infty} m(\,{:}q^\mu(m)p_\mu(-m){:} + {:}q^*_\mu(m)p_\mu(-m){:}\,), \tag{7.8.19}$$

where double dots indicate normal ordering with respect to frequency. This ensures that $H\big|0\big\rangle = 0$. The classical phase space $H^0_{cl}(\delta)$ is thus the the space of fields $\phi^\alpha(x)$ which solve $\mathcal{E}_\alpha(x) = 0$, and trajectories $q^\mu(t) = u^\mu t + a^\mu$, where $u^2 = 1$. After quantization, the fields and trajectories become operators which act on the physical Hilbert space $\mathcal{H} = H^0_{qm}(Q_{KT})$, which is the space of functions of the positive-energy modes of the classical phase space variables.

This construction differs technically from conventional canonical quantization, but there is also a physical difference. Consider the state $\big|\phi^\alpha(x)\big\rangle = \phi^\alpha(x)\big|0\big\rangle$ which excites one ϕ quantum from the vacuum. The Hamiltonian yields

$$\begin{aligned} H\big|\phi^\alpha(x)\big\rangle &= -i\dot{q}^\mu(t)\partial_\mu\phi^\alpha(x)\big|0\big\rangle \\ &= -i\big|\dot{q}^\mu(t)\partial_\mu\phi^\alpha(x)\big\rangle \\ &= -i\big|u^\mu\partial_\mu\phi^\alpha(x)\big\rangle. \end{aligned} \tag{7.8.20}$$

If u^μ were a classical variable, the state $\big|\phi^\alpha(x)\big\rangle$ would be a superposition of energy eigenstates:

$$H\big|\phi^\alpha(x)\big\rangle = -iu^\mu\partial_\mu\big|\phi^\alpha(x)\big\rangle. \tag{7.8.21}$$

In particular, let $u^\mu = (1,0,0,0)$ be a unit vector in the x^0 direction and $\phi^\alpha(x) = \exp(ik\cdot x)$ be a plane wave. We then define the state $\big|0;u,a\big\rangle$ by

$$q^\mu(t)\big|0;u,a\big\rangle = (u^\mu t + a^\mu)\big|0;u,a\big\rangle. \tag{7.8.22}$$

Now write $|k; u, a\rangle = \exp(ik \cdot x)|0; u, a\rangle$ for the single-quantum energy eigenstate.

$$H|k; u, a\rangle = k_\mu u^\mu |k; u, a\rangle, \tag{7.8.23}$$

so the eigenvalue of the Hamiltonian is $k_\mu u^\mu = k_0$, as expected. Moreover, the lowest-energy condition ensures that only quanta with positive energy will be excited; if $k_\mu u^\mu < 0$ then $|k; u, a\rangle = 0$.

However, the present analysis shows that it is in principle wrong to consider u^μ and a^μ as classical variables. The definition (7.8.22) means that the reference state $|0; u, a\rangle$ is a very complicated, mixed, macroscopic state where the observer moves along a well-defined, classical trajectory. This is of course an excellent approximation in practice, but in principle wrong.

7.9 Scalar field II: covariant quantization

Following the prescription in Section 7.8, we make the replacement $\phi(x) \to \phi(x, t)$, where $t \in \mathbb{R}$ is a parameter. The EL equation (7.7.1) becomes

$$\mathcal{E}(x, t) \equiv \Box\phi(x, t) + \omega^2\phi(x, t) = 0. \tag{7.9.1}$$

To remove this condition in cohomology we introduce antifields $\phi^*(x, t)$. But there is an extra condition

$$\partial_t\phi(x, t) \equiv \frac{\partial\phi(x, t)}{\partial t} = 0. \tag{7.9.2}$$

We can implement this condition by introducing new antifields $\bar{\phi}(x, t)$. However, the identities $\partial_t\mathcal{E}(x, t) \equiv 0$ give rise to unwanted cohomology. To kill this condition, we must introduce a second-order antifield $\bar{\phi}^*(x, t)$. After passage to jet space, the equation of motion and the time-independence condition translate into

$$\begin{aligned} \sum_\mu \phi_{,\mathbf{m}+2\mu}(t) + \omega^2\phi_{,\mathbf{m}}(t) &= 0, \\ D_t\phi_{,\mathbf{m}}(t) \equiv \dot{\phi}_{,\mathbf{m}}(t) - \sum_\mu \dot{q}^\mu(t)\phi_{,\mathbf{m}+\mu}(t) &= 0. \end{aligned} \tag{7.9.3}$$

We introduce anti-jets $\phi^*_{,\mathbf{m}}(t)$, $\bar{\phi}_{,\mathbf{m}}(t)$ and $\bar{\phi}^*_{,\mathbf{m}}(t)$ and the KT differential δ to implement these conditions:

$$\begin{aligned} \delta\phi_{,\mathbf{m}}(t) &= 0, \\ \delta\phi^*_{,\mathbf{m}}(t) &= \sum_\mu \phi_{,\mathbf{m}+2\mu}(t) + \omega^2\phi_{,\mathbf{m}}(t), \\ \delta\bar{\phi}_{,\mathbf{m}}(t) &= D_t\phi_{,\mathbf{m}}(t), \\ \delta\bar{\phi}^*_{,\mathbf{m}}(t) &= D_t\phi^*(t) - (\sum_\mu \bar{\phi}_{,\mathbf{m}+2\mu}(t) + \omega^2\bar{\phi}_{,\mathbf{m}}(t)). \end{aligned} \tag{7.9.4}$$

The classical cohomology group $H^0_{cl}(\delta)$ is spanned by of linear combinations of jets satisfying

$$\phi_{,\mathbf{m}}(t) = \mathrm{e}^{ik\cdot q(t)}(ik)^{\mathbf{m}} \tag{7.9.5}$$

where $k^2 = \omega^2$, $k \cdot q = k_\mu q^\mu$ and the power $k^{\mathbf{m}}$ is defined in analogy with (7.5.6). It is hardly surprising that the Taylor series can be summed, giving

$$\begin{aligned}\phi(x,t) &= \mathrm{e}^{ik\cdot q(t)} \sum_{\mathbf{m}} \frac{1}{\mathbf{m}!}(ik)^{\mathbf{m}}(x - q(t))^{\mathbf{m}} \\ &= \mathrm{e}^{ik\cdot q(t)}\mathrm{e}^{ik\cdot(x-q(t))} \\ &= \mathrm{e}^{ik\cdot x}.\end{aligned} \tag{7.9.6}$$

The physical Hamiltonian H is defined as in Equation (7.8.19). The classical phase space $H^0_{cl}(\delta)$ is thus the the space of plane waves $\mathrm{e}^{ik\cdot x}$, cf (7.9.6), and trajectories $q^\mu(t) = u^\mu t + a^\mu$. The energy is given by

$$\begin{aligned}[H, \mathrm{e}^{ik\cdot x}] &= k_\mu \dot{q}^\mu(t)\mathrm{e}^{ik\cdot x} = k_\mu u^\mu \mathrm{e}^{ik\cdot x}, \\ [H, q^\mu(t)] &= i\dot{q}^\mu(t).\end{aligned} \tag{7.9.7}$$

This is a covariant description of phase space, because the energy $k_\mu u^\mu$ is Poincaré invariant.

We now quantize the theory before imposing dynamics. To this end, we introduce the canonical momenta $\pi^{,\mathbf{m}}(t)$, $\pi_*^{;\mathbf{m}}(t)$, $\bar\pi^{,\mathbf{m}}(t)$, $\bar\pi_*^{;\mathbf{m}}(t)$ for the jets and antijets, and $p_\mu(t)$ and $p_*^\mu(t)$ for the observer's trajectory and its antifield. Since the jets also depend on the parameter t, we can define their Fourier components as in (7.5.13). The the Fock vacuum $\big|0\big\rangle$ is defined to be annihilated by the negative frequency modes of the jets and antijets, and the quantum Hamiltonian is still defined by (7.8.19), where double dots indicate normal ordering with respect to frequency, ensuring that $H\big|0\big\rangle = 0$.

The rest proceeds as in the end of Section 7.8. We can consider the one-quantum state with momentum k over the true Fock vacuum, $\big|k\big\rangle = \exp(ik\cdot x)\big|0\big\rangle$. This state is not an energy eigenstate, because the Hamiltonian excites a quantum of the observers trajectory: $H\big|k\big\rangle = k_\mu u^\mu \big|k\big\rangle$. We may think of the observer's trajectory as a classical variable and introduce the macroscopic reference state $\big|0; u, a\big\rangle$, on which $q^\mu(t)\big|0; u, a\big\rangle = (u^\mu t + a^\mu)\big|0; u, a\big\rangle$. We can then consider a state $\big|k; u, a\big\rangle = \exp(ik\cdot x)\big|k; u, a\big\rangle$ with one quantum over the reference state. The Hamiltonian gives $H\big|k; u, a\big\rangle = k_\mu u^\mu \big|k; u, a\big\rangle$. In particular, if $u^\mu = (1, 0, 0, 0)$, then the eigenvalue of the Hamiltonian is $k_\mu u^\mu = k_0$, as expected. Moreover, the lowest-energy condition (7.5.15) ensures that only quanta with positive energy will be excited; if $k_\mu u^\mu < 0$ then $\big|k; u, a\big\rangle = 0$.

7.10 MCCQ: Gauge symmetries

In the previous sections MCCQ was applied to the free scalar field. However, it is mainly useful for theories with gauge symmetries, due to its connection with the representation theory of gauge algebras developed earlier. We now come to this case, and assume that there are some relations between the EL equations (7.6.1). In other words, let there be identities of the form

$$r_a^\alpha \mathcal{E}_\alpha \equiv 0, \tag{7.10.1}$$

where the r_a^α are some functionals of ϕ^α. The zeroth cohomology group $H^0_{cl}(\delta) = C(\mathcal{Q})/\mathcal{N} = C(\Sigma)$ is not changed, but the higher cohomology groups no longer vanish, since $\delta(r_a^\alpha \phi^*_\alpha) = r_a^\alpha \mathcal{E}_\alpha \equiv 0$. The standard method to kill this unwanted cohomology is to introduce a bosonic second-order antifield ζ_a, so that $r_a^\alpha \phi^*_\alpha = \delta\zeta_a$ is KT exact. The differential (7.6.4) is thus modified to read

$$\begin{aligned} \delta\phi^\alpha &= 0, \\ \delta\phi^*_\alpha &= \mathcal{E}_\alpha, \\ \delta\zeta_a &= r_a^\alpha \phi^*_\alpha. \end{aligned} \tag{7.10.2}$$

By introducing canonical momenta $\chi^a = \delta/\delta\zeta_a$ for the second-order antifields, we can write the KT differential as a bracket, $\delta F = [Q_{KT}, F]$, where the full KT operator is

$$Q_{KT} = \mathcal{E}_\alpha \pi^\alpha_* + r_a^\alpha \phi^*_\alpha \chi^a. \tag{7.10.3}$$

Q_{KT} is an operator in the extended phase space $\mathcal{P}^*$ with basis $(\phi^\alpha, \pi_\beta, \phi^*_\alpha, \pi^\beta_*, \zeta_a, \chi^b)$, and $\{Q_{KT}, Q_{KT}\} = 0$.

The identity (7.10.1) implies that $J_a = r_a^\alpha \pi_\alpha$ generate a Lie algebra under the Poisson bracket. Namely, all J_a's preserve the action, because

$$[J_a, S] = r_a^\alpha [\pi_\alpha, S] = r_a^\alpha \mathcal{E}_\alpha \equiv 0, \tag{7.10.4}$$

and the bracket of two operators which preserve some structure also preserves the same structure. We will only consider the case that the J_a's generate a proper Lie algebra $\mathfrak{g}$ as in (7.2.6). The formalism extends without too much extra work to the more general case of structure functions $f_{ab}{}^c(\phi)$, but we will not need this complication here. It follows that the functions r_a^α satisfy the identity

$$\partial_\beta r_b^\alpha r_a^\beta - \partial_\beta r_a^\alpha r_b^\beta = f_{ab}{}^c r_c^\alpha. \tag{7.10.5}$$

The Lie algebra $\mathfrak{g}$ also acts on the antifields:

$$\begin{aligned} [J_a, \phi^\alpha] &= r_a^\alpha, \\ [J_a, \phi^*_\alpha] &= -\partial_\alpha r_a^\beta \phi^*_\beta \\ [J_a, \zeta_b] &= f_{ab}{}^c \zeta_c. \end{aligned} \tag{7.10.6}$$

In particular, it follows that ϕ^*_α carries a $\mathfrak{g}$ representation because it transforms in the same way as π_α does.

Classically, it is always possible to reduce the phase space further, by identifying points on $\mathfrak{g}$ orbits. To implement this additional reduction, we introduce ghosts c^a with anti-field number $\operatorname{afn} c^a = -1$, and ghost momenta b_a satisfying $\{b_a, c^b\} = \delta^b_a$. The Lie algebra $\mathfrak{g}$ acts on the ghosts as $[J_a, c^b] = -f_{ac}{}^b c^c$. The full extended phase space, still denoted by $\mathcal{P}^*$, is spanned by $(\phi^\alpha, \pi_\beta, \phi^*_\alpha, \pi^\beta_*, \zeta_a, \chi^b, c^a, b_b)$. The generators of $\mathfrak{g}$ are thus identified with the following vector fields in $\mathcal{P}^*$:

$$\begin{aligned} J_a &= r^\alpha_a \pi_\alpha - \partial_\alpha r^\beta_a \phi^*_\beta \pi^\alpha_* + f_{ab}{}^c \zeta_c \chi^b - f_{ab}{}^c c^b b_c \\ &= J^{field}_a + J^{ghost}_a, \end{aligned} \tag{7.10.7}$$

where $J^{ghost}_a = -f_{ab}{}^c c^b b_c$ and J^{field}_a is the rest.

Now define the longitudinal derivative d by

$$\begin{aligned} dc^a &= -\frac{1}{2} f_{bc}{}^a c^b c^c, \\ d\phi^\alpha &= r^\alpha_a c^a, \\ d\phi^*_\alpha &= \partial_\alpha r^\beta_a \phi^*_\beta c^a, \\ d\zeta_a &= -f_{ab}{}^c \zeta_c c^b. \end{aligned} \tag{7.10.8}$$

The longitudinal derivative can be written as $dF = [Q_{Long}, F]$ for every $F \in C(\mathcal{Q}^*)$, where

$$Q_{Long} = J^{field}_a c^a - \frac{1}{2} f_{ab}{}^c c^a c^b b_c = J^{field}_a c^a + \frac{1}{2} J^{ghost}_a c^a. \tag{7.10.9}$$

We note that Q_{Long} can be considered as smeared gauge generators, $\mathcal{J}_X = X^a J_a$, where the smearing function X^a is the fermonic ghost c^a:

$$Q_{Long} = \mathcal{J}^{field}_c + \frac{1}{2} \mathcal{J}^{ghost}_c. \tag{7.10.10}$$

One verifies that $d^2 = 0$ when acting on the fields and antifields by means of the identify (7.10.5) and the Jacobi identities for $\mathfrak{g}$. Moreover, it is straightforward to show that d anticommutes with the KT differential, $d\delta = -\delta d$; the proof is again done by checking the action on the fields. Hence we may define the nilpotent *BRST derivative* $s = \delta + d$,

$$\begin{aligned} sc^a &= -\frac{1}{2} f_{bc}{}^a c^b c^c, \\ s\phi^\alpha &= r^\alpha_a c^a, \\ s\phi^*_\alpha &= \mathcal{E}_\alpha + \partial_\alpha r^\beta_a \phi^*_\beta c^a, \\ s\zeta_a &= r^\alpha_a \phi^*_\alpha - f_{ab}{}^c \zeta_c c^b. \end{aligned} \tag{7.10.11}$$

Nilpotency immediately follows because $s^2 = \delta^2 + \delta d + d\delta + d^2 = 0$. The BRST operator can be written in the form $sF = [Q_{BRST}, F]$ with

$$\begin{aligned} Q_{BRST} &= Q_{KT} + Q_{Long} \\ &= \mathcal{E}_\alpha \pi_*^\alpha + r_a^\alpha \phi_\alpha^* \chi^a + J_a^{field} c^a + \frac{1}{2} J_a^{ghost} c^a \\ &= -\frac{1}{2} f_{ab}{}^c c^a c^b b_c + r_a^\alpha c^a \pi_\alpha + (\mathcal{E}_\alpha + \partial_\alpha r_a^\beta \phi_\beta^* c^a)\pi_*^\alpha \\ &\quad + (r_a^\alpha \phi_\alpha^* - f_{ab}{}^c \zeta_c c^b)\chi^a. \end{aligned} \tag{7.10.12}$$

In non-covariant quantization, we single out a privileged variable t among the α's, and declare it to be time. In the absense of gauge symmetries, the BRST operator reduces to the KT operator (7.10.3), which is already normal ordered and hence nilpotent on the quantum level. The question is whether the full BRST operator also has this property. The dangerous part is the longitudinal operator

$$Q_{Long} = \int dt \left\{ :J_a^{field}(t): c^a(t) + \frac{1}{2} :J_a^{ghost}(t) c^a(t): \right\}. \tag{7.10.13}$$

which ceases to be nilpotent unless the normal-ordered gauge generators $J_a(t) = :J_a^{field}(t): + :J_a^{ghost}(t):$ generate the algebra (7.2.6) without additional quantum corrections. If such an extension arises, the BRST operator ceases to be nilpotent. However, the situation is even worse. Not only do quantum effects generically ruin nilpotency of the BRST operator, but they make the gauge generators ill defined. However, it is possible to regularize the theory formulated in terms of Taylor data, in such a way that the full gauge symmetry of the original model is preserved, and the regularized gauge generators are well-defined operators. The price to pay is the appearance of an anomaly.

The next step in Section 7.8 was to introduce the observer's trajectory, expand all fields in a Taylor series around it, and quantize in the space of Taylor data histories. The motivation was mainly aesthetic; by adding the observer's trajectory, it is possible to write down a covariant expression (7.8.16) for the Hamiltonian, namely as the operator which translates the fields relative to the observer. However, it is in the presence of gauge symmetries that this construction becomes indispensable.

Thus, we reformulate the classical theory in jet coordinates. To the fields $c^a(x)$, $\phi^\alpha(x)$, $\phi^*_\alpha(x)$ and $\zeta_a(x)$ we associates p-jets $c^a_{,\mathbf{m}}(t)$, $\phi^\alpha_{,\mathbf{m}}(t)$, $\phi^*_{\alpha,\mathbf{m}}(t)$ and $\zeta_{a,\mathbf{m}}(t)$, with canonical momenta $b_a^{,\mathbf{m}}(t)$, $\pi_\alpha^{,\mathbf{m}}(t)$, $\pi_*^{\alpha,\mathbf{m}}(t)$ and $\chi^{a,\mathbf{m}}(t)$. We also introduce extra antifields $\bar\phi^\alpha_{,\mathbf{m}}(t)$ etc. to eliminate the t-dependence, but as in [19], they will not be written down explicitly.

To be concrete, consider the case that the symmetry is the DGRO algebra (7.4.1).

To each symmetry, we assign ghosts as in the following table:

	Gen	Smear	Ghost	Momentum	Q_{Long}
Diffeomorphisms	$\mathcal{L}_\xi$	$\xi^\mu(x)$	$c^\mu_{diff}(x,t)$	$b^{diff}_\mu(x,t)$	Q^{diff}_{Long}
Gauge	$\mathcal{J}_X$	$X^a(x)$	$c^a_{gauge}(x,t)$	$b^{gauge}_a(x,t)$	Q^{gauge}_{Long}
Reparametrizations	L_f	$f(t)$	$c_{rep}(t)$	$b^{rep}(t)$	Q^{rep}_{Long}

The BRST operator is $Q_{BRST} = Q_{Long} + Q_{KT}$, where the longitudinal operator is given by the prescription (7.10.10). For brevity, we only write down the formulas for the fields $\phi^\alpha(x,t)$ and the ghosts; the antifields do of course give rise to additional terms.

$$\begin{aligned}
Q^{diff}_{Long} &= -\int d^N x \int dt\, \Big\{ (c^\mu_{diff}(x,t)\partial_\mu\phi^\alpha(x,t) \\
&\quad + \partial_\nu c^\mu_{diff}(x,t) T^{\alpha\nu}_{\beta\mu}\phi^\beta(x,t))\pi_\alpha(x,t) \\
&\quad + c^\mu_{diff}(x,t)\partial_\mu c^\nu_{diff}(x,t) b^{diff}_\nu(x,t) \Big\}, \\
Q^{gauge}_{Long} &= -\int d^N x \int dt\, \Big\{ c^a_{gauge}(x,t) J^\alpha_{\beta a}\phi^\beta(x,t)\pi_\alpha(x,t) \\
&\quad + \frac{1}{2} f_{ab}{}^c c^a_{gauge}(x,t) c^b_{gauge}(x,t) b^{gauge}_c(x,t) \Big\}, \\
Q^{rep}_{Long} &= -\int d^N x \int dt\, \Big\{ (c_{rep}(t)\partial_t\phi^\alpha(x,t) \\
&\quad + \lambda(\dot{c}_{rep}(t) - i c_{rep}(t))\phi^\alpha(x,t))\pi_\alpha(x,t) \Big\} \\
&\quad - \int dt\; c_{rep}(t)\dot{c}_{rep}(t) b^{rep}(t).
\end{aligned} \tag{7.10.14}$$

These formulas assume that the field $\phi^\alpha(x)$ transforms as a tensor field. There is an additional term if the field is a connection, but this terms does not lead to any complications. After passage to jet space and normal ordering, we use the prescription (7.10.10) to find the longitudinal derivative, i.e. $\partial_{\mathbf{m}}\xi^\mu \to c^\mu_{diff,\mathbf{m}}$, $\partial_{\mathbf{m}} X^a \to c^a_{gauge,\mathbf{m}}$, and $f \to c_{rep}$:

$$
\begin{aligned}
Q^{diff}_{Long} &= \int dt \, \Big\{ c^{\mu}_{diff,\mathbf{0}} p_\mu(t) - \sum_{|\mathbf{n}| \leqslant |\mathbf{m}| \leqslant p} T^{\alpha\mathbf{n}}_{\beta\mathbf{m}}(c_{diff}(t)) :\phi^{\beta}_{,\mathbf{n}}(t)\pi^{;\mathbf{m}}_{\alpha}(t): \\
&\quad - \sum_{|\mathbf{n}| \leqslant |\mathbf{m}| \leqslant p} :T^{\mu\mathbf{n}}_{\nu\mathbf{m}}(c_{diff}(t)) c^{\nu}_{diff,\mathbf{n}}(t) b^{diff,\mathbf{m}}_{\mu}(t): \Big\}, \\
Q^{gauge}_{Long} &= -\int dt \, \Big\{ \sum_{|\mathbf{n}| \leqslant |\mathbf{m}| \leqslant p} J^{\alpha\mathbf{n}}_{\beta\mathbf{m}}(c_{gauge}(t)) :\phi^{\beta}_{,\mathbf{n}}(t)\pi^{;\mathbf{m}}_{\alpha}(t): \\
&\quad -\frac{1}{2} \sum_{|\mathbf{n}| \leqslant |\mathbf{m}| \leqslant p} :J^{a\mathbf{n}}_{b\mathbf{m}}(c_{gauge}(t)) c^{b}_{gauge,\mathbf{n}}(t) b^{gauge,\mathbf{m}}_{a}(t): \Big\}, \qquad (7.10.15) \\
Q^{rep}_{Long} &= -\int dt \, \Big\{ \sum_{|\mathbf{m}| \leqslant p} c_{rep}(t) :\dot{\phi}^{\alpha}_{,\mathbf{m}}(t)\pi^{;\mathbf{m}}_{\alpha}(t): \\
&\quad +\lambda \sum_{|\mathbf{m}| \leqslant p} (\dot{c}_{rep}(t) - i c_{rep}(t)) :\phi^{\alpha}_{,\mathbf{m}}(t)\pi^{;\mathbf{m}}_{\alpha}(t): \\
&\quad + :c_{rep}(t)\dot{c}_{rep}(t) b^{rep}(t): \Big\}.
\end{aligned}
$$

The matrices are given by (cf. (7.5.10))

$$
\begin{aligned}
T^{\mathbf{m}}_{\mathbf{n}}(c_{diff}(t)) &\equiv (T^{\alpha\mathbf{m}}_{\beta\mathbf{n}}(c_{diff}(t))) = \sum_{\mu\nu} \binom{\mathbf{n}}{\mathbf{m}} c^{\mu}_{diff,\mathbf{n}-\mathbf{m}+\nu}(t) T^{\nu}_{\mu} \\
&\quad + \sum_{\mu} \binom{\mathbf{n}}{\mathbf{m}-\mu} c^{\mu}_{diff,\mathbf{n}-\mathbf{m}+\mu}(t) - \sum_{\mu} \delta^{\mathbf{m}-\mu}_{\mathbf{n}} c^{\mu}_{diff,\mathbf{0}}(t), \\
J^{\mathbf{m}}_{\mathbf{n}}(c_{gauge}(t)) &\equiv (J^{\alpha\mathbf{m}}_{\beta\mathbf{n}}(c_{gauge}(t))) = \binom{\mathbf{n}}{\mathbf{m}} c^{a}_{gauge,\mathbf{n}-\mathbf{m}}(t) J_a. \qquad (7.10.16)
\end{aligned}
$$

$T^{\mu\mathbf{m}}_{\nu\mathbf{n}}$ and $J^{a\mathbf{m}}_{b\mathbf{n}}$ denote the specializations of $T^{\alpha\mathbf{m}}_{\beta\mathbf{n}}$ and $J^{\alpha\mathbf{m}}_{\beta\mathbf{n}}$ to the adjoint representations; $\sum_{\mathbf{m}} T^{\mu\mathbf{m}}_{\nu\mathbf{n}}(c) c^{\nu}_{,\mathbf{m}} = (c^{\nu} c^{\mu}_{,\nu})_{,\mathbf{n}}$ and $\sum_{\mathbf{m}} J^{a\mathbf{m}}_{b\mathbf{n}}(c) c^{b}_{,\mathbf{m}} = (c^{a} c^{b})_{,\mathbf{n}}$

The condition for $Q^2_{Long} = 0$, and thus $Q^2_{BRST} = 0$, is that the algebra generated by the normal-ordered gauge generators is anomaly free. However, even if this condition fails, which is the typical situation, everything is not lost. The KT operator is still nilpotent, and we can implement dynamics as the KT cohomology in the extended phase space without ghosts. The physical phase space now grows, because some gauge degrees of freedom become physical upon quantization.

In the next two sections, we apply this formalism to some well-known theories.

7.11 The free Maxwell field

The Maxwell field $A_\mu(x)$ transforms as a vector field under the Poincaré group and as a connection under the gauge algebra $\mathfrak{map}(N, \mathfrak{u}(1))$, whose smeared generators are denoted by $\mathcal{J}_X = \int d^4x\; X(x)J(x)$:

$$[\mathcal{J}_X, A_\mu(x)] = \partial_\mu X(x). \tag{7.11.1}$$

We use the Minkowski metric $\eta_{\mu\nu}$ and its inverse $\eta^{\mu\nu}$ to freely raise and lower indices, e.g. $F^{\mu\nu} = \eta^{\mu\rho}\eta^{\nu\sigma}F_{\rho\sigma}$. The field strength $F_{\mu\nu}(x) = \partial_\mu A_\nu(x) - \partial_\nu A_\mu(x)$ transforms in the adjoint representation, i.e. trivially. The action

$$S = \frac{1}{4}\int d^4x\; F_{\mu\nu}(x)F^{\mu\nu}(x) \tag{7.11.2}$$

leads to the equations of motion

$$\mathcal{E}^\mu(x) \equiv -\frac{\delta S}{\delta A_\mu(x)} = \partial_\nu F^{\mu\nu}(x) = 0. \tag{7.11.3}$$

The Maxwell equations are not all independent, because of the constraints

$$\partial_\mu \mathcal{E}^\mu(x) = \partial_\mu\partial_\nu F^{\mu\nu}(x) \equiv 0. \tag{7.11.4}$$

We are thus instructed to introduce the following fields: the first-order antifield $A^\mu_*(x)$ for the EL equation $\partial_\nu F^{\mu\nu}(x) = 0$; the second-order antifield $\zeta(x)$ for the identity $\partial_\mu\partial_\nu F^{\mu\nu}(x) \equiv 0$; and the ghost $c(x)$ to identify fields related by a gauge transformation of the form (7.11.1).

The BRST operator s acts as

$$\begin{aligned} sc(x) &= 0, \\ sA_\mu(x) &= \partial_\mu c(x), \\ sA^\mu_*(x) &= \partial_\nu F^{\mu\nu}(x), \\ s\zeta(x) &= \partial_\mu A^\mu_*(x), \end{aligned} \tag{7.11.5}$$

We check that $s^2 = 0$ and $sF_{\mu\nu} = s\partial_\mu A^\mu_* = 0$, so the kernel of s is spanned by c, the field strengths $F_{\mu\nu}$, and $\partial_\mu A^\mu_*$. im s is generated by the ideals $\partial_\mu c$, $\partial_\nu F^{\mu\nu}$, and $\partial_\mu A^\mu_*$. Hence $H^0_{cl}(s)$ consists of the gauge-invariant parts of A_μ (i.e. $F_{\mu\nu}$) which solve the Maxwell equations, as expected.

Introduce canonical momenta $E^\mu(x)$, $E^*_\mu(x)$, $\chi(x)$ and $b(x)$, defined by the following non-zero brackets:

$$\begin{aligned} [E^\mu(x), A_\nu(x')] &= \delta^\mu_\nu\delta(x - x'), \\ \{E^*_\mu(x), A^\nu_*(x')\} &= \delta^\nu_\mu\delta(x - x'), \\ [\chi(x), \zeta(x')] &= \delta(x - x'), \\ \{b(x), c(x')\} &= \delta(x - x'). \end{aligned} \tag{7.11.6}$$

It should be emphasized that $E^\mu = \delta/\delta A_\mu$ is the conjugate of the gauge potential in history space, and not yet related to the electric field $F^{\mu 0}$. We could introduce the condition $E^\mu \approx F^{\mu 0}$ as a constraint in the history phase space, turning the Maxwell equations into second class constraints. By keeping dynamics as a first-class constraint no such condition, which would ruin covariance, is necessary. The BRST operator can explicitly be written as

$$Q_{BRST} = \int d^4x \left\{ \partial_\mu c(x) E^\mu(x) + \partial_\nu F^{\mu\nu}(x) E^*_\mu(x) + \partial_\mu A^\mu_*(x)\chi(x) \right\}. \quad (7.11.7)$$

The physical content of the theory is clearer in Fourier space. The BRST operator

$$\begin{aligned} Q_{BRST} \quad = \quad & \int d^N k \left\{ k_\mu c(k) E^\mu(-k) + (k^\mu k_\nu A^\nu(k) - k^\nu k_\nu A^\mu(k)) E^*_\mu(-k) \right. \\ & \left. + k_\mu A^\mu_*(k)\chi(-k) \right\}, \end{aligned} \quad (7.11.8)$$

acts on the Fourier modes as

$$\begin{aligned} sc(k) &= 0, \\ sA_\mu(k) &= k_\mu c(k), \\ sA^\mu_*(k) &= k^\mu k_\nu A^\nu(k) - k^\nu k_\nu A^\mu(k), \\ s\zeta(k) &= k_\mu A^\mu_*(k). \end{aligned} \quad (7.11.9)$$

We distinguish between two cases:

1. $k^2 = \omega^2 \neq 0$, say $k = (\omega, 0, 0, 0)$. Then $sc = 0$, $sA_0 = \omega c$, $sA_1 = sA_2 = sA_3 = 0$, $sA^0_* = \omega^2 A_0 - \omega\omega A_0 = 0$, $sA^1_* = \omega^2 A_1$ $sA^2_* = \omega^2 A_2$, $sA^3_* = \omega^2 A_3$ and $s\zeta = \omega A^0_*$. The kernel is thus spanned by c, A_1, A_2, A_3 and A^0_*, and the image is spanned by c, A_1, A_2, A_3 and A^0_*. Since $\ker s = \operatorname{im} s$ there is no cohomology.

2. $k^2 = 0$, say $k = (k_0, 0, 0, k_0)$. Then $sc = 0$, $sA_0 = sA_3 = k_0 c$, $sA_1 = sA_2 = 0$, $sA^0_* = sA^3_* = k^0 k_\nu A^\nu$, $sA^1_* = sA^2_* = 0$ and $s\zeta = k_\mu A^\mu_*$. The kernel is thus spanned by c, A_1, A_2, $k^\mu A_\mu = A_0 - A_3$, A^1_*, A^2_*, and $k_\mu A^\mu_* = A^0_* - A^3_*$. The image is spanned by c, $k^\mu A_\mu$ and $k_\mu A^\mu_*$, which factor out in cohomology. We are left with two physical polarizations A_1 and A_2.

We here assumed that the momenta factor out in cohomology. As for the scalar field in Section 7.7 and the harmonic oscillator in [18], this is not quite true. There is unwanted cohomology because the Hessian is singular. However, this problem has nothing to do with gauge invariance.

We now quantize in the history phase space before introducing dynamics by passing to the BRST cohomology. We single out one direction x^0 as time, and take the Hamiltonian to be the generator of rigid time translations,

$$
\begin{aligned}
H &= -i\int d^4x \,\Big\{\partial_0 A_\mu(x)E^\mu(x) + \partial_0 A_*^\mu(x)E_\mu^*(x) \\
&\qquad +\partial_0\zeta(x)\chi(x) + \partial_0 c(x)b(x)\Big\} \\
&= \int d^Nk\; k_0\Big\{A_\mu(k)E^\mu(-k) + A_*^\mu(k)E_\mu^*(-k) \\
&\qquad +\zeta(k)\chi(-k) + c(k)b(-k)\Big\}.
\end{aligned}
\tag{7.11.10}
$$

Note that at this stage we break Poincaré invariance, since the Hamiltonian treats the x^0 coordinate differently from the other x^μ. Quantize by introducing a Fock vacuum $|0\rangle$ satisfying

$$
\begin{aligned}
&A_\mu(k)|0\rangle = E^\mu(k)|0\rangle = A_*^\mu(k)|0\rangle = E_\mu^*(k)|0\rangle = \\
&\zeta(k)|0\rangle = \chi(k)|0\rangle = c(k)|0\rangle = b(k)|0\rangle = 0,
\end{aligned}
\tag{7.11.11}
$$

for all k such that $k_0 < 0$.

At this point we want to pass to BRST cohomology. There might be problems with normal ordering, but in fact the BRST operator (7.11.8) is already normal ordered. This is because the generator of $\mathfrak{u}(1)$ gauge transformations

$$
\mathcal{J}_X = -\int d^4x \; X(x)\partial_\mu E^\mu(x) \tag{7.11.12}
$$

is itself already normal ordered. There are thus no anomalies, and the BRST operator (7.11.8) remains nilpotent. We define the BRST state cohomology as the space of physical states, where a state is physical if it is BRST closed, $Q_{BRST}|phys\rangle = 0$, and two physical states are equivalent if they differ by a BRST exact state, $|phys\rangle \sim |phys'\rangle$ if $|phys\rangle - |phys'\rangle = Q_{BRST}|\rangle$.

The rest proceeds as for the harmonic oscillator [18] or the free scalar field in Section 7.7. After adding a small perturbation to make the Hessian invertible, all momenta vanish in cohomology, and only the transverse polarizations $\epsilon^\mu A_\mu(k) = 0$ with $\epsilon^\mu k_\mu = 0$ and $\epsilon^0 = 0$ survive. A basis for the history Hilbert space consists of multi-quanta states

$$
\epsilon_1^\mu A_\mu(k^{(1)})\ldots\epsilon_n^\mu A_\mu(k^{(n)})|0\rangle \tag{7.11.13}
$$

where $k_\mu^{(j)}k^{(j)\mu} = 0$ and $k_0^{(j)} > 0$. The energy is given by $H = k_0^{(1)} + \ldots + k_0^{(n)}$. The gauge generators (7.11.12) act in a well-defined manner, in fact trivially, on the Hilbert space, because $\epsilon_j^\mu k_\mu^{(j)} = 0$.

As in Section 7.8, we want to give a completely covariant description of the Hamiltonian. Therefore we pass to jet data, e.g.

$$
A_\mu(x) = \sum_{|\mathbf{m}|\leqslant p} \frac{1}{\mathbf{m}!} A_{\mu,\mathbf{m}}(t)(x - q(t))^{\mathbf{m}}. \tag{7.11.14}
$$

The equations of motion (7.11.3) translate into

$$\sum_\nu F^{\mu\nu}_{,\mathbf{m}+\nu}(t) = 0, \tag{7.11.15}$$

and the constraint (7.11.4) becomes

$$\sum_\mu \mathcal{E}^\mu_{\mu,\mathbf{m}}(t) = \sum_{\mu\nu} F^{\mu\nu}_{,\mathbf{m}+\mu+\nu}(t) \equiv 0, \tag{7.11.16}$$

where the field strength is

$$F_{\mu\nu,\mathbf{m}}(t) = A_{\mu,\mathbf{m}+\nu}(t) - A_{\nu,\mathbf{m}+\mu}(t) = 0. \tag{7.11.17}$$

We introduce jets also for the antifields and for the ghost, denoted by $A^\mu_{*,\mathbf{m}}(t)$, $\zeta_{,\mathbf{m}}(t)$, and $c_{,\mathbf{m}}(t) \equiv c^{gauge}_{,\mathbf{m}}(t)$. The BRST differential s which implements all these conditions is defined by

$$\begin{aligned} sc_{,\mathbf{m}}(t) &= 0, \\ sA_{\mu,\mathbf{m}}(t) &= c_{,\mathbf{m}+\mu}(t), \\ sA^\mu_{*,\mathbf{m}}(t) &= \sum_\nu F^{\mu\nu}_{,\mathbf{m}+\nu}(t), \\ s\zeta_{,\mathbf{m}}(t) &= \sum_\mu A^\mu_{*,\mathbf{m}+\mu}(t). \end{aligned} \tag{7.11.18}$$

Moreover, we demand that the Taylor series does not depend on the parameter t, which gives rise to conditions of the type

$$D_t A_{\mu,\mathbf{m}}(t) \equiv \dot{A}_{\mu,\mathbf{m}}(t) - \sum_\nu \dot{q}^\nu(t) A_{\mu,\mathbf{m}+\nu}(t) = 0. \tag{7.11.19}$$

As in (7.8.6), we need to double the number of antifields and introduce an additional differential σ to remove these conditions in cohomology. Thus we introduce antifields $\bar{c}_{,\mathbf{m}}(t)$, $\bar{A}_{\mu,\mathbf{m}}(t)$, $\bar{A}^\mu_{*,\mathbf{m}}(t)$, $\bar{\zeta}_{,\mathbf{m}}(t)$ and set

$$\begin{aligned} \sigma\bar{c}_{,\mathbf{m}}(t) &= \dot{c}_{,\mathbf{m}}(t), \\ \sigma\bar{A}_{\mu,\mathbf{m}}(t) &= \dot{A}_{\mu,\mathbf{m}}(t), \\ \sigma\bar{A}^\mu_{*,\mathbf{m}}(t) &= \dot{A}^\mu_{*,\mathbf{m}}(t), \\ \sigma\bar{\zeta}_{,\mathbf{m}}(t) &= \dot{\zeta}_{,\mathbf{m}}(t) \\ \sigma c_{,\mathbf{m}}(t) &= \sigma A_{\mu,\mathbf{m}}(t) = \sigma A^\mu_{*,\mathbf{m}}(t) = \sigma\zeta_{,\mathbf{m}}(t) = 0. \end{aligned} \tag{7.11.20}$$

Clearly, $\sigma^2 = 0$. We also extend the definition of the BRST differential s to the barred antifields:

$$\begin{array}{rcl} s\bar{c}_{,\mathbf{m}}(t) & = & 0, \\ s\bar{A}_{\mu,\mathbf{m}}(t) & = & -\bar{c}_{,\mathbf{m}+\mu}(t), \\ s\bar{A}^{\mu}_{*,\mathbf{m}}(t) & = & -\sum_{\nu}(\bar{A}_{\nu,\mathbf{m}+\mu}(t) - \bar{A}_{\mu,\mathbf{m}+\nu}(t)), \\ s\bar{\zeta}_{,\mathbf{m}}(t) & = & -\sum_{\mu}\bar{A}^{\mu}_{*,\mathbf{m}+\mu}(t). \end{array} \tag{7.11.21}$$

That $s^2 = 0$ follows in the same way as for (7.11.18). Moreover, we verify that $s\sigma = -\sigma s$, and hence $s + \sigma$ is nilpotent.

The classical cohomology group $H^0_{cl}(s+\sigma)$ consists of linear combinations of jets

$$A_{\mu,\mathbf{m}}(t) = \epsilon_\mu(t)\mathrm{e}^{ik\cdot q(t)}(ik)^{\mathbf{m}} \tag{7.11.22}$$

where $k^2 = 0$ and the polarization vector $\epsilon_\mu(t)$ is perpendicular both to the photon momentum and the observer's trajectory:

$$\epsilon_\mu(t)k^\mu = \epsilon_\mu(t)\dot{q}^\mu(t) = 0. \tag{7.11.23}$$

The latter is evidently equivalent to the non-covariant condition $\epsilon^0 = 0$. Moreover, $k \cdot q = k_\mu q^\mu$. The Taylor series (7.11.14) can be summed in the same way as for the scalar field (7.9.6).

We now quantize the theory before imposing dynamics. To this end, we introduce the canonical momenta for all jets and antijets, and $p_\mu(t)$ and $p^\mu_*(t)$ for the observer's trajectory and its antifield. The defining relations are

$$\begin{array}{rcl} [E^{\mu,\mathbf{m}}(t), A_{\nu,\mathbf{n}}(t')] & = & \delta^\mu_\nu \delta^{\mathbf{m}}_{\mathbf{n}}\delta(t-t'), \\ \{E^{*,\mathbf{m}}_{\mu}(t), A^{\nu}_{*,\mathbf{n}}(t')\} & = & \delta^\nu_\mu \delta^{\mathbf{m}}_{\mathbf{n}}\delta(t-t'), \\ [\chi^{,\mathbf{m}}(t), \zeta_{,\mathbf{n}}(t')] & = & \delta^{\mathbf{m}}_{\mathbf{n}}\delta(t-t'), \\ \{b^{,\mathbf{m}}(t), c_{,\mathbf{n}}(t')\} & = & \delta^{\mathbf{m}}_{\mathbf{n}}\delta(t-t'), \\ [p_\nu(t), q^\mu(t')] & = & \delta^\mu_\nu\delta(t-t'). \end{array} \tag{7.11.24}$$

Since the jets also depend on the parameter t, we can define their Fourier components as in (7.5.13). The Fock vacuum (7.11.11) is replaced by a new vacuum, also denoted by $|0\rangle$, which is defined to be annihilated by the negative frequency modes. The quantum Hamiltonian is still defined by (7.8.19), where double dots indicate normal ordering with respect to frequency, ensuring that $H|0\rangle = 0$.

It remains to check that the algebra of $\mathfrak{u}(1)$ gauge transformations acts in a well-defined manner before we can pass to the BRST cohomology. Since a gauge potential transforms as

$$[\mathcal{J}_X, A_{\mu,\mathbf{m}}(t)] = \partial_{\mathbf{m}+\mu}X(q(t)) \tag{7.11.25}$$

we have

$$\mathcal{J}_X = \sum_{|\mathbf{m}|\leqslant p} \sum_{\mu} \int dt\ \partial_{\mathbf{m}+\mu} X(q(t)) E^{\mu,\mathbf{m}}(t). \tag{7.11.26}$$

There are no contributions from the antifields, since A^μ_*, ζ and c all transform trivially under $\mathfrak{map}(N, \mathfrak{u}(1))$. The prescription (7.10.10) gives

$$Q_{Long} = \sum_{|\mathbf{m}|\leqslant p} \sum_{\mu} \int dt\ c_{,\mathbf{m}+\mu}(t) E^{\mu,\mathbf{m}}(t). \tag{7.11.27}$$

The expressions (7.11.26) and (7.11.27) are evidently normal ordered as they stand, and consequently there are no gauge anomalies.

The rest proceeds as for the scalar field.

7.12 Gravity

Finally we are ready to apply the MCCQ formalism to general relativity. For simplicity we consider only pure gravity. The only field is the symmetric metric $g_{\mu\nu}(x)$. The inverse $g^{\mu\nu}$, the determinant $g = \det(g_{\mu\nu})$, the Levi-Cività connection $\Gamma^\mu_{\nu\rho}$, Riemann's curvature tensor $R^\rho{}_{\sigma\mu\nu}$, the Ricci tensor $R_{\mu\nu}$, the scalar curvature $R = g^{\mu\nu} R_{\mu\nu}$ and the Einstein tensor $G^{\mu\nu} = R^{\mu\nu} - \frac{1}{2} g^{\mu\nu} R$ are defined as usual. The covariant derivative is

$$\nabla_\mu = \partial_\mu + \Gamma^\rho_{\nu\mu} T^\nu_\rho, \tag{7.12.1}$$

where T^μ_ν are finite-dimensional matrices satisfying $gl(N)$ (7.5.2).

The Einstein action

$$S_E = \frac{1}{16\pi} \int d^4x\ \sqrt{g(x)} R(x). \tag{7.12.2}$$

leads to Einstein's equation of motion

$$G^{\mu\nu}(x) = 0, \tag{7.12.3}$$

which is subject to the identity

$$\nabla_\nu G^{\mu\nu}(x) \equiv 0 \tag{7.12.4}$$

We introduce a fermionic antifield $g^{\mu\nu}_*(x)$ for (7.12.3), a bosonic second-order antifield $\zeta^\mu(x)$ for (7.12.4), and a ghost $c^\mu_{diff}(x)$ to eliminate diffeomorphisms. The total field content in the extended history phase space is thus

afn	Field	Momentum	Parity
-1	$c^\mu_{diff}(x)$	$b^{diff}_\mu(x)$	F
0	$g_{\mu\nu}(x)$	$\pi^{\mu\nu}(x)$	B
1	$g^{\mu\nu}_*(x)$	$\pi^*_{\mu\nu}$	F
2	$\zeta^\mu(x)$	$\chi_\mu(x)$	B

The KT differential δ is defined by

$$\begin{aligned}
\delta c^\mu_{diff}(x) &= 0, \\
\delta g_{\mu\nu}(x) &= 0, \\
\delta g_*^{\mu\nu}(x) &= G^{\mu\nu}(x) \\
\delta \zeta_\mu(x) &= \nabla_\nu G^{\mu\nu}(x),
\end{aligned} \tag{7.12.5}$$

i.e. the KT operator is

$$Q_{KT} = \int d^4x \ \ (G^{\mu\nu}(x)\pi^*_{\mu\nu}(x) + \nabla_\nu G^{\mu\nu}(x)\chi_\mu(x)). \tag{7.12.6}$$

The longitudinal operator was written down in (7.10.14), $Q_{Long} = Q^{diff}_{Long}$, and the BRST operator is the sum of the KT and the longitudinal operators, as usual. Of the originally ten degrees of freedom $g_{\mu\nu}(x)$, the antifield $g_*^{\mu\nu}(x)$ eliminates four and the ghost $c^\mu_{diff}(x)$ another four, leaving two graviton polarizations in the BRST cohomology.

We now quantize as usual by passing to jet space, introducing a Fock vacuum that is annihilated by the negative frequency modes of all fields and antifields, and normal ordering. The fields are symmetric tensor fields, i.e. they correspond to the symmetric $gl(N)$ modules S_ℓ and S^ℓ in (7.5.18). The values of the parameters in (7.5.20) are

Field	ϱ	u	v	w	x	p
$c^\mu_{diff}(x)$	S^1	1	0	-1	N	$p+1$
$g_{\mu\nu}(x)$	S_2	$-(N+2)$	-1	$-(N+1)$	$-N(N+1)/2$	p
$g_*^{\mu\nu}(x)$	S^2	$(N+2)$	1	$-(N+1)$	$N(N+1)/2$	$p-2$
$\zeta^\mu(x)$	S^1	-1	0	1	$-N$	$p-3$

The parameters were written down in arbitrary dimension N for generality, although we are primarily interested in the physical case $N = 4$. The last column is the truncation order for the corresponding jets.

The diffeomorphism anomalies are now read off from (7.5.19); for definiteness, we only consider c_1. Depending on whether we exclude the ghost $c^\mu_{diff}(x)$ or not, the abelian charge $c_1 = c_1^{tot}$ becomes $c_1^{tot} = c_1^{field} \equiv 1 + c_1^g + c_1^{g*} + c_1^\zeta$ or $c_1^{tot} = c_1^{field} + c_1^{ghost}$, where

$$\begin{aligned}
c_1^{ghost} &= -\binom{N+p+1}{N} - N\binom{N+p+2}{N+2}, \\
c_1^g &= (N+2)\binom{N+p}{N} + \frac{N(N+1)}{2}\binom{N+p+1}{N+2}, \\
c_1^{g*} &= -(N+2)\binom{N+p-2}{N} - \frac{N(N+1)}{2}\binom{N+p-1}{N+2}, \\
c_1^\zeta &= \binom{N+p-3}{N} + N\binom{N+p-2}{N+2}.
\end{aligned} \tag{7.12.7}$$

It is clear that c_1^{tot} does not vanish for generic p, with or without the ghost contribution (the leading term proportional to $p^{N+2}/(N+2)!$ does however vanish in the $p \to \infty$ limit). The longitudinal operator (7.10.15) thus acquires an anomaly, and we can only implement the KT cohomology. Hence the ghost plays no role and should be discarded.

The quantum KT operator becomes

$$\begin{aligned} Q_{KT} &= \int dt \, \Big\{ \sum_{|\mathbf{m}|\leqslant p-2} :G^{\mu\nu}_{,\mathbf{m}}(t)\pi^{*,\mathbf{m}}_{\mu\nu}(t): \\ &+ \sum_{|\mathbf{m}|\leqslant p-3} :(\nabla_\nu G^{\mu\nu})_{,\mathbf{m}}(t)\chi^{,\mathbf{m}}_{\mu}(t): \Big\}, \end{aligned} \tag{7.12.8}$$

where $G^{\mu\nu}_{,\mathbf{m}}(t)$ and $(\nabla_\nu G^{\mu\nu})_{,\mathbf{m}}(t)$ are the corresponding jets. Note that the sums run up to $|\mathbf{m}| = p-2$ and $p-3$, respectively, because Einstein's equation is second order and the identity (7.12.4) is third order.

One difference compared to Minkowski space is that the geodesic equation depends on a dynamical field. In order to make the geodesic operator $\mathcal{G}_\mu(t)$ transform tensorially under reparametrizations as well, we need to add an extra term. We can construct the following quantites from the metric and the observer's trajectory:

1. The Levi-Cività connection $\Gamma^\nu_{\sigma\tau}(x,t) = \frac{1}{2}g^{\nu\rho}(x,t)(\partial_\sigma g_{\rho\tau}(x,t) + \partial_\tau g_{\sigma\rho}(x,t) - \partial_\rho g_{\sigma\tau}(x,t))$.
2. The einbein $e(t) = \sqrt{g_{\mu\nu,\mathbf{0}}(t)\dot{q}^\mu(t)\dot{q}^\nu(t)}$.
3. The reparametrization connection $\Gamma(t) = -e^{-1}(t)\dot{e}(t)$.

The geodesic operator reads [16]

$$\mathcal{G}_\mu(t) = e^{-1}(t)g_{\mu\nu,\mathbf{0}}(t)(\ddot{q}^\nu(t) + \Gamma(t)\dot{q}^\nu(t) + \Gamma^\nu_{\sigma\tau,\mathbf{0}}(t)\dot{q}^\sigma(t)\dot{q}^\tau(t)), \tag{7.12.9}$$

where $g_{\mu\nu,\mathbf{0}}(t)$ and $\Gamma^\nu_{\sigma\tau,\mathbf{0}}(t)$ are the zero-jets corresponding to the metric and Levi-Cività connection, respectively. It is straightforward to check that (7.12.9) transforms nicely under the full DGRO algebra,

$$\begin{aligned} [\mathcal{L}_\xi, \mathcal{G}_\nu(t)] &= -\partial_\nu \xi^\mu(q(t))\mathcal{G}_\mu(t), \\ [L_f, \mathcal{G}_\nu(t)] &= -f(t)\dot{\mathcal{G}}_\nu(t) - \dot{f}(t)\mathcal{G}_\nu(t). \end{aligned} \tag{7.12.10}$$

The contribution to the KT operator is thus $Q_{KT} = \int dt \; \mathcal{G}_\mu(t)p^\mu_*(t)$, which eliminates the observer's trajectory $q^\mu(t)$ in cohomology.

7.13 Finiteness conditions

In the previous section we applied the MCCQ formalism to gravity, and found a well-defined but anomalous action of the DGRO algebra. However, the passage to the space of p-jets amounts to a regularization. The regularization is unique in that it preserves the full constraint algebra, but it must nevertheless be removed in the end. In order to reconstruct the original field by means of the Taylor series (7.5.5), we must take the limit $p \to \infty$. A necessary condition for taking this limit is that the abelian charges have a finite limit.

Taken at face value, the prospects for succeeding appear bleak. When p is large, $\binom{m+p}{n} \approx p^n/n!$, so the abelian charges (7.5.19) diverge; the worst case is $c_1 \approx c_2 \approx p^{N+2}/(N+2)!$, which diverges in all dimensions $N > -2$. In [15] a way out of this problem was devised: consider a more general realization by taking the direct sum of operators corresponding to different values of the jet order p. Take the sum of $r+1$ terms like those in (7.5.12), with p replaced by p, $p-1$, ..., $p-r$, respectively, and with ϱ and M replaced by $\varrho^{(i)}$ and $M^{(i)}$ in the $p-i$ term.

Such a sum of contributions arises naturally from the KT complex, because the antifields are only defined up to an order smaller than p (e.g. $p - o_\alpha$ or $p - \varsigma_a$). Denote the numbers u, v, w, x, y in the modules $\varrho^{(i)}$ and $M^{(i)}$, defined as in (7.5.20), by u_i, v_i, w_i, x_i, y_i, respectively. Of course, there is only one contribution from the observer's trajectory. Then it was shown in [15], Theorem 3, that

$$
\begin{aligned}
c_1 &= -U\binom{N+p-r}{N-r}, & c_2 &= -V\binom{N+p-r}{N-r}, \\
c_3 &= W\binom{N+p-r}{N-r}, & c_4 &= -X\binom{N+p-r}{N-r}, \\
c_5 &= Y\binom{N+p-r}{N-r},
\end{aligned}
\tag{7.13.1}
$$

where $u_0 = U$, $v_0 = V$, $w_0 = W$, $x_0 = X$ and $y_0 = Y$, provided that the following conditions hold:

$$
\begin{aligned}
& u_i + (-)^i\binom{r-2}{i-2}X = (-)^i\binom{r}{i}U, \\
& v_i - 2(-)^i\binom{r-1}{i-1}W - (-)^i\binom{r-2}{i-2}X = (-)^i\binom{r}{i}V, \\
& w_i - (-)^i\binom{r-1}{i-1}X = (-)^i\binom{r}{i}W, \\
& x_i = (-)^i\binom{r}{i}X, \\
& y_i = (-)^i\binom{r}{i}Y.
\end{aligned}
\tag{7.13.2}
$$

The contributions from the observer's trajectory have also been eliminated by antifields coming from the geodesic equation; this is not important in the sequel because these contributions were finite anyway.

Let us now consider the solutions to (7.13.2) for the numbers x_i, which can be interpreted as the number of fields and anti-fields. First assume that the field $\phi^\alpha_{,\mathbf{m}}(t)$ is fermionic with x_F components, which gives $x_0 = x_F$. We may assume, by the spin-statistics theorem, that the EL equations are first order, so the bosonic antifields $\phi^*_{\alpha,\mathbf{m}}(t)$ contribute $-x_F$ to x_1. The barred antifields $\bar{\phi}^\alpha_{,\mathbf{m}}(t)$ are also defined up to order $p-1$, and so give $x_1 = -x_F$, and the barred second-order antifields $\bar{\phi}^*_{\alpha,\mathbf{m}}(t)$ give $x_2 = x_F$. Further assume that the fermionic EL equations have x_S gauge symmetries, i.e. the second-order antifields $\zeta_{a,\mathbf{m}}(t)$ give $x_2 = x_S$. In established theories, $x_S = 0$, but we will need a non-zero value for x_S. Finally, the corresponding barred antifields give $x_3 = -x_S$.

For bosons the situation is analogous, with two exceptions: all signs are reversed, and the EL equations are assumed to be second order. Hence $\phi^*_{\alpha,\mathbf{m}}(t)$ yields $x_2 = x_B$ and the gauge antifields $\zeta_{a,\mathbf{m}}(t)$ give $x_3 = -x_G$. Accordingly, the barred antifields are one order higher.

The situation is summarized in the following tables, where the upper half is valid if the original field is fermionic and the lower half if it is bosonic:

afn	Jet	Order	x
0	$\phi^\alpha_{,\mathbf{m}}(t)$	p	x_F
1	$\bar{\phi}^\alpha_{,\mathbf{m}}(t)$	$p-1$	$-x_F$
1	$\phi^*_{\alpha,\mathbf{m}}(t)$	$p-1$	$-x_F$
2	$\bar{\phi}^*_{\alpha,\mathbf{m}}(t)$	$p-2$	x_F
2	$\zeta_{a,\mathbf{m}}(t)$	$p-2$	x_S
3	$\bar{\zeta}_{a,\mathbf{m}}(t)$	$p-3$	$-x_S$
0	$\phi^\alpha_{,\mathbf{m}}(t)$	p	$-x_B$
1	$\bar{\phi}^\alpha_{,\mathbf{m}}(t)$	$p-1$	x_B
1	$\phi^*_{\alpha,\mathbf{m}}(t)$	$p-2$	x_B
2	$\bar{\phi}^*_{\alpha,\mathbf{m}}(t)$	$p-3$	$-x_B$
2	$\zeta_{a,\mathbf{m}}(t)$	$p-3$	$-x_G$
3	$\bar{\zeta}_{a,\mathbf{m}}(t)$	$p-4$	x_G

(7.13.3)

If we add all contributions of the same order, we see that fourth relation in (7.13.2) can only be satisfied provided that

$$\begin{aligned}
p: &\quad x_F - x_B = X \\
p-1: &\quad -2x_F + x_B = -rX, \\
p-2: &\quad x_B + x_F + x_S = \binom{r}{2} X, \\
p-3: &\quad -x_B - x_S - x_G = -\binom{r}{3} X, \\
p-4: &\quad x_G = \binom{r}{4} X, \\
p-5: &\quad 0 = -\binom{r}{5} X, \ldots
\end{aligned} \tag{7.13.4}$$

The last equation holds only if $r \leqslant 4$ (or trivially if $X = 0$). On the other hand, if we demand that there is at least one bosonic gauge condition, the $p-4$ equation yields $r \geqslant 4$. Such a demand is natural, because both the Maxwell/Yang-Mills and the Einstein equations have this property. Therefore, we are unambigiously guided to consider $r = 4$ (and thus $N = 4$). The specialization of (7.13.4) to four dimensions reads

$$\begin{aligned}
p: &\quad x_F - x_B = X \\
p-1: &\quad -2x_F + x_B = -4X, \\
p-2: &\quad x_B + x_F + x_S = 6X, \\
p-3: &\quad -x_B - x_S - x_G = -4X, \\
p-4: &\quad x_G = X.
\end{aligned} \tag{7.13.5}$$

Clearly, the unique solution to these equations is

$$x_F = 3X, \qquad x_B = 2X, \qquad x_S = X, \qquad x_G = X. \tag{7.13.6}$$

The solutions to the remaining equations in (7.13.2) are found by analogous reasoning. The result is

$$\begin{aligned}
u_B &= 2U & v_B &= 2V + 2W \\
u_F &= 3U & v_F &= 3V + 2W \\
u_S &= U - X & v_S &= V + 2W + X \\
u_G &= U - X & v_G &= V + 2W + X \\
\\
w_B &= 2W + X & y_B &= 2Y \\
w_F &= 3W + X & y_F &= 3Y \\
w_S &= W + X & y_S &= Y \\
w_G &= W + X & y_G &= Y
\end{aligned} \tag{7.13.7}$$

This result expresses the twenty parameters $x_B - w_G$ in terms of the five parameters X, Y, U, V, W. For this particular choice of parameters, the abelian charges in (7.13.1) are given by

$$c_1 = -U, \qquad c_2 = -V, \qquad c_3 = W, \qquad c_4 = -X, \qquad c_5 = Y, \tag{7.13.8}$$

independent of p. Hence there is no manifest obstruction to the limit $p \to \infty$.

The prediction that spacetime has $N = 4$ dimensions is of course very nice. Unfortunately, at closer scrutiny the situation appears less appealing. In particular, the need for fermionic gauge symmetries ($x_S \neq 0$) is apparently in disagreement with observation. It was also found in [19] that the gauge anomaly c_5 does not have a finite $p \to \infty$ limit for reasonable choices of field content.

Hence it is presently unclear how to remove the regulator and take the field limit, and this is of course a major unsolved problem. Nevertheless, it should be emphasized that already the regularized theories carry representations of the *full* gauge and diffeomorphism algebras.

7.14 Conceptual issues

One of the most important tasks of any putative quantum theory of gravity is to shed light on the various conceptual difficulties which arise when the principles of quantum mechanics are combined with general covariance [6, 25]. These issues include:

1. In conventional canonical quantization, the canonical commutation relations are defined on a "spacelike" surface. However, a surface is spacelike w.r.t. some particular spacetime metric $g_{\mu\nu}$, which is itself a quantum operator.

2. Microcausality requires that the field variables defined in spacelike separated regions commute. Again, it is unclear what this means when the notion of spacelikeness is dynamical.

3. Different choices of foliation lead to a priori different quantum theories, and it by no means clear that these are unitarily equivalent.

4. The problem of time: The Hamiltonian of general relativity is a first class constraint, hence it vanishes on the reduced phase space. This means that there is no notion of time evolution among diffeomorphism-invariant degrees of freedom.

5. The notion of time as a causal order is lost. This is not really a problem in the classical theory, where one can solve the equations of motion first, but in quantum theory causality is needed from the outset.

6. QFT rests on two pillars: quantum mechanics and locality. However, locality is at odds with diffeomorphism invariance underlying gravity; "there are no local observables in quantum gravity".

Let us see how MCCQ addresses these conceptual issues.

1. The canonical commutation relations are defined throughout the history phase space $\mathcal{P}$, and hence not restricted to variables living on a spacelike surface. Dynamics is implemented as a first class constraint in $\mathcal{P}$. Only if we solve this constraint prior to quantization need we restrict quantization to a spacelike surface.

2. By passing to p-jet space, we eliminate the notion of spacelikeness altogher. The p-jets live on the observer's trajectory, and the observer moves along a timelike curve. It might seem strange to dismiss the notion of spacelike separation, but distant events can never be directly observed, and a physical theory only needs to describe directly observable events. What can be observed are indirect effects of distant events. E.g., a terrestial detector does not directly observe the sun, but only photons emanating from the sun eight light-minutes ago. The detector signals are of course compatible with the existence of the sun, but a physical theory only needs to deal with directly observed events, i.e. the absorbtion of photons in the detector.

3. In MCCQ there is no foliation, but rather an explicit observer, or detector. The theory is unique since the observer's trajectory is a quantum object; we do not deal with a family of theories parametrized by the choice of observer, but instead the observer's trajectory is represented on the Hilbert space in the same way as the quantum fields.

4. By introducing an explicit observer, we can define a genuine energy operator (7.8.19) which translates the fields relative to the observer, or vice versa. In contrast, there is also a Hamiltonian constraint, which translates both the observer and the fields the same amount. This constraint is killed in KT cohomology and is thus identically zero on physical observables.

5. The p-jets live on the observer's trajectory $q(t)$ and are thus causally related; causal order is defined by the parameter t. The relation between this order and the fields is encoded in the geodesic equation (7.12.9).

6. As we saw in Section 7.2, locality is compatible with infinite-dimensional spacetime symmetries, but only in the presence of an anomaly. This is the key lesson from CFT.

It is gratifying that the MCCQ formalism yields natural explanations of many of the conceptual problems that plague quantum gravity.

7.15 Conclusion

The key insight underlying the present work is that the process of observation must be localized in spacetime in order to be compatible with the philosophy of QFT. The innocent-looking introduction of the observer's trajectory leads to dramatic consequences, because new gauge and diffeomorphism anomalies arise. On the mathematical side, this construction leads to well-defined realizations of the constraint algebra generators as operators on a linear space, as least for the regularized theory.

We have also developed a manifestly covariant canonical quantization method, based on the form of the DGRO algebra modules. This formalism is convenient due to its relation to representation theory, but it is presumably possible to repeat the analysis in any sensible quantization scheme, at the cost of additional work. In contrast, the introduction of the observer's trajectory is absolutely crucial, because the new anomalies can not be formulated without it. Anomalies matter!

Four critical problems remain to be solved. As was discussed in Section 7.13, the original fields must be reconstructed from the p-jets, i.e. we must take the limit $p \to \infty$. This limit is problematic because the abelian charges diverge. Second, the issue of unitarity needs to be understood. So far we only noted that an extension is necessary for unitarity by restriction to Virasoro subalgebras, and then we proceeded to construct anomalous representations. The main problem is to find an invariant inner product. Third, perturbation theory and renormalization must be transcribed to ths formalism, to make contact with numerical predictions of ordinary QFT. Finally, we know from CFT that reducibility conditions analogous to Kac' formula [8] are needed in physically interesting situations. Unfortunately, none of these problems appears to be easy.

References

[1] L. Bonora, P. Pasti and M. Tonin, The anomaly structure of theories with external gravity, *J. Math. Phys.* **27** (1986) 2259–2270.

[2] S. Berman and Y. Billig, Irreducible representations for toroidal Lie algebras, *J. Algebra* **221** (1999) 188–231.

[3] S. Berman, Y. Billig and J. Szmigielski; Vertex operator algebras and the representation theory of toroidal algebras, `math.QA/0101094` (2001)

[4] Y. Billig, Principal vertex operator representations for toroidal Lie algebras, *J. Math. Phys.* **7** (1998) 3844–3864.

[5] Y. Billig and K. Zhao, Weight modules over exp-polynomial Lie algebras, `math.RT/0305293` (2003)

[6] S. Carlip, Quantum gravity: a progress report, `gr-qc/0108040`

[7] A. Dzhumadildaev, Virasoro type Lie algebras and deformations, *Z. Phys. C* **72** (1996) 509–517.

[8] P. Di Francesco, P. Mathieu, and D. Sénéchal, Conformal field theory, New York: Springer-Verlag, 1996

[9] P. Goddard and D. Olive, Kac-Moody and Virasoro algebras in relation to quantum physics. *Int. J. Mod. Phys.* **1** (1986) 303–414.

[10] M. Henneaux and C. Teitelboim, Quantization of gauge systems, Princeton Univ. Press (1992)

[11] C. J. Isham and N. Linden, Continuous histories and the history group in generalised quantum theory, *J. Math. Phys.* **36** (1995) 5392.

[12] C. Kassel, Kahler differentials and coverings of complex simple Lie algebras extended over a commutative algebra, *J. Pure and Appl. Algebra* **34** (1985) 256–275.

[13] T.A. Larsson, Central and non-central extensions of multi-graded Lie algebras, *J. Phys. A.* **25** (1992) 1177–1184.

[14] T.A. Larsson, Extended diffeomorphism algebras and trajectories in jet space. *Comm. Math. Phys.* **214** (2000) 469–491. `math-ph/9810003`

[15] T.A. Larsson, Multi-dimensional diffeomorphism and current algebras from Virasoro and Kac-Moody Currents, `math-ph/0101007` (2001)

[16] T.A. Larsson, Koszul-Tate cohomology as lowest-energy modules of non-centrally extended diffeomorphism algebras, `math-ph/0210023` (2002)

[17] T.A. Larsson, Multi-dimensional Virasoro algebra and quantum gravity, in *Mathematical physics research on the leading edge*, ed. C. V. Benton, New York: Nova Science Publishers, 2003.

[18] T.A. Larsson, Manifestly covariant canonical quantization I: the free scalar field, `hep-th/0411028` (2004)

[19] T.A. Larsson, Manifestly covariant canonical quantization II: Gauge theory and anomalies, `hep-th/0501043` (2005)

[20] J. Mickelsson, Current algebras and groups, *Plenum Monographs in Nonlinear Physics*, London: Plenum Press, 1989.

[21] P. Nelson and L. Alvarez-Gaumé, Hamiltonian interpretation of anomalies, *Comm. Math. Phys.* **99** (1985) 103–114.

[22] D. Pickrell, On the Mickelsson-Faddeev extensions and unitary representations, *Comm. Math. Phys.* **123** (1989) 617.

[23] S.E. Rao and R.V. Moody, Vertex representations for N-toroidal Lie algebras and a generalization of the Virasoro algebra. *Comm. Math. Phys.* **159** (1994) 239–264.

[24] K. Savvidou, The action operator in continuous time histories, *J. Math. Phys.* **40** (1999) 5657.

[25] N. Savvidou, General relativity histories theory, `gr-qc/0412059` (2004)

In: Focus on Quantum Gravity Research
Editor: David C. Moore, pp. 311-326

ISBN 1-59454-660-6

Chapter 8

Singularity and Freezing of 5th Dimension

V. Dzhunushaliev[1]
Dept. Phys. and Microel. Engineer., Kyrgyz-Russian
Slavic University, Bishkek, Kievskaya Str. 44, 720021,
Kyrgyz Republic
E-mail: dzhun@hotmail.kg
R. Myrzakulov
Institute of Physics and Technology, 480082, Almaty-82,
Kazakhstan
E-mail: cnlpmyra@satsun.sci.kz

Some reasonings are presented that the problem of a singularity in general relativity with the problem of freezing of the 5th dimension can be connected. It is shown that some solutions in the 5D Kaluza-Klein gravity with the cross section in the Planck region have a region ($\approx l_{Pl}$) where the metric signature changes from $\{+,-,-,-,-\}$ to $\{-,-,-,-,+\}$. The idea is discussed that such switching can not occur following some general point of view that the Planck length is the minimal length in the nature and consequently the physical quantities can not change very quickly in the course of this length. In this case the dynamic of the G_{55} metric component should be frozen.

8.1 Introduction

One of the principal problems in physics is the problem of a singularity. This problem exists both in classical and quantum physics. In classical physics it is, for example, an infinite energy of a point-like charge in electrodynamics or singularities in the

[1]Senior Associate of the Abdus Salam ICTP

Schwarzschield and Reissner-Nordström solutions in general relativity. In quantum physics it is well known singularities connected with the loops of Feynman diagrams. The origin of these singularities is that the corresponding particles are point like. In classical electrodynamics it result in the fact that near to the point-like particle the electric field tends to infinity by such a manner that full energy is infinite. In general relativity exist general theorems which tell us that in some general sense the singularities are unavoidable. In this sense general relativity can not solve the problem of an infinite energy of point-like particle : Reissner-Nordström solution has a singularity at $r = 0$.

The modern string theory approach smooths this problem since the string is an extended object. Roughly speaking one can say that near to string the fields are not so singular in comparison with point-like particle. But string point of view has another disagreeable peculiarity : remaining within the framework of this theory we can not ask a question on an internal structure of string. In string theory the string is a fundamental structureless object and the question on an internal structure is the nonsensical question.

Thus we can think that the problem of the point-like singularity is connected with an internal structure of elementary particles. Probably the resolution of this problem can be received on the framework of a quantum theory of Everything including gravity. The problem here is that this quantum theory can not be based on the Feynman diagram technique since this approach from the very beginning on default assumes that quantum fields consist of point-like *free* particles, *i.e.* it is necessary to have a non-perturbative quantization method for strongly interacting fields (especially for gravity). Unfortunately now we do not have such method.

In this notice we would like to present some qualitative reasonings and quantitative calculations that the singularity problem can be closely connected with the problem of freezing (splitting off) extra dimensions. The idea is that all over the space the metric on the extra dimensions is not variable in the sense that we have not any equations for it. But near to a point-like singularity the gravitational field becomes so strong that excites the additional degrees of freedom : the metric on the extra dimensions. Such idea is not new but ordinary it is supposed that such compactification of the extra dimensions is a dynamical process which can be described on the language of field equations (classical or quantum, more probably that quantum). The idea presented here is that this process can not be described on the field equations level. The matter is that we suppose that on the level of Planck sizes fluctuate not only fields but field equations also.

In this connection we would like to show that one can build a wormhole-like object with a very long 5-dimensional throat (and with the cross section in Planck region) and two asymptotically flat tails. The main point of view is that the wormhole is 5-dimensional object but on the tails G_{55} metric component is frozen in the sense that the corresponding equation is absent but on the throat this equation is switched on. The main question is why this equation is switched on on the throat and switched off on the tails. Here we would like to show that such switching can not occur following some general point of view that the Planck length is the minimal length in the nature and consequently the physical quantities can not change very quickly in the course of this length.

We will use this statement for the metric signature. For example, the next dynamic is impossible: the metric signature by $r < r_H - l_{Pl}$ is $\{+,-,-,-,-\}$ and by $r > r_H + l_{Pl}$ is $\{-,-,-,-,+\}$ (r_H is some constant). According to the above-mentioned statement in the region $|r - r_H| < l_{Pl}$ have to be some quantum gravitational effects which will prevent such dynamic. In this notice we do not give any microscopical description how it happens but we consider only the consequences of such prohibition. Such approach is similar to the *macroscopical* thermodynamical investigation of a physical process when we do not interesting for the *microscopical* description of the process.

8.2 Wormhole-like flux tubes

The 5-dimensional wormhole-like flux tube metric is

$$\begin{aligned} ds^2 &= \frac{dt^2}{\Delta(r)} - \Delta(r) e^{2\psi(r)} \left[d\chi + \omega(r) dt + Q\cos\theta d\varphi\right]^2 \\ &- dr^2 - a(r)(d\theta^2 + \sin^2\theta d\varphi^2), \end{aligned} \tag{8.2.1}$$

where χ is the 5th extra coordinate; r, θ, φ are $3D$ spherical-polar coordinates; $r \in \{-\infty, +\infty\}$ is the longitudinal coordinate; Q is the magnetic charge.

According the correspondence between 5-dimensional gravity and 4-dimensional gravity + electrodynamics + scalar field the metric (8.2.1) has the following components of the 4-dimensional electromagnetic potential A_μ

$$A_t = \omega(r) \quad \text{and} \quad A_\varphi = Q\cos\theta. \tag{8.2.2}$$

For this potential the Maxwell tensor is

$$F_{rt} = \omega'(r) \quad \text{and} \quad F_{\theta\varphi} = -Q\sin\theta. \tag{8.2.3}$$

This means that we have radial Kaluza-Klein electric $E_r \propto F_{tr}$ and magnetic $H_r \propto F_{\theta\varphi}$ fields.

Substituting this ansatz into the 5-dimensional Einstein vacuum equations

$$R_{AB} - \frac{1}{2}\eta_{AB} R = 0 \tag{8.2.4}$$

$A, B = 0, 1, 2, 3, 5$ and η_{AB} is the metric signature, gives us

$$\frac{\Delta''}{\Delta} - \frac{\Delta'^2}{\Delta^2} + \frac{\Delta' a'}{\Delta a} + \frac{\Delta'\psi'}{\Delta} + \frac{q^2}{a^2\Delta^2}e^{-4\psi} = 0, \quad (8.2.5)$$

$$\frac{a''}{a} + \frac{a'\psi'}{a} - \frac{2}{a} + \frac{Q^2}{a^2}\Delta e^{2\psi} = 0, \quad (8.2.6)$$

$$\psi'' + \psi'^2 + \frac{a'\psi'}{a} - \frac{Q^2}{2a^2}\Delta e^{2\psi} = 0, \quad (8.2.7)$$

$$-\frac{\Delta'^2}{\Delta^2} + \frac{a'^2}{a^2} - 2\frac{\Delta'\psi'}{\Delta} - \frac{4}{a} + 4\frac{a'\psi'}{a} + \frac{q^2}{a^2\Delta^2}e^{-4\psi} + \frac{Q^2}{a^2}\Delta e^{2\psi} = 0 \quad (8.2.8)$$

q is the electric charge. These equations are derived after substitution the expression (8.2.9) for the electric field in the initial Einstein's equations (8.2.4). The 5-dimensional $\binom{\chi}{t}$-Einstein's equation (4-dimensional Maxwell equation) is taken as having the following solution

$$\omega' = \frac{q}{a\Delta^2}e^{-3\psi}. \quad (8.2.9)$$

For the determination of the physical sense of the constant q let us write the $\binom{\chi}{t}$-Einstein's equation (from the 4-dimensional viewpoint it is the Maxwell equations) in the following way :

$$\left(\omega'\Delta^2 e^{3\psi} 4\pi a\right)' = 0. \quad (8.2.10)$$

After the dimensional reduction of the 5-dimensional Kaluza - Klein gravity the Maxwell's tensor is

$$F_{\mu\nu} = \partial_\mu A_\nu - \partial_\nu A_\mu. \quad (8.2.11)$$

That allows us to write the electric field as $E_r = \omega'$. Eq.(8.2.10) with the electric field defined by (8.2.11) can be compared with the Maxwell's equations in a continuous medium

$$\mathrm{div}\vec{\mathcal{D}} = 0 \quad (8.2.12)$$

where $\vec{\mathcal{D}} = \epsilon\vec{E}$ is an electric displacement and ϵ is a dielectric permeability. Comparing Eq. (8.2.12) with Eq. (8.2.10) we see that the magnitude $q/a = \omega'\Delta^2 e^{3\psi}$ is like to the electric displacement and the dielectric permeability is $\epsilon = \Delta^2 e^{3\psi}$. It means that q can be taken as the Kaluza-Klein electric charge because the flux of electric field is

$$\mathbf{\Phi} = 4\pi a\mathcal{D} = 4\pi q. \quad (8.2.13)$$

As the electric q and magnetic Q charges are varied it was found [1] that the solutions to the metric in Eq. (8.2.5)-(8.2.8) evolve in the following way :

1. $0 \leqslant Q < q$. The solution is *a regular gravitational flux tube.* The solution is filled with both electric and magnetic fields. The longitudinal distance between the $\pm r_H$ surfaces increases, and the cross-sectional size does not increase as rapidly as $r \to r_H$ with $q \to Q$. The values $r = \pm r_H$ are defined by the following way $\Delta(\pm r_H) = 0$.

2. $Q = q$. In this case the solution is *an infinite flux tube* filled with constant electric and magnetic fields. The cross-sectional size of this solution is constant ($a = const.$).

3. $0 \leqslant q < Q$. In this case we have *a singular gravitational gravitational flux tube* located between two (+) and (-) electric and magnetic charges at $\pm r_{sing}$. At $r = \pm r_{sing}$ this solution has real singularities where are the charges.

We will consider the case with $q \approx Q$ but $q > Q$. In this case there is a region $|r| \leqslant r_H$ where the solution is like to a tube filled with almost equal electric and magnetic fields. The length $L = 2r_H$ of the throat of the flux tube (by $|r| < r_H$) depends on relation $\delta = 1 - Q/q$ by the following manner $L \stackrel{\delta \to 0}{\longrightarrow} \infty$ but $\delta > 0$. The numerical analysis [2] shows that the spatial cross section of the tube ($t, \chi, r = \text{const}$) does not change significantly, for example, $a(r_H) \approx 2a(0)$. The cross section of the tube at the center $a(0)$ is arbitrary and we choose it as $a(0) \approx l_{Pl}^2$. This allows us to say that we have a super-thin and super-long gravitational flux tube: $L \gg \sqrt{a(0)}$, $a(0) \approx l_{Pl}$.

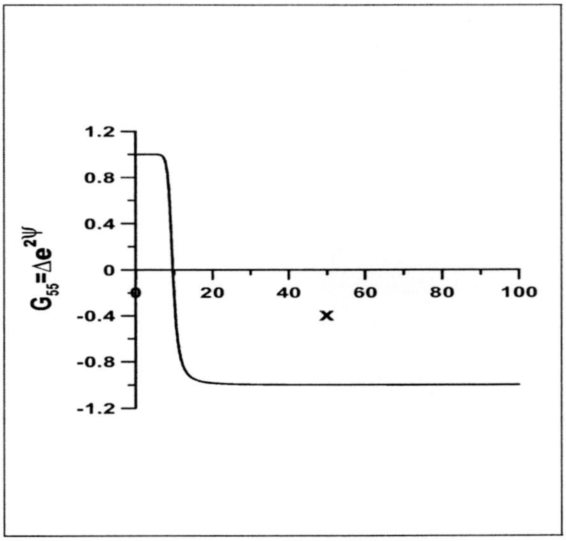

Figure 8.2.1: The metric component $G_{55} = \Delta(x)e^{2\psi(x)}$, $x = r/\sqrt{a(0)}$ is the dimensionless radius.

On Fig. (8.2.1) the profile of the function $\Delta e^{2\psi}$ (which is equal to G_{55}) from Ref. [2] is presented. We see that nearby the value $r = r_H$ this function changes drastically from the value $G_{55} \approx +1$ by $r \gtrsim +r_H - l_0$ to $G_{55} \approx -1$ by $r \lesssim +r_H + l_0$. The same occur near $r = -r_H$.

In order to present the metric on the throat by $|r| < r_H - l_0$ we present the exact solution with $q = Q$

$$\begin{aligned} a &= a(0) = \frac{Q_0^2}{2} = const, && (8.2.14)\\ e^{2\psi} &= \frac{1}{\Delta} = \cosh^2 \frac{r}{\sqrt{a(0)}}, && (8.2.15)\\ \omega &= \sqrt{2}\sinh \frac{r}{\sqrt{a(0)}}, && (8.2.16)\\ G_{55} &= \Delta \mathrm{e}^{2\psi} = 1 && (8.2.17) \end{aligned}$$

here we have parallel electric E and magnetic H fields with equal electric q_0 and magnetic Q_0 charges $q_0 = Q_0 = \sqrt{2a(0)}$. On the throat by $|r| < r_H - l_0$ (the solution with $q \approx Q$ but $q > Q$) all equals signs of Eq's (8.2.14)-(8.2.17) should be changed on the approximate equality signs

$$\begin{aligned} a &\approx \frac{Q_0^2}{2} = const, && (8.2.18)\\ e^{2\psi} &\approx \frac{1}{\Delta} = \cosh \frac{r}{\sqrt{a(0)}}, && (8.2.19)\\ \omega &\approx \sqrt{2}\sinh \frac{r}{\sqrt{a(0)}}, && (8.2.20)\\ G_{55} &= \Delta \mathrm{e}^{2\psi} \approx 1 && (8.2.21) \end{aligned}$$

Such approximation is valid only by $|r| \lesssim r_H - l_0$ where l_0 is some small quantity $l_0 \ll r_H$.

Now we would like to estimate the length l_0 of region where the change of the metric component $G_{55} = \Delta \mathrm{e}^{2\psi}$ is

$$\Delta \mathrm{e}^{2\psi}\Big|_{r \approx r_H + l_0} - \Delta \mathrm{e}^{2\psi}\Big|_{r \approx r_H - l_0} \approx 2. \tag{8.2.22}$$

For this estimation will be used Eq. (8.2.6). On the throat by $|r| < r_H - l_0$ this equation approximately is

$$-\frac{2}{a} + \frac{Q^2}{a^2}\Delta e^{2\psi} \approx 0. \tag{8.2.23}$$

We can estimate l_0 by solving Einsten's equations (8.2.5)-(8.2.8) nearby $r = +r_H$ (for $r = -r_H$ the analysis is the same) and define $r = r_H - l_0$ where the last two terms in Eq. (8.2.6) will have the same order

$$\left(\frac{2}{a}\right)\Big|_{r = r_H - l_0} \approx \left(\frac{Q^2}{a^2}\Delta e^{2\psi}\right)\Big|_{r = r_H - l_0} \tag{8.2.24}$$

The solution close to $r = +r_H$ we search in the following form

$$\Delta(r) = \Delta_1 (r_H - r) + \Delta_2 (r_H - r)^2 + \cdots . \tag{8.2.25}$$

The substitution in Eq's (8.2.5) gives us the following solution

$$\Delta_1 = \frac{qe^{-2\psi_H}}{a_H}. \tag{8.2.26}$$

After the substitution into Eq. (8.2.24) we have

$$l_0 \approx \sqrt{a(0)} = l_{Pl} \tag{8.2.27}$$

here we took into account that the numerical analysis [2] shows that $a_H \approx 2a(0)$. It means that the change of macroscopical dimensionless function $G_{55} = \Delta e^{2\psi}$ as in Eq. (8.2.22) occurs during the Planck length. The metric (8.2.1) by $|r| \approx r_H - l_{Pl}$ approximately is

$$ds^2 \approx e^{2\psi_H} dt^2 - dr^2 - a(r_H)\left(d\theta^2 + \sin\theta d\varphi^2\right) - (d\chi + \omega dt + Q\cos\theta d\phi)^2 \tag{8.2.28}$$

by $|r| \approx r_H + l_{Pl}$ the metric approximately is

$$ds^2 \approx -e^{2\psi_H} dt^2 - dr^2 - a(r_H)\left(d\theta^2 + \sin\theta d\varphi^2\right) + (d\chi + \omega dt + Q\cos\theta d\phi)^2 \tag{8.2.29}$$

here we took into account that numerical calculations [2] show that $\psi \approx \psi_H = \text{const}$ by $|r| \gtrsim r_H$. We see that during the Planck length the metric signature changes from $\{+,-,-,-,-\}$ to $\{-,-,-,-,+\}$. Simultaneously it is necessary to mention that the metric (8.2.1) is non-singular by $|r| = r_H$ and approximately is [2]

$$\begin{aligned} ds^2 \approx\ & e^{2\psi_H} dt^2 - e^{\psi_H} dt\,(d\chi + Q\cos\theta d\varphi) - dr^2 - a(r_H)\left(d\theta^2 + \sin^2\theta d\varphi^2\right) = \\ & \left[e^{\psi_H} dt - \frac{1}{2}(d\chi + Q\cos\theta d\varphi)\right]^2 - dr^2 - a(r_H)\left(d\theta^2 + \sin^2\theta d\varphi^2\right) - \\ & \frac{1}{4}(d\chi + Q\cos\theta d\varphi)^2 \end{aligned} \tag{8.2.30}$$

where ψ_H is some constant.

If we write the metric (8.2.1) in the 5-bein formalism

$$\begin{aligned} ds^2 &= \omega^A \omega^B \eta_{AB}, \\ \omega^A &= e^A_\mu dx^\mu, \quad x^\mu = t, r, \theta, \varphi, \chi \end{aligned} \tag{8.2.31}$$

then we see that

$$\eta_{AB} = \{+1,-1,-1,-1,-1\} \quad \text{by} \quad |r| \lesssim r_H - l_{Pl} \tag{8.2.32}$$

$$\eta_{AB} = \{-1,-1,-1,-1,+1\} \quad \text{by} \quad |r| \gtrsim r_H + l_{Pl} \tag{8.2.33}$$

It is necessary to note that for the mechanism presented here the change of the sign of two components η_{00} and η_{55} is very important. The reason is that the G_{55} metric component can be made dimensional by such a manner

$$G_{55} d\chi^2 = \left(l_0^2 G_{55}\right)\left(\frac{d\chi}{l_0}\right)^2 \tag{8.2.34}$$

where $l_0 \approx l_{Pl}$ is the characteristic length of the 5th dimension. In this case the quantity $\sqrt{l_{Pl} G_{55}}$ changes

$$l_{Pl}\sqrt{G_{55}}\Big|_{r\approx r_H+l_{Pl}} - l_{Pl}\sqrt{G_{55}}\Big|_{r\approx r_H-l_{Pl}} \approx l_{Pl} \tag{8.2.35}$$

by changing the radial coordinate

$$\Delta r \approx l_{Pl}. \tag{8.2.36}$$

Such variation of G_{55} is possible but simultaneously the dimensionless quantity $\eta_{00}(e_t^0)^2$ changes

$$\eta_{00}(e_t^0)^2\big|_{r\approx r_H+l_{Pl}} - \eta_{00}(e_t^0)^2\big|_{r\approx r_H-l_{Pl}} \approx 2e^{2\psi_H} \gg 1 \tag{8.2.37}$$

during the Planck length.

But this is not the whole of history.

8.2.1 Soft singularity

Let us consider more carefully what occurs at the place $r = \pm r_H$ with the G_{55} metric component. For this more convenient to present the metric (8.2.1) in the form

$$ds^2 = \frac{e^{2\psi(r)}}{\tilde{\Delta}(r)}dt^2 - \tilde{\Delta}(r)\left[d\chi + \omega(r)dt + Q\cos\theta d\varphi\right]^2 - dr^2 - a(r)(d\theta^2 + \sin^2\theta d\varphi^2), \tag{8.2.38}$$

comparing the metrics (8.2.1) and (8.2.38) we see the following connection between functions $\Delta(r)$ and $\tilde{\Delta}(r)$

$$G_{55} = \tilde{\Delta}(r) = \Delta(r)e^{2\psi(r)}. \tag{8.2.39}$$

By definition the derivative of the G_{55} is equal

$$\tilde{\Delta}'(r)\Big|_{r=r_H} = \lim_{\Delta r\to 0}\frac{\tilde{\Delta}\left(r_H+\Delta r\right) - \tilde{\Delta}\left(r_H - \Delta r\right)}{2\Delta r}. \tag{8.2.40}$$

Taking into account quantum gravity one can see that here there is one subtlety. The point is that Δr can not converge to zero since there is the minimal length : Planck length. Ordinary it is not important but in our case we have a *soft* singularity in the following sense. Equations (8.2.22) and (8.2.40) give us

$$\begin{aligned}\tilde{\Delta}'(r)\Big|_{r=r_H} &\approx \tfrac{\tilde{\Delta}(r_H+l_{Pl})-\tilde{\Delta}(r_H-l_{Pl})}{2l_{Pl}} = \tfrac{\Theta(r_H+l_{Pl})-\Theta(r_H-l_{Pl})}{2l_{Pl}} \\ &\approx \Theta'(r)\big|_{r=r_H} \approx \delta\left(r_H\right) = \infty\end{aligned} \tag{8.2.41}$$

where

$$\Theta\left(x\right) = \begin{cases} +1, & \text{if } x > 0 \\ -1, & \text{if } x < 0 \end{cases} \tag{8.2.42}$$

is the step function, $\delta(r)$ is the Dirac's delta function. It means that near to $r = \pm r_H$

$$\tilde{\Delta}'(r) \approx \delta\left(r\right). \tag{8.2.43}$$

It is not very bad if everywhere we would have only $\tilde{\Delta}'(r)$ not $\tilde{\Delta}'^2(r)$. But it is not the case. For example the Ricci scalar is

$$\begin{aligned}\frac{1}{2}R =& \psi'^2(r) - \frac{\psi'(r)\tilde{\Delta}'(r)}{\Delta(r)} + \frac{1}{2}\frac{\tilde{\Delta}'^2(r)}{\tilde{\Delta}^2(r)} + 2\psi''(r) - \frac{1}{2}\frac{\tilde{\Delta}^2(r)\omega^2(r)}{e^{2\psi(r)}} + \\ & 2\frac{a'(r)\psi'(r)}{a(r)} - \frac{1}{2}\frac{a'^2(r)}{a^2(r)} + 2\frac{a''(r)}{a(r)} - \frac{2}{a(r)} + \frac{1}{2}\frac{Q^2\tilde{\Delta}(r)}{a^2(r)}.\end{aligned} \tag{8.2.44}$$

We see that we have $\tilde{\Delta}'^2(r)$. It means that by $r = \pm r_H$ we have a singularity - *a soft* singularity. The word soft signifies that in this case singularity differs from (*hard*) singularity, for example, by $r = 0$ in the Schwarzschield black hole.

8.2.2 Pure quantum freezing of the 5th dimension

From the previous sections we saw that the gravitational flux tube solutions in the 5-dimensional Kaluza-Klein gravity have two regions where G_{55} metric component just has such fast change. One of the basic paradigm of quantum gravity tell us that the Planck length is the minimal length in the nature and consequently not any physical fields and quantities can change in the course of the Planck length. Consequently we can draw a conclusion that such classical dynamic of the metric signature is *impossible*. We suppose that only one way is to avoid such dynamical behavior of this quantity: *any* dynamic of G_{55} field variable should be forbidden with the conservation of its last value. One can say that it is *a pure quantum freezing of the dynamic of* 5th *dimension.* Mathematically it means that G_{55} becomes non-dynamical quantity and should not be varied by the derivation of Einstein's equations from the corresponding Lagrangian.

Thus we can say that the analysis of the classical dynamic of the metric signature for the super-thin and super-long gravitational flux tube shows that there is a region where this quantity changes too much quickly from the quantum gravity viewpoint. It leads to the fact that some pure quantum gravity effects have to happen in order to avoid such variation of the metric signature. We do not consider the mechanism of these effects but the authors point of view is that such mechanism can not be based on any field-theoretical consideration. This pure quantum freezing of extra dimension is like to a trigger which has only two states: in one state the dynamic of G_{55} is switched on in another one is switched off. One can say that such quantum dynamic which can be realized only in the Planck region is a non-differentiable dynamic. The examples of such hypothesized non-differentiable dynamic can be above-mentioned freezing of extra dimensions, the change of metric signature and may be other phenomena [3], [4].

The next question arising in this context is the dynamic of residuary degrees of freedom. Since $G_{55} = \text{const}$ we have Kaluza-Klein gravity in its initial interpretation when 5-dimensional Kaluza-Klein theory with $G_{55} = \text{const}$ is equivalent to 4-dimensional electrogravity. It prompts an idea that the spacetime with ($r \gtrsim r_H + l_{Pl}$) (the same for $r \lesssim -r_H - l_{Pl}$) will be the Reissner-Nordström solution with the corresponding electric and magnetic fields.

8.3 Continuity of electric and magnetic fields

In this section we would like to discuss the problems concerning to joining the electric and magnetic fields of the gravitational flux tube considered above and the extreme Reissner-Nordström solution. For this we will compare the fluxes of electric and magnetic fields on the throat and the Reissner-Nordström spacetime.

8.3.1 Continuity of electric field

At first we consider the electric field. The Maxwell equations for the electric field on the curve space are

$$\frac{1}{\sqrt{\gamma}} \frac{\partial}{\partial x^\alpha} \left(\sqrt{\gamma} \overset{(4)}{D^\alpha} \right) = 0 \tag{8.3.1}$$

where $\sqrt{\gamma}$ is the determinant of the 3D spatial metric and $\overset{(4)}{D^\alpha}$ is an analog of the electric displacement in the electrodynamics of continua medium in the presence of the gravitational field

$$\overset{(4)}{D^\alpha} = -\sqrt{g_{00}} \overset{(4)}{F^{0\alpha}}. \tag{8.3.2}$$

For the Reissner-Nordström spacetime the corresponding metric is

$$\begin{aligned} \overset{(4)}{ds^2} &= \left(1 - \frac{r_g}{R} + \frac{r_{q,Q}}{R^2} \right) \overset{(4)}{dt^2} + \frac{dR^2}{1 - \frac{r_g}{R} + \frac{r_{q,Q}}{R^2}} - R^2 \left(d\theta^2 + \cos^2\theta d\varphi^2 \right), \\ r_g &= \frac{2Gm}{c^2}, \quad r_{q,Q} = \sqrt{\frac{G}{c^4} \left(q^2 + Q^2 \right)} \end{aligned} \tag{8.3.3}$$

where G is the gravitational constant; $\overset{(4)}{(\,)}$ means that the corresponding quantity is 4-dimensional one; R is the 4-dimensional radial coordinate. Therefore Eq. (8.3.1) is

$$\frac{d}{dR} \left(R^2 \overset{(4)}{F_{tR}} \right) = 0, \quad \overset{(4)}{F_{tR}} = \frac{d\phi}{dR} \tag{8.3.4}$$

here ϕ is the scalar potential. This equation shows that the flux $\overset{(4)}{\Phi_e}$ of the electric field $\overset{(4)}{E_R} = \phi'$ does conserve

$$\overset{(4)}{\Phi_e} = 4\pi R^2 \phi' = 4\pi \overset{(4)}{q} = \text{const}. \tag{8.3.5}$$

The corresponding $\binom{5}{1}$ Einstein equation on the throat of the 5-dimensional flux tube is

$$\frac{d}{dr}\left[a\left(\omega'\Delta^2 e^{3\psi}\right)\right] = 0. \tag{8.3.6}$$

In the consequence of Eq. (8.3.6) we have also the conserved flux of an analog of electric displacement

$$\overset{(5)}{\Phi}_e = 4\pi a \overset{(5)}{D^r} = 4\pi \overset{(5)}{q} = \text{const} \tag{8.3.7}$$

where

$$\overset{(5)}{D^r} = \omega'\Delta^2 e^{3\psi} = \omega' e^{-\psi} G_{55}. \tag{8.3.8}$$

The simplest assumption about two electric fields on the throat and Reissner-Nordström spacetimes is to join the fluxes (8.3.5) and (8.3.7)

$$\overset{(5)}{\Phi}_e = \overset{(4)}{\Phi}_e \tag{8.3.9}$$

here $\overset{(5)}{\Phi}_e$ have to be calculated at the ends of throat where freezing of 5th coordinate happens, i.e. by $r \approx r_H - l_{Pl}$ on the right side of the throat and $r \approx -r_H + l_{Pl}$ on the left side of the throat. There $G_{55} \approx 1$ and

$$\phi' \approx \omega' e^{-\psi_H} \tag{8.3.10}$$

that results to

$$\overset{(5)}{q} = \overset{(4)}{q} = q \tag{8.3.11}$$

here we have used the conditions

$$\left.\overset{(4)}{g}_{\theta\theta}\right|_{R=r_0} = r_0^2 = a(0) = \left.\overset{(5)}{G}_{\theta\theta}\right|_{r=r_H-l_{Pl}} \tag{8.3.12}$$

$$\psi_H \approx \psi|_{r=r_H-l_{Pl}} = \psi|_{r=-r_H+l_{Pl}} \tag{8.3.13}$$

8.3.2 Continuity of magnetic field

Now we will repeat the similar calculations for the magnetic field. The Maxwell equation for the magnetic field having the information about the flux of magnetic field is the same for the throat and the tails (= Reissner-Nordström spacetimes)

$$\frac{1}{\sqrt{\gamma}}\frac{\partial}{\partial x^\alpha}\left[\sqrt{\gamma}\left(\frac{-1}{2\sqrt{\gamma}}\epsilon^{\alpha\beta\delta}F_{\beta\delta}\right)\right] = 0 \tag{8.3.14}$$

here we introduce the magnetic field B^α

$$B^\alpha = \frac{-1}{2\sqrt{\gamma}}\epsilon^{\alpha\beta\delta}F_{\beta\delta}. \tag{8.3.15}$$

More concretely Eq. (8.3.15) has the following form for the Reissner-Nordström spacetime

$$\frac{d}{dR}\left[R^2\left(\frac{\overset{(4)}{Q}}{R^2}\right)\right] = 0 \tag{8.3.16}$$

and similar for the 5-dimensional case

$$\frac{d}{dr}\left[r^2\left(\frac{\overset{(5)}{Q}}{r^2}\right)\right] = 0 \tag{8.3.17}$$

here r, R are the radial coordinates on the throat and the tails correspondingly. It allows us to introduce the flux of magnetic field for the throat and the tails which will be equal

$$\overset{(4)}{\Phi}_m = 4\pi R^2 \frac{\overset{(4)}{Q}}{R^2} = 4\pi a \frac{\overset{(5)}{Q}}{a} = \overset{(5)}{\Phi}_m. \tag{8.3.18}$$

Therefore one can introduce correctly the magnetic charge

$$\overset{(4)}{Q} = \overset{(5)}{Q} = Q. \tag{8.3.19}$$

8.3.3 Continuity of metric

After quantum freezing of the 5th dimension 5-dimensional metric (8.2.1) will be

$$\overset{(5)}{ds^2} = A(r)dt^2 - B(r)dr^2 - a(r)(d\theta^2 + \sin^2\theta d\varphi^2) - [d\chi + \omega(r)dt + Q\cos\theta d\varphi]^2 . \tag{8.3.20}$$

The interpretation of this metric in the Kaluza-Klein gravity tells us that we have 4-dimensional metric

$$\overset{(4)}{ds^2} = A(r)dt^2 - B(r)dr^2 - a(r)(d\theta^2 + \sin^2\theta d\varphi^2) \tag{8.3.21}$$

and the electromagnetic potential

$$A_\mu = \{\omega, 0, 0, Q\cos\theta\} . \tag{8.3.22}$$

In the previous subsections we have joined the electric and magnetic fields. Now we have to consider the metric components. We will interpret the metric (8.3.21) as Reissner-Nordström metric. After freezing of 5th coordinate the corresponding equations become 5-dimensional Einstein's equations but without $\binom{5}{5}$ Einstein equation, i.e. we have the ordinary 4-dimensional electro-gravity. Taking this into account the metric (8.3.20) will be Reissner-Nordström metric (8.3.3). We will join $g_{\theta\theta}$ and g_{tt} components of the metrics (8.2.1) and (8.3.21). It is necessary to note that joining the g_{rr} components of these metrics is not necessary as they measure the distance on the transversal direction to the surface of joining. We do not join (as usually) the first derivatives of metric components since on the 5-dimensional throat and the tails there are the different equations set: on the throat - 5-dimensional Einstein's equations but on the tails 4-dimensional electro-gravity equations : let us remember that according to the initial interpretation of the Kaluza-Klein gravity the 4-dimensional electro-gravity spacetime can be considered as 5-dimensional spacetime with frozen 5th dimension (G_{55} = const).

Joining of $g_{\theta\theta}$ gives

$$\overset{(4)}{g}_{rr}(r_0) = a|_{\pm r_H \mp l_{Pl}} \quad \Rightarrow \quad r_0^2 = a(0) = l_{Pl}^2. \tag{8.3.23}$$

For the g_{tt} components

$$\left(1 - \frac{r_g}{r_0} + \frac{r_{q,Q}^2}{r_0^2}\right) \overset{(4)}{dt^2} = e^{2\psi_H} \overset{(5)}{dt^2} \tag{8.3.24}$$

here $\overset{(5)}{t}$ and $\overset{(4)}{t}$ are the time coordinates on the 5-dimensional throat and the 4-dimensional tails correspondingly. One can rewrite this relation as

$$\frac{\overset{(5)}{dt}}{\overset{(4)}{dt}} = e^{-\psi_H} \sqrt{1 - \frac{r_g}{l_{Pl}} + \frac{r_{q,Q}^2}{l_{Pl}^2}}. \tag{8.3.25}$$

Only by such relation between the time coordinates on the 5-dimensional throat and the 4-dimensional tails the time will pass equally on the hypersurface of joining.

It is necessary to mention that the analysis presented in this section is very simple and, for example, do not give us the possibility to determine the mass m for the Reissner-Nordström solution. Probably more exact calculations could be made on the quantum field-theoretical language.

8.4 Conclusions and discussion

Thus in this notice we wanted to show that in quantum gravity there are some reasonings speaking about that some classical dynamic of G_{55} is forbidden. We have shown that the super-thin and super-long gravitational flux tube solutions in 5-dimensional Kaluza-Klein gravity have two regions where the classical description can not be applied. Some metric components change too quickly: the metric signature changes from $\eta_{ab} = \{+,-,-,-,-\}$ to $\eta_{ab} = \{-,-,-,-,+\}$ and consequently $\Delta G_{55} \approx 2$ during the Planck length. To avoid such variation the dynamic of the metric component G_{55} must be frozen. Then the metric on the throat can be extended via the right and left ends to the Reissner-Nordström spacetimes. If we additionally will freeze 4-dimensional metric $g_{\mu\nu}$ then we will have only Maxwell electrodynamic on the flat space which is similar to the AdS/S correspondence idea. At the joint places takes place *a pure quantum freezing of the dynamic of* G_{55} *metric component.* Such object looks as two extremal Reissner-Nordström spacetimes ($r_g^2/4 < r_{q,Q}^2$) connected with a super-thin and super-long flux tube filled with the electric and magnetic fields. Let us note that the presented here point of view is based only on the statement which does not depend on the details of one or another quantum gravity theory.

Such model allow us successfully to resolve one of the hardest problems of general relativity: the avoidance of a singularity. The construction presented here shows that for an extremal Reissner-Nordström solution (at least for $q > Q$) the gravitational field near to singularity becomes so strong that the dynamic on the 5th dimension becomes excited and a flux tube from one singularity to another one appears. The force lines do not convergence in a pointlike singularity but leave our universe to another one (or to a remote part of our universe) through the flux tube and there appear near to another singularity. Similar idea in the 4-dimensional case is presented in Ref. [6] as a model of an elementary particle.

There is another argument for presented here the pure quantum freezing mechanism. We see that by $|r| \lesssim r_H - l_{Pl}$ the metric component $G_{55} \approx 1$ is almost frozen. It means that with big accuracy we have *dynamical* freezing of the 5th coordinate on the throat but near to the surface of the change of metric signature the quantity G_{55} becomes dynamical. One can suppose that there is a quantum mechanism which keep this freezing from dynamical one to pure quantum one.

It is necessary to note that such mechanism does work only for some special extremal Reissner-Nordström solutions having the electric and magnetic charges with the relation $q > Q$. Certainly the question arises: is it possible to extend the result presented here about avoidance of a singularity to other Reissner-Nordström solutions ? In this connection we have to emphasize that the analysis carried out in tis notice is very simple and more careful analysis with the quantization of corresponding field quantities can give more exact results. In this connection one can mention the results of Ref. [5] where it is shown that the super-thin and super-long flux tube solutions without freezing 5th dimension have interesting properties on the tails: the magnetic fields decreases faster the electric field and there is a rotation connected with the magnetic field on the throat. One can presuppose that these properties will remain the same by switching on freezing 5th coordinate (may be in some weaker form).

The construction presented here (two Reissner-Nordström spacetimes connected by a flux tube) can be considered as a model of a spacetime with frozen 5th dimension but with piecewise impregnations where the 5th dimension is unfrozen. The similar idea about spacetime regions with compactified and uncompactified phases was considered by Guendelman in Ref. [7]. The construction presented in [7] is also based on the flux tube (Levi-Civita - Bertotti - Robison solution [8]-[10]) inserted between two 4-dimensional black holes. Similarly in Ref. [6] the model of classical electron with Levi-Civita - Bertotti - Robison flux tube between two Reissnaer-Nordström black holes is considered.

References

[1] V. Dzhunushaliev and D. Singleton, *Phys. Rev.* **D59**, 064018 (1999).

[2] V. Dzhunushaliev, "Strings and Branes under Microscope", gr-qc/0312038; "Wormhole solutions in 5D Kaluza-Klein theory as string-like objects", *New Developments in Quantum Cosmology Research. Series: Horizons in World Physicss*, Vol. 247, Ed. A. Reimer, Vol. 247, 113-140 (2005), (New-York, Nova Science Publishers, 2005), gr-qc/0405017.

[3] G. Bogdanoff and I. Bogdanoff, *Class. Quant. Grav.* **18**, 4341(2001); V. Dzhunushaliev and D. Singleton, *Class. Quant. Grav.* **18**, 1787(2001).

[4] C. Castro and M. Pavsic, *Int. J. Theor. Phys.* **42**, 1693 (2003).

[5] V. Dzhunushaliev, *Phys. Lett. B.*, **600**, 171 (2004).

[6] O. B. Zaslavskii, *Phys. Rev. D*, **70**, 104017 (2004).

[7] E. I. Guendelman, *Gen. Rel. Grav.* **23** (1991) 1415.

[8] T. Levi-Civita, *Atti Acad. Naz. Lincei*, **26**, 519 (1917).

[9] B. Bertotti, *Phys. Rev.*, **116**, 1331 (1959).

[10] I. Robinson, *Bull. Akad. Pol.*, **7**, 351 (1959).

In: Focus on Quantum Gravity Research
Editor: David C. Moore, pp. 327-349

ISBN 1-59454-660-6

Chapter 9

An $\mathrm{SL}(2,\mathbb{R})$-covariant, First Order, κ-supersymmetric Action for the $D5$-brane

DFTT - 41/2002

Leonardo Modesto
Dipartimento di Fisica Teorica, Universitá di Torino,
& INFN - Sezione di Torino
via P. Giuria 1, I-10125 Torino, Italy

This work is supported in part by the European Union RTN contracts HPRN-CT-2000-00122 and HPRN-CT-2000-00131

The new first order, rheonomic, κ–supersymmetric formalism recently introduced by us for the world-volume action of the $D3$ brane is extended to the case of $D5$ branes. This extension requires the dual formulation of the Free Differential Algebra of type IIB supergravity in terms of 6–form gauge potentials which was so far missing and is given here. Furthermore relying on our new approach we are able to write the $D5$ world volume action in a manifestly $\mathrm{SL}(2,\mathbb{R})$ covariant form. This is important in order to solve the outstanding problem of finding the appropriate boundary actions of $D3$–branes on smooth ALE manifolds with twisted fields. The application of our results to this problem is however postponed to a subsequent publication.

9.1 Introduction

In a recent paper [1], the present authors introduced a **new first order formalism** that is able of generating second order world volume actions of the Born–Infeld type for p-branes. This construction allows for a description within the standard framework of rheonomy of those κ-supersymmetric boundary actions that function as sources for supergravity p–solutions in the bulk. Furthermore in this approach the SL(2, R) duality symmetry can be manifest at all levels. These features are very much valuable and quite essential in order to discuss various questions relative to the microscopic interpretation of classical p–solutions like the $D3$–brane with flux on smooth ALE manifolds [2] and other issues in the general quest of the gauge/gravity correspondence. The new formulation which, in our opinion, is particularly compact and elegant is based on the introduction of an additional auxiliary field, besides the world volume vielbein, and on the enlargement of the local symmetry from the Lorentz group to the general linear group:

$$\mathrm{SO}(1, \mathrm{d}-1) \overset{\text{enlarged}}{\Longrightarrow} \mathrm{GL}(\mathrm{d}, \mathbb{R}) \tag{9.1.1}$$

Within this new formalism in [1] we constructed the first order version of the κ–supersymmetric action for a $D3$–brane. This choice was not random rather it was rooted in the main motivations to undertake such a new construction. Indeed we needed a suitable formulation of the $D3$–brane action holding true on a generic background in order to apply it to the aforementioned $D3$–brane bulk solution with flux [2] or to more complicated ones. However, as for the the solution in [2], the task was not exhausted once we had the $D3$-brane action. Indeed this latter accounts only for the source of the Ramond-Ramond $A^{[4]}$–form with self dual field strength but not for the source of the doublet $\{B^{[2]}, C^{[2]}\}$ of 2–forms whose flux is trapped on the homology 2–cycles of the transverse ALE manifold. The source of such fields can only be accounted for by the presence of 5–branes whose world-volume sweeps both the world volume of the $D3$–brane and the ALE 2–cycles. Hence we need to extend our first order formalism also to 5–branes and this is what we do in the present paper.

It was noted already in [3, 2, 4] that the a $D5$ brane wrapped on a 2–cycle of the ALE manifold cannot be the direct source of the smooth, fractional brane like solution of [2]. The reason is simple. In the solution of [2] the dilaton field is constant and therefore there is no source for its field equation. On the other hand the $D5$–brane couples to the dilaton and contributes such a source. It is only in the orbifold limit that such a coupling vanishes together with the vanishing of the homology cycle trapping the flux of the 2–forms. In order to reproduce the correct boundary action for the smooth supersymmetric boundary solution of [2] we need a 5–brane that has vanishing coupling to the dilaton field. As we are going to show in a next coming publication [5] such a brane exists and it is a mixture of $D5$–branes on which closed **fundamental strings** can annihilate and of $D5'$–branes where such annihilation occurs for closed D**-strings**. To describe the world–volume action of such an object we need an explicitly SL(2, ℝ)-covariant formulation of the 5 brane that depends on two vectors q_Λ and p_Λ that transform in the fundamental representation of SL(2, ℝ). In this paper using the first order rheonomic formalism of [1] and relying on a dual formulation of type IIB supergravity we construct the required SL(2, ℝ) covariant 5–brane action.

9.2 Type IIB supergravity with dual 6–form potentials

In the appendix B of [1] we gave a short but comprehensive summary of type IIB supergravity in the rheonomic approach. This summary was entirely based on the original papers by Castellani and Pesando [6, 7] but it also contained some new useful results, in particular the transcription of the curvatures from the complex SU(1, 1) basis to the real SL(2, ℝ) basis and a comparison of the supergravity field equations as written in the rheonomy approach and as written in string text–books like Polchinsky's [11].

In order to deal with 5–branes we need to include also the 7–form field strengths that are dual to the 3–form field strengths originally used in [6, 7] and summarized in [1]. This amounts to extending the original Free Differential Algebra of type IIB supergravity with the 6–form generators that are associated with the new cohomology classes of such an algebra. This is the exact analogue of the procedure which leads to the extended Free Differential Algebra of M-theory [8] containing both the 3–form (source of $M2$-branes) and the 6–form (source of the $M5$-branes). This latter arises by including the new generator associated with the cohomology 7–cocycle of the original Free Differential Algebra of M–theory in its 3–form formulation [9].

In our case the new complete form of the curvatures is the following one:

$$
\begin{aligned}
R^{\underline{a}} &= \mathcal{D}V^{\underline{a}} - i\bar{\psi}\wedge\Gamma^{\underline{a}}\psi && (9.2.1)\\
R^{\underline{ab}} &= d\omega^{\underline{ab}} - \omega^{\underline{ac}}\wedge\omega^{\underline{db}}\eta_{\underline{cd}} \\
\rho &= \mathcal{D}\psi \equiv d\psi - \tfrac{1}{4}\omega^{\underline{ab}}\wedge\Gamma_{ab}\psi - \tfrac{1}{2}\mathrm{i}Q\wedge\psi && (9.2.2)\\
\mathcal{H}^{\alpha}_{[3]} &= \sqrt{2}\,dA^{\alpha}_{[2]} + 2i\Lambda^{\alpha}_{+}\bar{\psi}\wedge\Gamma_{\underline{a}}\psi^{*}\wedge V^{\underline{a}} + 2\mathrm{i}\Lambda^{\alpha}_{-}\bar{\psi}^{*}\wedge\Gamma_{\underline{a}}\psi\wedge V^{\underline{a}} && (9.2.3)\\
\mathcal{F}_{[5]} &= dC_{[4]} + \tfrac{1}{16}\,\mathrm{i}\,\epsilon_{\alpha\beta}\sqrt{2}\,A^{\alpha}_{[2]}\wedge\mathcal{H}^{\beta}_{[3]} + \tfrac{1}{6}\,\bar{\psi}\wedge\Gamma_{\underline{abc}}\psi\wedge V^{\underline{a}}\wedge V^{\underline{b}}\wedge V^{\underline{c}} \\
&\quad + \tfrac{1}{8}\,\epsilon_{\alpha\beta}\sqrt{2}\,A^{\alpha}_{[2]}\wedge\left(\Lambda^{\beta}_{+}\bar{\psi}\Gamma_{\underline{a}}\psi^{*} + \Lambda^{\beta}_{-}\bar{\psi}^{*}\Gamma_{\underline{a}}\psi\right)\wedge V^{\underline{a}} && (9.2.4)\\
\mathcal{D}\lambda &= d\lambda - \frac{1}{4}\omega^{\underline{ab}}\Gamma_{\underline{ab}}\lambda - \mathrm{i}\tfrac{3}{2}\,Q\lambda && (9.2.5)\\
\mathcal{D}\Lambda^{\alpha}_{\pm} &= d\Lambda^{\alpha}_{\pm} \mp \mathrm{i}\,Q\,\Lambda^{\alpha}_{\pm}. && (9.2.6)\\
\mathcal{H}^{\alpha}_{[7]} &= dC^{\alpha}_{[6]} + \Lambda^{\alpha}_{+}\,\bar{\psi}\,\Gamma_{a_1a_2a_3a_4a_5}\,\psi^{*}\,V^{a_1a_2a_3a_4a_5} - \Lambda^{\alpha}_{-}\,\bar{\psi}^{*}\,\Gamma_{a_1a_2a_3a_4a_5}\,\psi\,V^{a_1a_2a_3a_4a_5} \\
&\quad + \frac{10}{3}\,\bar{\psi}\,\Gamma_{a_1a_2a_3}\,\psi\,V^{a_1a_2a_3}\,\sqrt{2}\,A^{\alpha}_{[2]} \\
&\quad + 40\,\sqrt{2}\,dA^{\alpha}_{[2]}\,C_{[4]} && (9.2.7)
\end{aligned}
$$

(where $\psi^* = C\bar{\psi}^T$) the novelty being the expression of the $\mathcal{H}^{\alpha}_{[7]}$ curvature. Its definition is a direct consequence of the following Fierz identities:

$$\begin{aligned}
\bar{\psi}\Gamma^a\psi^*\bar{\psi}\Gamma^a &= 0 \\
\Gamma^a\psi\bar{\psi}^*\Gamma^a\psi &= 0 \\
\bar{\psi}\Gamma_{a_1a_2a_3}\psi\bar{\psi}\Gamma^{a_3}\psi &= \bar{\psi}^*\Gamma_{[a_1}\psi\bar{\psi}_{a_2]}\psi^* \\
\bar{\psi}\Gamma_{a_1a_2a_3a_4a_5}\psi^*\bar{\psi}\Gamma^{a_1}\psi &= -4\bar{\psi}\Gamma_{[a_2a_3a_4}\psi\bar{\psi}\gamma_{a_5]}\psi^* \\
\bar{\psi}^*\Gamma_{a_1a_2a_3a_4a_5}\psi\bar{\psi}\Gamma^{a_1}\psi &= 4\bar{\psi}\Gamma_{[a_2a_3a_4}\psi\bar{\psi}^*\gamma_{a_5]}\psi
\end{aligned} \tag{9.2.8}$$

From eq.s (9.2.7) and (9.2.4) one obtains the supersymmetry transformation rules of the background fields relevant in establishing the κ-supersymmetry transformation rules against which the 5-brane action is invariant.

9.3 Establishing the κ-supersymmetric action of the 5–brane

Next, following the general scheme outlined in [1] we apply the new first order formalism to the case $d = 6$ in order to derive the κ-supersymmetric action of a 5–brane. As extensively discussed in [1], the κ–supersymmetry transformations just follow, via a suitable projection, from the bulk supersymmetries as derived from supergravity, type IIB theory, in this case. The latter has a duality symmetry with respect to an SL(2, $\mathbb{R}$) group of transformations that acts non linearly on the two scalars of massless spectrum, the dilaton ϕ and the Ramond scalar C_0. Indeed these two parametrize the coset manifold SL(2, $\mathbb{R}$)/O(2) and actually correspond to its solvable parametrization. Because of the special relevance of this issue in the context of our problem, we recall below such a solvable parametrization. This also helps to fix the notations.

<u>SL(2, $\mathbb{R}$) *Lie algebra*</u>

$$[L_0\,,\,L_\pm] = \pm\,L_\pm \quad ; \quad [L_+\,,\,L_-] = 2\,L_0 \tag{9.3.1}$$

with explicit 2–dimensional representation:

$$L_0 = \frac{1}{2}\begin{pmatrix} 1 & 0 \\ 0 & -1 \end{pmatrix} \quad ; \quad L_+ = \begin{pmatrix} 0 & 1 \\ 0 & 0 \end{pmatrix} \quad ; \quad L_- = \begin{pmatrix} 0 & 0 \\ 1 & 0 \end{pmatrix} \quad ; \tag{9.3.2}$$

<u>*Coset representative of* SL(2, $\mathbb{R}$)/O(2) *in the solvable parametrization*</u>

$$\mathbb{L}\left(\varphi, C_{[0]}\right) = \exp\left[\varphi\,L_0\right]\,\exp\left[C_{[0]}e^{\varphi}\,L_-\right] = \begin{pmatrix} \exp[\varphi/2] & 0 \\ C_{[0]}e^{\varphi/2} & \exp[-\varphi/2] \end{pmatrix} \tag{9.3.3}$$

where $\varphi(x)$ and $C_{[0]}$ are respectively identified with the dilaton and with the Ramond-Ramond 0-form of the superstring massless spectrum. The isomorphism of SL(2, ℝ) with SU(1, 1) is realized by conjugation with the Cayley matrix:

$$\mathcal{C} = \frac{1}{\sqrt{2}} \begin{pmatrix} 1 & -\mathrm{i} \\ 1 & \mathrm{i} \end{pmatrix} \tag{9.3.4}$$

Introducing the SU(1, 1) coset representative

$$\mathrm{SU}(1,1) \ni \Lambda = \mathcal{C}\,\mathbb{L}\,\mathcal{C}^{-1} \tag{9.3.5}$$

from the left invariant 1–form $\Lambda^{-1}\, d\Lambda$ we can extract the 1-forms corresponding to the scalar vielbein P and the U(1) connection Q

<u>*The* SU(1, 1)/U(1) *vielbein and connection*</u>

$$\Lambda^{-1}\, d\Lambda = \begin{pmatrix} -\mathrm{i}\,Q & P \\ P^{\star} & \mathrm{i}\,Q \end{pmatrix} \tag{9.3.6}$$

Explicitly

$$\begin{array}{rcll} P & = & \frac{1}{2}\left(d\varphi - \mathrm{i}\, e^{\varphi}\, dC_{[0]}\right) & \text{scalar vielbein} \\ Q & = & \frac{1}{2}\exp[\varphi]\, dC_{[0]} & \text{U(1)-connection} \end{array} \tag{9.3.7}$$

As stressed in the introduction the 5–brane action we want to write, not only should be cast into first order formalism, but should also display manifest covariance with respect to SL(2, ℝ). This covariance relies on introducing two vectors with two components q_α and p_α that transform in the fundamental representation of SU(1, 1) and expresse the charges carried by the $D5$ brane with respect to the 2–forms $A^\alpha_{[2]}$ of bulk supergravity (both the Neveu Schwarz $B_{[2]}$ and Ramond–Ramond $C_{[2]}$). According to the geometrical formulation of type IIB supergravity summarized above we set:

$$\begin{array}{rclcrcl} A^\Lambda & = & (B_{[2]}\,,\, C_{[2]}) & ; & A^\alpha & = & \mathcal{C}^\alpha{}_\Lambda\, A^\Lambda \\ A^{\alpha=1} & = & \frac{1}{\sqrt{2}}\,(B_{[2]} \,-\, i\,C_{[2]}) & ; & A^{\alpha=2} & = & \frac{1}{\sqrt{2}}\,(B_{[2]} \,+\, i\,C_{[2]}) \end{array} \tag{9.3.8}$$

In terms of these objects and using the new first order formalism that we introduced in [1] we write down the complete action of the $D5$–brane as follows:

$$\begin{aligned} \mathcal{L} \;=\; & \\ & \rho\,\Pi^{\underline{a}}_i\, V^{\underline{b}}\,\eta_{\underline{ab}}\,\eta^{i\ell_1} \wedge e^{\ell_2} \wedge \cdots \wedge e^{\ell_6}\,\epsilon_{\ell_1\ldots\ell_6} + \rho\, a_1\, \Pi^{\underline{a}}_i\, \Pi^{\underline{b}}_j\, \eta_{\underline{ab}}\, h^{ij}\, e^{\ell_1} \wedge \cdots \wedge e^{\ell_6}\,\epsilon_{\ell_1\ldots\ell_6} \\ & +\rho\, a_2\left[\det\left(h^{-1} + \mu\mathcal{F}\right)\right]^\alpha e^{\ell_1} \wedge \cdots \wedge e^{\ell_6}\,\epsilon_{\ell_1\ldots\ell_6} \\ & +\rho\, a_3 \mathcal{F}^{ij}\, F^{[2]} \wedge e^{\ell_3} \wedge e^{\ell_6}\,\epsilon_{ij\ell_3\ldots\ell_6} \\ & +\nu\, F \wedge F \wedge F + a_5\, p_\alpha\, A^\alpha \wedge F \wedge F + a_6\, C_{[4]} \wedge F + a_7\, \widetilde{q}_\alpha C^\alpha_{[6]} \end{aligned} \tag{9.3.9}$$

where the vectors $q_\alpha = \frac{1}{\sqrt{2}}(q_1, q_1^\star)$, $p_\alpha = \frac{1}{\sqrt{2}}(p_1, p_1^\star)$ satisfy the constraint (see appendix 9.4):

$$q_\alpha\, p_\beta\, \epsilon^{\alpha\beta} = 1 \tag{9.3.10}$$

$\rho = \rho(\phi)$ is a function of the dilaton to be determined, $C_{[4]}$ is the 4–form potential, and $C^{\alpha}_{[6]}$ are the dual potentials to the $A^{\alpha}_{[2]}$ forms. The coefficients

$$\alpha = \frac{1}{4} \qquad a_1 = -\frac{1}{12} \qquad a_2 = -\frac{1}{3} \qquad a_3 = \frac{5}{4} \tag{9.3.11}$$

where already determined in paper [1]. The coefficients a_5, a_6, ν are new and they must be fixed by to be κ–supersymmetry. The first two are numerical, while ν is also a function of the bulk scalars to be determined through κ-supersymmetry.
In the action (9.3.9)

$$F \equiv F^{[2]} \equiv F_{[G]} + q_\alpha A^\alpha \tag{9.3.12}$$

is the field strength of the world–volume gauge field and depends on the charge vector q_α. The physical interpretation of $F^{[2]}$ is as follows. By definition a *Dp*–brane is a locus in space–time where open strings can end or, in the dual picture, boundaries for closed string world–volumes can be located. The type IIB theory contains two kind of strings, the fundamental strings and the *D*-strings which are rotated one into the other by the $\mathrm{SL}(2,\mathbb{Z}) \subset \mathrm{SL}(2,\mathbb{R})$ group. Correspondingly a *D*5 brane can be a boundary either for fundamental or for *D*–strings or for a mixture of the two. The charge vectors q_α and p_α just express this fact. Furthermore the definition (9.3.12) of $F^{[2]}$ encodes the following idea: the world–volume gauge 1-form $A^{[1]}$ is just the parameter of a gauge transformation for the 2–form $q_\alpha A^\alpha$, which in a space–time with boundaries can be reabsorbed everywhere except on the boundary itself. Note that if we take $q_\alpha = \frac{1}{\sqrt{2}}(1\,,\,1)$ and $p_\alpha = \frac{i}{\sqrt{2}}(1\,,\,-1)$ we obtain :

$$q_\alpha A^\alpha = B_{[2]} \quad ; \quad p_\beta A^\beta = C_{[2]} \tag{9.3.13}$$

9.3.1 κ–supersymmetry

Next we want to prove that with an appropriate choice of ν, a_5, a_6 and a_7 the action (9.3.9) is invariant against bulk supersymmetries characterized by a projected spinor parameter. For simplicity we do this in the case of the choice $\widetilde{q}_\alpha = q_\alpha = \frac{1}{\sqrt{2}}(1\,,\,1)$ and $p_\alpha = \frac{i}{\sqrt{2}}(1\,,\,-1)$. For other choices of the charge type the modifications needed in the prove will be obvious from its details.

To accomplish our goal we begin by writing the supersymmetry transformations of the bulk differential forms $V^{\underline{a}}$, $B_{[2]}$, $C_{[2]}$, $C_{[4]}$ and $C_{[6]}$ which appear in the action. From the rheonomic parametrizations of the curvatures recalled in [1] we immediately obtain:

$$\begin{aligned}
\delta V^{\underline{a}} &= \mathrm{i}\,\tfrac{1}{2}\,(\bar{\epsilon}\,\Gamma^{\underline{a}}\,\psi + \bar{\epsilon}^*\,\Gamma^{\underline{a}}\,\psi^*)\\
\delta B_{[2]} &= -\,2\,i\,[(\Lambda^1_+ + \Lambda^2_+)\,\bar{\epsilon}\,\Gamma_{\underline{a}}\,\psi^*\,V^{\underline{a}} + (\Lambda^1_- + \Lambda^2_-)\,\bar{\epsilon}^*\,\Gamma_{\underline{a}}\,\psi\,V^{\underline{a}}]\\
\delta C_{[2]} &= 2\,[(\Lambda^1_+ - \Lambda^2_+)\,\bar{\epsilon}\,\Gamma_{\underline{a}}\,\psi^*\,V^{\underline{a}} + (\Lambda^1_- - \Lambda^2_-)\,\bar{\epsilon}^*\,\Gamma_{\underline{a}}\,\psi\,V^{\underline{a}}]\\
\delta C_{[4]} &= \underbrace{-\frac{1}{6}(\bar{\epsilon}\,\Gamma_{\underline{abc}}\,\psi \;-\; \bar{\epsilon}^*\,\Gamma_{\underline{abc}}\,\psi^*)\,V^{\underline{abc}}}_{\delta C'_{[4]}} + \frac{1}{8}\,[B_{[2]}\,\delta C_{[2]} \;-\; C_{[2]}\,\delta B_{[2]}]
\end{aligned} \tag{9.3.14}$$

and :

$$\begin{aligned} \delta C_{[6]} &= g_1\, C_{[4]}\, \delta F_{[2]} + g_2\, F_{[2]} \delta C'_{[4]} + \delta C'_{[6]} \\ &= \frac{1}{\sqrt{2}} (\delta C^1_{[6]} + \delta C^2_{[6]}) \\ &= -\frac{2}{\sqrt{2}} (\Lambda^1_+ + \Lambda^2_+) \bar{\epsilon}\, \Gamma_{\underline{a_1 \ldots a_5}}\, \psi^* + \frac{2}{\sqrt{2}} (\Lambda^1_- + \Lambda^2_-) \bar{\epsilon}^*\, \Gamma_{\underline{a_1 \ldots a_5}}\, \psi\, V^{\underline{a_1 \ldots a_5}} + \\ &\quad +20\sqrt{2}\, \delta C'_{[4]}\, F_{[2]} - 40\sqrt{2}\, C_{[4]}\, \delta F_{[2]} \end{aligned} \tag{9.3.15}$$

Note that in writing the above transformations we have neglected all terms involving the dilatino field. This is appropriate since the background value of all fermion fields is zero. The gravitino 1–form ψ is instead what we need to keep track of. Proving κ–supersymmetry is identical with showing that all ψ terms cancel against each other in the variation of the action. Relying on (9.3.14) the variation of the W.Z.T term is as follows:

$$\begin{aligned} &\delta(\nu\, F \wedge F \wedge F + a_5\, C_{[2]} \wedge F \wedge F + a_6\, C_{[4]} \wedge F + a_7\, C_{[6]}) = \\ &= 3\,\nu\, F\, F\, \delta F - \frac{3 g_1 a_7}{16} F\, F\, \delta C_{[2]} + (g_2 - g_1) a_7\, F\, \delta C'_{[4]} + a_7\, \delta C'_{[6]} \end{aligned} \tag{9.3.16}$$

for $a_5 = -\frac{g_1 a_7}{16}$, $a_6 = -g_1 a_7$, $g_1 = -40\sqrt{2}$ and $g_2 = 20\sqrt{2}$.
And with such a choice the complete variation of the Lagrangian under a supersymmetry transformation of arbitrary parameter is:

$$\begin{aligned} \delta\mathcal{L} &= \delta\mathcal{L}_\psi + \delta\mathcal{L}_{\psi^*} \\ \delta\mathcal{L}_\psi &= [-5!\, i\, (h^{-1})^{ij}\, (\bar{\epsilon}\, \gamma_i\, \psi) + (\mu_3\, \mathcal{F}^{ij} + \mu_4\, \widetilde{F\,F}^{ij})\, (\bar{\epsilon}^*\, \gamma_i\, \psi) + \\ &\quad +(g_1 - g_2) \frac{a_7}{3}\, \widetilde{F}^{klmj}\, (\bar{\epsilon}\, \gamma_{klm} \psi) + a_7\, \mu_6\, (\bar{\epsilon}^*\, \widetilde{\gamma}^j\, \psi)]\, \Omega^{[5]}_j \\ \delta\mathcal{L}_{\psi^*} &= [-5!\, i\, (h^{-1})^{ij}\, (\bar{\epsilon}^*\, \gamma_i\, \psi^*) + (\mu_1\, \mathcal{F}^{ij} + \mu_2\, \widetilde{F\,F}^{ij})\, (\bar{\epsilon}\, \gamma_i\, \psi^*) + \\ &\quad -(g_1 - g_2) \frac{a_7}{3}\, \widetilde{F}^{klmj}\, (\bar{\epsilon}^*\, \gamma_{klm} \psi^*) + a_7\, \mu_5\, (\bar{\epsilon}\, \widetilde{\gamma}^j\, \psi^*)]\, \Omega^{[5]}_j \end{aligned} \tag{9.3.17}$$

where :

$$\begin{aligned} \mu_1 &= -4\, 4!\, i\, a_3\, (\Lambda^1_+ + \Lambda^2_+) \\ \mu_2 &= 4\, [-3\,\nu\, i\, (\Lambda^1_+ + \Lambda^2_+) - \tfrac{3}{16}\, g_1\, a_7\, (\Lambda^1_+ - \Lambda^2_+)] \\ \mu_3 &= -4\, 4!\, i\, a_3\, (\Lambda^1_- + \Lambda^2_-) \\ \mu_4 &= 4\, [-3\,\nu\, i\, (\Lambda^1_- + \Lambda^2_-) - \tfrac{3}{16}\, g_1\, a_7\, (\Lambda^1_- - \Lambda^2_-)] \end{aligned} \tag{9.3.18}$$

Recalling eq.s (9.3.5) and (9.3.3) the above eq.s (9.3.18) become:

$$\begin{aligned} \mu_1 &= -5!\, i\, e^{\phi/2} \quad ; \quad \mu_2 = \tfrac{3}{4}\, g_1\, a_7\, e^{-\frac{\phi}{2}} \\ \mu_3 &= -5!\, i\, e^{\phi/2} \quad ; \quad \mu_4 = -\tfrac{3}{4}\, g_1\, a_7\, e^{-\frac{\phi}{2}} \end{aligned} \tag{9.3.19}$$

where we have chosen:

$$\nu = \frac{g_1\, a_7}{16}\, C_{[0]} = \frac{g_1\, a_7}{16} \mathrm{Re}\, \mathcal{N} \tag{9.3.20}$$

In the above equation we have introduced the complex kinetic matrix which would appear in a $d = 4$ gauge theory with scalars sitting in SU(1, 1)/U(1) and determined by the classical Gaillard–Zumino general formula [1] applied to the specific coset:

$$\mathcal{N} = \mathrm{i}\, \frac{\Lambda^1_- - \Lambda^2_-}{\Lambda^1_- + \Lambda^2_-} \quad \Rightarrow \quad \begin{cases} \mathrm{Re}\, \mathcal{N} = C_0 \\ \mathrm{Im}\, \mathcal{N} = e^{-\phi} \end{cases} \tag{9.3.21}$$

It is convenient to rewrite the full variation (9.3.17) of the Lagrangian in matrix form in the 2–dimensional space spanned by the fermion parameters $(\epsilon\,,\, \epsilon^*)$:

$$\delta\mathcal{L} = \delta\mathcal{L}_\psi + \delta\mathcal{L}_{\psi^*} = (\bar{\epsilon}\,,\, \bar{\epsilon}^*)\, A \begin{pmatrix} \psi \\ \psi^* \end{pmatrix} \tag{9.3.22}$$

$$A_p = \begin{pmatrix} -5!\, i\, \gamma_p - 20\sqrt{2} a_7\, \widetilde{F}^{klmj}\, h_{jp}\, \gamma_{klm} & (-5!\, i\, e^{\frac{\phi}{2}}\, \mathcal{F}^{ij}\, \gamma_i - 30\sqrt{2}\, a_7\, e^{-\frac{\phi}{2}}\, \widetilde{F\, F}^{ij}\, \gamma_i - \frac{2}{\sqrt{2}}\, a_7\, e^{\frac{\phi}{2}}\, \widetilde{\gamma}^j)\, h_{jp} \\ (-5!\, i\, e^{\frac{\phi}{2}}\, \mathcal{F}^{ij} \gamma_i + 30\sqrt{2} a_7 e^{-\frac{\phi}{2}}\, \widetilde{F\, F}^{ij} \gamma_i + \frac{2}{\sqrt{2}} a_7\, e^{\frac{\phi}{2}}\, \widetilde{\gamma}^j)\, h_{jp} & -5!\, i\, \gamma_p + 20\sqrt{2}\, a_7\, \widetilde{F}^{klmj}\, h_{jp}\, \gamma_{klm} \end{pmatrix} \tag{9.3.23}$$

where $A = A_p\, \Omega^p_{[5]}$, and $e^i \wedge e^j \wedge e^k \wedge e^l \wedge e^m \equiv \epsilon^{ijklmp} \Omega^{[5]}_p$ denotes the quadruplet of five–volume forms

The matrix A_p is a tensor product of a matrices in spinor space and 2×2 matrices in the space spanned by $(\epsilon\,,\, \epsilon^*)$. It is convenient to spell out this tensor product structure which is achieved by the following rewriting:

$$\begin{aligned} A_p &= f_1 \gamma_p \otimes 1\!\!1 + f_2\, \widetilde{F}^{klmj}\, h_{jp}\, \gamma_{klm} \otimes \sigma_3 \\ &+ f_3\, e^{\frac{\phi}{2}} \mathcal{F}^{ij} h_{jp}\, \gamma_i \otimes \sigma_1 + (f_4\, \widetilde{\gamma}^j h_{jp} + f_5\, e^{-\frac{\phi}{2}}\, \widetilde{F\, F}^{ij} \gamma_i\, h_{jp}) \otimes \sigma_2 \end{aligned} \tag{9.3.24}$$

where :

$$\begin{aligned} f_1 &= -5!\, i\, \rho\ ;\ f_2 = -20\sqrt{2}\, a_7 \quad ; \\ f_3 &= -5!\, i\, \rho\ ;\ f_4 = -\tfrac{2}{\sqrt{2}}\, i\, a_7\, e^{\frac{\phi}{2}}\ ;\ f_5 = -30\, i\, \sqrt{2}\, a_7 \end{aligned} \tag{9.3.25}$$

and:

$$\begin{aligned} \widetilde{\gamma}^i &\equiv \epsilon^{lmnpqi}\, \gamma_{lmnpq} \quad ; \\ \widetilde{F}^{klmp} &\equiv \tfrac{1}{2}\, F_{ij}\, \epsilon^{ijklmp} \quad ; \quad (\widetilde{F\, F})_{ij} \equiv \tfrac{1}{2}\, \epsilon_{ijlmpq}\, F^{lm}\, F^{pq} \end{aligned} \tag{9.3.26}$$

now using the solution of the first order equations already determined in [1], namely

$$\begin{cases} \widehat{G} = \eta \\ \mathcal{F} = \frac{1}{\mu^2} h^{-1} F\, \eta \\ h\eta = (1\!\!1 - \frac{1}{\mu^2}\, F\, \eta\, F\, \eta) \left[\det\left(\eta - \frac{1}{\mu}\, F\right)\right]^{-1/2} \end{cases} \tag{9.3.27}$$

[1]For a general discussion of the Gaillard-Zumino formula see for instance [12]

or, in more compact form:

$$h = (\eta - \frac{1}{\mu^2}F^2)\left[\det\left(\eta - \frac{1}{\mu}F\right)\right]^{-1/2} \tag{9.3.28}$$

we can set

$$\begin{aligned} \frac{1}{\mu} &= e^{-\phi/2} = \sqrt{\mathrm{Im}\,\mathcal{N}} \\ \hat{F} &\equiv \sqrt{\mathrm{Im}\,\mathcal{N}}F \\ \rho &= e^{\frac{\phi}{2}} \end{aligned} \tag{9.3.29}$$

and we obtain:

$$\begin{aligned} & e^{\phi/2}\mathcal{F}^{ij}\,h_{jp} = e^{\phi/2}e^{-\phi}(Fh^{-1})^{ij}h_{jp} \equiv \hat{F}^{i}{}_{p} \\ & \widetilde{F}^{klmp} = e^{\frac{\phi}{2}}\,\widetilde{\hat{F}}^{klmp} \\ & e^{-\frac{\phi}{2}}(\widetilde{F\,F})_{ij} = e^{\frac{\phi}{2}}\,(\widetilde{\hat{F}\,\hat{F}})_{ij} \end{aligned} \tag{9.3.30}$$

This observation further simplifies the expression of A_k which can be rewritten as:

$$\begin{aligned} A_p &= e^{\frac{\phi}{2}}[z_1\gamma_p \otimes 1\!\!1 + z_2\,\widetilde{\hat{F}}^{klmj}\,h_{jp}\,\gamma_{klm}\otimes\sigma_3 \\ &+ z_3\,(\widetilde{\hat{F}\,\hat{F}})^{ij}\gamma_i\,h_{jp}\otimes\sigma_2 + z_4\hat{F}^i_p\,\gamma_i\otimes\sigma_1 + z_5\widetilde{\gamma}^j\,h_{jp}\otimes\sigma_2] \end{aligned} \tag{9.3.31}$$

where :

$$z_1 = f_1 \quad ; \quad z_2 = f_2 \quad ; \quad z_3 = f_5 \quad ; \quad z_4 = f_3 \quad ; \quad z_5 = f_4\,. \tag{9.3.32}$$

We introduce :

$$\begin{aligned} \Pi^p_{(1)} &= \widetilde{\hat{F}}^{klmp}\,\gamma_{klm} \equiv \frac{1}{2}\widetilde{\gamma}^{pij}\,F_{ij} \\ \Pi^p_{(2)} &= (\widetilde{\hat{F}\,\hat{F}})^{ip}\,\gamma_i \\ \Pi^p_{(3)} &= F^i{}_p\,\gamma_i \end{aligned} \tag{9.3.33}$$

and we obtain :

$$A_p = e^{\frac{\phi}{2}}[z_1\,\gamma_p\otimes 1\!\!1 + z_2\Pi^{(1)j}\,h_{jp}\otimes\sigma_3 + z_3\Pi^{(2)j}\,h_{jp} + z_4\Pi^{(3)}_p\otimes\sigma_1 + z_5\widetilde{\gamma}^j\,h_{jp}\otimes\sigma_2] \tag{9.3.34}$$

Just as in the case of the *D*3–brane discussed in [1], the proof of κ–supersymmetry can now be reduced to the following simple computation. Assume we have a matrix operator Γ with the following properties:

$$\begin{aligned} &[a] \qquad \Gamma^2 = 1\!\!1 \\ &[b] \qquad \Gamma\,A_k = A_k \end{aligned} \tag{9.3.35}$$

It follows that

$$P = \frac{1}{2}(1\!\!1 - \Gamma) \tag{9.3.36}$$

is a projector since $P^2 = P$ and that

$$PA_k = \frac{1}{N}(1\!\!1 - \Gamma)\, A_k = 0 \tag{9.3.37}$$

Therefore if we use supersymmetry parameters $(\overline{\kappa}, \overline{\kappa}^*) = (\overline{\epsilon}, \overline{\epsilon}^*)\ P$ projected with this P, then the action is invariant and this is just the proof of κ–supersymmetry.

The appropriate Γ is the following:

$$\Gamma = \frac{1}{N}[(\omega_{[6]} + \omega_{[2]}) \otimes \sigma_2 + (\omega_{[4]} + \omega_{[0]}) \otimes \sigma_3] \tag{9.3.38}$$

where:

$$\begin{aligned}
\omega_{[6]} &= \alpha_6\, \epsilon^{ijklmn}\, \gamma_{ijklmn} \\
\omega_{[4]} &= \alpha_4\, \epsilon^{ijklmn}\, \hat{F}_{ij}\, \gamma_{klmn} \\
\omega_{[0]} &= \alpha_0\, \epsilon^{ijklmn}\, \hat{F}_{ij}\, \hat{F}_{kl}\, \hat{F}_{mn} \\
\omega_{[2]} &= \alpha_2\, \epsilon^{ijklmn}\, \hat{F}_{ij}\, \hat{F}_{kl} \gamma_{mn} \\
N &= \left[\det\left(1\!\!1 \pm \hat{F}\right)\right]^{1/2}
\end{aligned} \tag{9.3.39}$$

and the coefficients are fixed to:

$$\alpha_6 = \frac{i}{6!} \quad ; \quad \alpha_4 = \frac{1}{2\,4!} \quad ; \quad \alpha_0 = \frac{1}{48} \quad ; \quad \alpha_2 = \frac{i}{16} \tag{9.3.40}$$

This choice suffices to guarantee property [a] in the above list. An outline of the details of the calculation leading to such a result is given below. We have:

$$\begin{aligned}
\Gamma^2 &= \frac{1}{N^2}\left[\omega_{[6]}^2 + \omega_{[2]}^2 + \omega_{[6]}\,\omega_{[2]} + \omega_{[2]}\,\omega_{[6]} + \omega_{[4]}^2\omega_{[0]}^2 + 2\omega_{[4]}\omega_{[0]}\right] \otimes 1\!\!1_{2\times 2} \\
&\quad + \left[\omega_{[6]}\omega_{[4]} - \omega_{[4]}\omega_{[6]} + \omega_{[6]}\omega_{[0]} - \omega_{[0]}\omega_{[6]}\right] \otimes \sigma_2\,\sigma_3 = \\
&= \frac{1}{N^2}\left[\omega_{[6]}^2 + \omega_{[2]}^2 + 2\omega_{[6]}\,\omega_{[2]} + \omega_{[4]}^2\omega_{[0]}^2 + 2\omega_{[4]}\omega_{[0]}\right] \otimes 1\!\!1_{2\times 2}
\end{aligned} \tag{9.3.41}$$

and by straightforward manipulations we obtain

$$\begin{aligned}
\omega_{[6]}^2 &= -(\alpha_6)^2(6!)^2 \\
\omega_{[4]}^2 &= -(4!)^2(\alpha_4)^2\, \hat{F}_{a_1a_2}\, \hat{F}_{a_3a_4}\, \gamma^{a_1a_2a_3a_4} - 2(4!)^2(\alpha_4)^2 Tr(\hat{F}^2) \\
\omega_{[2]}^2 &= \frac{4(\alpha_2)^2}{9\alpha_0\alpha_4}\,\omega_{[4]}\omega_{[0]} - 32(\alpha_2)^2\left[Tr(\hat{F}^2)\,Tr(\hat{F}^2) - 2\,Tr(\hat{F}^4)\right] \\
\omega_{[6]}\omega_{[2]} &= -2\,(6!)\,\alpha_6\,\alpha_2\,\hat{F}_{a_1a_2}\hat{F}_{a_3a_4}\,\gamma^{a_1a_2a_3a_4}
\end{aligned} \tag{9.3.42}$$

Using the following identities on determinants

$$\det\left(\mathbb{1} \pm \hat{F}\right) = 1 - \frac{1}{2}Tr(\hat{F}^2) + \frac{1}{8}\left[Tr(\hat{F}^2)\right]^2 - \frac{1}{4}Tr(\hat{F}^4) + \det(\hat{F}) \quad (9.3.43)$$

and

$$\det(\hat{F}) = -\tfrac{1}{6}\,Tr(\hat{F}^6) - \tfrac{1}{48}\left[Tr(\hat{F}^2)\right]^3 + \tfrac{1}{8}\left[Tr(\hat{F}^2)\right]\left[Tr(\hat{F}^4)\right] \quad (9.3.44)$$

$$= \left(\tfrac{1}{48}\,\epsilon_{a_1a_2a_3a_4a_5a_6}\,\hat{F}^{a_1a_2}\,\hat{F}^{a_3a_4}\,\hat{F}^{a_5a_6}\right) = \omega_{[0]}^2 \quad (9.3.45)$$

$$\left(\hat{F}\,\widetilde{\hat{F}\,\hat{F}}\right)_{ab} = -\tfrac{1}{12}\left(\epsilon_{a_1a_2a_3a_4a_5a_6}\,\hat{F}^{a_1a_2}\,\hat{F}^{a_3a_4}\,\hat{F}^{a_5a_6}\right)\mathbb{1}_{ab}$$
$$= -4\,\omega_{[0]}\,\mathbb{1}_{ab} \quad (9.3.46)$$

that are just consequences of $\widehat{F} = -\widehat{F}^T$ being antisymmetric property $[a]$ is proved.

Let us now turn to supersymmetry, namely to property $[b]$:

$$\Gamma A_p = A_p \quad (9.3.47)$$

We need to calculate the products of the matrices $\omega_{[i]}$ ($i = 6, 4, 2$) defined above with the gamma matrix structures appearing in the matrix A_p. Explicitly for $\omega_{[6]}$ and omitting the simbol " ^ " we find:

$$\begin{aligned}
\omega_{[6]}\,\gamma_a &= \frac{i}{5!}\,\widetilde{\gamma}_a \\
\omega_{[6]}\,\Pi^{b_1}_{(1)} &= \frac{3!}{2}\,i\,F_{b_2b_3}\,\gamma^{b_1b_2b_3} \\
\omega_{[6]}\,\Pi^{b_2}_{(2)} &= \frac{i}{5!}\,(\widetilde{F\,F})^{b_1b_2}\,\widetilde{\gamma}_{b_1} \\
\omega_{[6]}\,\Pi^{b_2}_{(3)} &= \frac{i}{5!}\,F^{b_1b_2}\,\widetilde{\gamma}_{b_1} \\
\omega_{[6]}\,\widetilde{\gamma}^{b_1} &= -5!\,i\,\gamma^{b_1}
\end{aligned} \quad (9.3.48)$$

while for $\omega_{[4]}$ we obtain

$$\begin{aligned}
\omega_{[4]}\,\gamma_{b_1} &= \frac{1}{4!}\,\widetilde{F}^{a_1a_2a_3a_4}\,\gamma_{a_1a_2a_3a_4b_1} - \frac{1}{3!}\,\widetilde{F}_{b_1a_2a_3a_4}\,\gamma^{a_2a_3a_4} \\
\omega_{[4]}\,\Pi^{b_4}_{(1)} &= -\frac{3}{2}F_{a_3a_4}\,F_{a_5a_6}\,\gamma^{b_4a_3a_4a_5a_6} \\
&\quad -3\,F^{b_5b_6}\,F^{b_3b_4}\,\gamma_{b_5b_6b_3} - 3\,Tr(F^2)\gamma^{b_4} + 6\,(F^2)^{b_5}{}_{b_6}\,\gamma^{b_6} \\
\omega_{[4]}\,\Pi^{b_2}_{(2)} &= \frac{1}{4!}\widetilde{F}_{a_1a_2a_3a_4}\,(\widetilde{F\,F})_{b_1}{}^{b_2}\,\gamma^{a_1a_2a_3a_4b_1} - \frac{1}{6}\,\widetilde{F}_{a_1a_2a_3a_4}\,(\widetilde{F\,F})^{a_1b_2}\,\gamma^{a_2a_3a_4} \\
\omega_{[4]}\Pi^{b_2}_{(3)} &= \frac{1}{4!}\,\widetilde{F}_{a_1a_2a_3a_4}\,F_{b_1}{}^{b_2}\,\gamma^{a_1a_2a_3a_4b_1} + \frac{1}{12}F^{a_5a_6}\,F^{a_1b_2}\,\gamma_{a_1a_5a_6} \\
\omega_{[4]}\widetilde{\gamma}^{b_6} &= \frac{5!}{2}\,F_{b_4b_5}\,\gamma^{b_6b_4b_5} + 5!\,F^{b_5b_6}\,\gamma_{b_5}
\end{aligned} \quad (9.3.49)$$

and finally for $\omega_{[2]}$ we have:

$$\begin{aligned}
\omega_{[2]}\,\gamma_{b_1} &= \frac{i}{8}\,(\widetilde{F\,F})_{a_1a_2}\,\gamma_{b_1}^{a_1a_2} - \frac{i}{4}\,(\widetilde{F\,F})_{b_1a_2}\,\gamma^{a_2} \\
\omega_{[2]}\,\Pi^{b_4}_{(1)} &= \frac{i}{8}(\widetilde{FF})_{a_1a_2}\,\widetilde{F}^{b_1b_2b_3b_4}\,\gamma^{a_1a_2}_{b_1b_2b_3} \\
&\quad -\frac{3}{4}\,i\,(\widetilde{FF})_{a_1a_2}\,\widetilde{F}^{a_1b_2b_3b_4}\,\gamma^{a_2}_{b_2b_3} - \frac{3}{4}\,i\,(\widetilde{FF})_{a_1a_2}\,\widetilde{F}^{a_1a_2b_3b_4}\,\gamma_{b_3} \\
\omega_{[2]}\,\Pi^{b_2}_{(2)} &= \frac{i}{8}(\widetilde{FF})_{a_1a_2}\,F^{b_1b_2}\,\gamma^{a_1a_2}_{b_1} - \frac{i}{4}\,(\widetilde{FF})_{b_1a_2}\,(\widetilde{FF})^{b_1b_2}\,\gamma^{a_2} \\
\omega_{[2]}\,\Pi^{b_2}_{(3)} &= \frac{i}{8}\,(\widetilde{F\,F})_{a_1a_2}\,F^{b_1b_2}\,\gamma^{a_1a_2}_{b_1} + \frac{1}{4}[F\,(\widetilde{FF})]^{b_2}{}_{a_2}\,\gamma^{a_2} \\
\omega_{[2]}\,\widetilde{\gamma}^{b_6} &= -\frac{5}{2}\,i\,(\widetilde{F\,F})_{a_1a_2}\,\widetilde{\gamma}^{a_1a_2b_6}
\end{aligned} \tag{9.3.50}$$

With these relations we determine the values of the parameter $a_7 = \frac{\sqrt{2}}{2}\,i$ from the supersymmetry relation [b] (9.3.35) and the projector (9.3.38).

9.4 Outlook and conclusions

In this paper we have applied the new first order formalism for p–brane world volume actions introduced by us in [1] to the case of 5 branes. The motivations for such a calculation were already discussed in the introduction and are not repeated here. We recall the essential features of the new formalism that we have adopted and that allows to reproduce the Born–Infeld second order action via the elimination of a set composed by three auxiliary fields:

- Π^{a}_{i}
- h^{ij} (symmetric)
- $\mathcal{F}^{ij}$ (antisymmetric)

Distinctive properties are:

1. All fermion fields are implicitly hidden inside the definition of the p–form potentials of supergravity
2. κ–supersymmetry is easily proven from supergravity rheonomic parametrizations
3. The action is manifestly covariant with respect to the duality group $SL(2,\mathbb{R})$ of type IIB supergravity.
4. The action functional can be computed on any background which is an exact solution of the supergravity bulk equations.

Of specific interest in applications are precisely the last two properties. Putting together our result we can summarize the $D5$ brane action we have found as follows:

$$\begin{aligned}
\mathcal{L} &= \frac{1}{\sqrt{\mathrm{Im}\mathcal{N}}}\Pi_i^{\underline{a}} V^{\underline{b}} \eta_{\underline{ab}} \eta^{i\ell_1} \wedge e^{\ell_2} \wedge \cdots \wedge e^{\ell_6} \epsilon_{\ell_1 \ldots \ell_6} \\
&- \frac{1}{12\sqrt{\mathrm{Im}\mathcal{N}}}\Pi_i^{\underline{a}} \Pi_j^{\underline{b}} \eta_{\underline{ab}} h^{ij} e^{\ell_1} \wedge \cdots \wedge e^{\ell_6} \epsilon_{\ell_1 \ldots \ell_6} \\
&- \frac{1}{3\sqrt{\mathrm{Im}\mathcal{N}}}\left[\det\left(h^{-1} + \frac{1}{\sqrt{\mathrm{Im}\mathcal{N}}}\mathcal{F}\right)\right]^{\frac{1}{4}} e^{\ell_1} \wedge \cdots \wedge e^{\ell_6} \epsilon_{\ell_1 \ldots \ell_6} \\
&+ \frac{5}{4\sqrt{\mathrm{Im}\mathcal{N}}} \mathcal{F}^{ij} F^{[2]} \wedge e^{\ell_3} \wedge \cdots \wedge e^{\ell_6} \epsilon_{ij\ell_3 \ldots \ell_6} \\
&- \frac{5}{2} i \,\mathrm{Re}\mathcal{N}\, F \wedge F \wedge F + \frac{5}{2} i \, p_\beta A^\beta \wedge F \wedge F \\
&+ 40\, i\, C_{[4]} \wedge F + \frac{\sqrt{2}}{2} i\, q_\alpha C^\alpha_{[6]}
\end{aligned} \tag{9.4.1}$$

where:

$$\begin{aligned}
F &\equiv F_{[G]} + q_\alpha A^\alpha \\
\mathcal{N} &= \frac{p_\alpha \Lambda^{\alpha}_{-}}{q_\beta \Lambda^{\beta}_{-}} \\
A^\alpha &= \frac{1}{\sqrt{2}}\left(B_{[2]} - iC_{[2]} \, , \, B_{[2]} + iC_{[2]}\right) \\
q_\alpha &= \frac{1}{\sqrt{2}}(p + iq \, , \, p - iq) \\
p_\alpha &= \frac{1}{\sqrt{2}}(r + is \, , \, r - is) \\
1 &= q_\alpha \, p_\beta \, \epsilon^{\alpha\beta}
\end{aligned} \tag{9.4.2}$$

In the particular case of $q_\Lambda = \{1, 0\}$ that implies $q_\alpha = \frac{1}{\sqrt{2}}(1, 1)$ and $p_\Lambda = \{0, 1\}$ that implies $p_\alpha = \frac{i}{\sqrt{2}}(1, -1)$ we have:

$$\mathcal{N} = \mathrm{i}\,\frac{\Lambda^1_- - \Lambda^2_-}{\Lambda^1_- + \Lambda^2_-} = i\, e^{-\phi} + C_{[0]} \tag{9.4.3}$$

As many times stressed, our specific interest in the above action is given by its evaluation on the background provided by the bulk solution found in [2] which describes a $D3$–brane with an $\mathbb{R}^2 \times ALE$ transverse manifold and a 2–form flux trapped on a homology 2–cycle of the ALE. Combining this with the world volume action of a $D3$–brane obtained in [1] we will finally obtain the appropriate source term of that exact solution which was so far missing. Alternatively by expanding (9.4.1) for small fluctuations around the same background we can use it as a token to explore the gauge/gravity correspondence. The key point is to reconcile this with the fact there is no running dilaton in the bulk solution of [1]. This will achieved by an appropriate choice of the charge vectors q_Λ and p_Λ, thanks to present manifestly $SL(2,R)$-covariant formulation. Such applications are postponed to a forthcoming paper.

A. Rheonomic parametrization of the type IIB supergravity curvatures

In order to obtain the supersymmetry transformation rules used in the text one needs the rheonomic parametrizations of the curvatures. For simplicity we write them only in the complex basis and we disregard the bilinear fermionic terms calculated by Castellani and Pesando. We have:

$$R^{\underline{a}} = 0 \tag{9.4.1}$$

$$\begin{aligned}\rho &= \rho_{\underline{ab}} V^{\underline{a}} \wedge V^{\underline{b}} + \frac{5}{16} \mathrm{i} \Gamma^{\underline{a_1 - a_4}} \psi V^{\underline{a_5}} \left(F_{\underline{a_1 - a_5}} + \frac{1}{5!} \epsilon_{\underline{a_1 - a_{10}}} F_{\underline{a_6 - a_{10}}} \right) \\ &+ \frac{1}{32} \left(-\Gamma^{\underline{a_1 - a_4}} \psi^* V_{\underline{a_1}} + 9 \Gamma^{\underline{a_2 a_3}} \psi^* V^{\underline{a_4}} \right) \Lambda^\alpha_+ \mathcal{H}^\beta_{\underline{a_2 - a_4}} \epsilon_{\alpha\beta} \\ &+ \text{fermion bilinears}\end{aligned} \tag{9.4.2}$$

$$\mathcal{H}^\alpha_{[3]} = \mathcal{H}^\alpha_{\underline{abc}} V^{\underline{a}} \wedge V^{\underline{b}} \wedge V^{\underline{c}} + \Lambda^\alpha_+ \bar{\psi}^* \Gamma_{\underline{ab}} \lambda^* V^{\underline{a}} \wedge V^{\underline{b}} + \Lambda^\alpha_- \bar{\psi} \Gamma_{\underline{ab}} \lambda V^{\underline{a}} \wedge V^{\underline{b}} \tag{9.4.3}$$

$$\mathcal{F}_{[5]} = F_{\underline{a_1 - a_5}} V^{\underline{a_1}} \wedge \cdots \wedge V^{\underline{a_5}} \tag{9.4.4}$$

$$\mathcal{D}\lambda = \mathcal{D}_{\underline{a}} \lambda V^{\underline{a}} + \mathrm{i} P_{\underline{a}} \Gamma^{\underline{a}} \psi^* - \frac{1}{8} \mathrm{i} \Gamma^{\underline{a_1 - a_3}} \psi \epsilon_{\alpha\beta} \Lambda^\alpha_+ \mathcal{H}^\beta_{\underline{a_1 - a_3}} \tag{9.4.5}$$

$$\mathcal{D}\Lambda^\alpha_+ = \Lambda^\alpha_- P_{\underline{a}} V^{\underline{a}} + \Lambda^\alpha_- \bar{\psi}^* \lambda \tag{9.4.6}$$

$$\mathcal{D}\Lambda^\alpha_- = \Lambda^\alpha_+ P^*_{\underline{a}} V^{\underline{a}} + \Lambda^\alpha_+ \bar{\psi} \lambda^* \tag{9.4.7}$$

$$R^{\underline{ab}} = R^{\underline{ab}}{}_{\underline{cd}} V^{\underline{c}} \wedge V^{\underline{d}} + \text{fermionic terms} \tag{9.4.8}$$

$$\mathcal{H}^\alpha_{[7]} = \frac{1}{3!\,7!} \epsilon^{\underline{abcd_1 \ldots d_7}} \mathcal{H}^\alpha_{\underline{abc}} V_{\underline{d_1}} \wedge \cdots \wedge V_{\underline{d_7}} + \ldots \tag{9.4.9}$$

(9.4.10)

B. SL(2, ℝ) and SU(1, 1) covariant formalism for D-branes

We define a two component vector for the fields $B_{[2]}$ and $C_{[2]}$:

$$A^{\Sigma} = (B_{[2]}, C_{[2]}) \tag{9.4.1}$$

The variation of this vector and $\mathcal{N}$ under $SL(2, R)$ are :

$$\begin{pmatrix} B'_{[2]} \\ C'_{[2]} \end{pmatrix} = \begin{pmatrix} p & q \\ r & s \end{pmatrix} \begin{pmatrix} B_{[2]} \\ C_{[2]} \end{pmatrix} = \begin{pmatrix} p\, B_{[2]} + q\, C_{[2]} \\ r\, B_{[2]} + s\, C_{[2]} \end{pmatrix} \tag{9.4.2}$$

$$\mathcal{N}' = \frac{s\mathcal{N} + r}{q\mathcal{N} + p} = \frac{sq\,(C_{[0]}^2 + e^{-2\phi}) + (sp + rq)\, C_{[0]} + rp + i\, e^{-\phi}}{q^2\,(C_{[0]}^2 + e^{-2\phi}) + 2qp\, C_{[0]} + p^2} \tag{9.4.3}$$

We can see that there is a subgroup of SL(2, ℝ) that leaves invariant the Lagrangian. If we take the element :

$$T = \begin{pmatrix} 1 & 0 \\ r & 1 \end{pmatrix} \tag{9.4.4}$$

we obtain for D3-brane (we have the same results for D5-brane) :

$$\begin{aligned} &B'_{[2]} = B_{[2]} \;\;,\;\; C'_{[2]} = rB_{[2]} + C_{[2]} \\ &\mathcal{N}' = \mathcal{N} + r \\ &Im(\mathcal{N}') = Im(\mathcal{N}) \;\;,\;\; Re(\mathcal{N}') = Re(\mathcal{N}) + r \end{aligned} \tag{9.4.5}$$

$$\begin{aligned} &\delta_T \mathcal{L}_{B.I} = 0 \\ &\mathcal{L}'_{W.Z.T} = \\ &-\tfrac{3}{4} i \,[\mathrm{Re}\mathcal{N} + r]\; B_{[2]} \wedge B_{[2]} + \tfrac{3}{4}\, i\, [C_{[2]} + rB_{[2]}] \wedge B_{[2]} + 6i\, C_{[4]} = \mathcal{L}_{W.Z.T} \end{aligned} \tag{9.4.6}$$

this is a one-parameter subgroup of SL(2, ℝ)
Now we introduce e covariant formalism for SL(2, ℝ) covariance; in particular we introduce two vectors q_Λ and p_Λ that transform in the fundamental representation of SL(2, ℝ). In analogy with $\varphi_i\, \hat{n}_i$ in field theory, where $\hat{n}_i$ is a fix vector and φ_i is a scalar field in the fundamental representation of SO(N), to do this we replace in the Lagrangian :

$$\begin{aligned} B_{[2]} &\rightarrow q_\Lambda A^\Lambda \\ C_{[2]} &\rightarrow p_\Lambda A^\Lambda \end{aligned} \tag{9.4.7}$$

where A^Λ is definite in (9.4.1).
The vectors q_Λ and p_Λ have this general form :

$$\begin{aligned} q_\Lambda &= (p, q) \\ p_\Lambda &= (r, s) \end{aligned} \tag{9.4.8}$$

with the constrain :

$$q_\Lambda p_\Sigma \epsilon^{\Lambda\Sigma} = 1 \tag{9.4.9}$$

This is actually the condition that the determinant of a general element of SL$(2,\mathbb{R})$ is equal to one.
We have in total only three real parameters but as we will see in the SU$(1,1)$ formalism one of those is connected to the invariance (9.4.7).
We must introduce also the covariant form of $\mathcal{N}$, but, to do this, is suitable to pass at the SU$(1,1)$ formalism. We pass to SU$(1,1)$ formalism using the Cayley matrix :

$$C^\alpha{}_\Lambda = \frac{1}{\sqrt{2}} \begin{pmatrix} 1 & -i \\ 1 & i \end{pmatrix} \tag{9.4.10}$$

and the inverse matrix :

$$\widetilde{C}_\alpha{}^\Lambda = (C^{-1})^\Lambda{}_\alpha = \frac{1}{\sqrt{2}} \begin{pmatrix} 1 & 1 \\ i & -i \end{pmatrix} \tag{9.4.11}$$

In this way we have that :

$$\begin{aligned} q_\alpha &= \widetilde{C}_\alpha{}^\Lambda q_\Lambda \\ p_\alpha &= \widetilde{C}_\alpha{}^\Lambda p_\Lambda \\ A^\alpha &= C^\alpha{}_\Lambda A^\Lambda \end{aligned} \tag{9.4.12}$$

and :

$$q_\alpha A^\alpha = \widetilde{C}_\alpha{}^\Sigma q_\Sigma C^\alpha{}_\Lambda A^\Lambda = (C^{-1})^\Sigma{}_\alpha C^\alpha{}_\Lambda A^\Lambda = q_\Lambda A^\Lambda \tag{9.4.13}$$

From (9.4.1), (9.4.8), (9.4.12)

$$\begin{aligned} A^\alpha &= C^\alpha{}_\Sigma A^\Sigma = \frac{1}{\sqrt{2}} (B_{[2]} - iC_{[2]} \,,\, B_{[2]} + iC_{[2]}) \\ q_\alpha &= \frac{1}{\sqrt{2}}(p + iq \,,\, p - iq) = \frac{1}{\sqrt{2}}(q_1, q_1^\star) \\ p_\alpha &= \frac{1}{\sqrt{2}}(r + is \,,\, r - is) = \frac{1}{\sqrt{2}}(p_1, p_1^\star) \end{aligned} \tag{9.4.14}$$

The vectors q_Λ and p_Λ transform in the fundamental representation of SL(2,R) as :

$$\begin{aligned} q'_\Lambda &= q_\Sigma U^\Sigma{}_\Lambda \\ p'_\Lambda &= p_\Sigma U^\Sigma{}_\Lambda \end{aligned} \tag{9.4.15}$$

and in the formalism of SU(1,1) the vectors q_α and p_α transform as :

$$q'_\alpha = q'_\Sigma (C^{-1})^\Sigma{}_\alpha = q_\Lambda U^\Lambda{}_\Sigma (C^{-1})^\Sigma{}_\alpha = q_\beta C^\beta{}_\Lambda U^\Lambda{}_\Sigma (C^{-1})^\Sigma{}_\alpha \tag{9.4.16}$$

so the relation between SU(1,1) and SL(2,R) is :

$$U^{\beta}{}_{\alpha} = C^{\beta}{}_{\Lambda}\, U^{\Lambda}{}_{\Sigma}\, (C^{-1})^{\Sigma}{}_{\alpha} = (C\, U\, C^{-1})^{\beta}{}_{\alpha} \tag{9.4.17}$$

$$\begin{aligned} U^{\alpha}{}_{\beta} &= \frac{1}{2}\begin{pmatrix} 1 & -i \\ 1 & i \end{pmatrix}\begin{pmatrix} p & q \\ r & s \end{pmatrix}\begin{pmatrix} 1 & 1 \\ i & -i \end{pmatrix} \\ &= \frac{1}{2}\begin{pmatrix} p - ir + iq + s & p - ir - iq - s \\ p + ir + iq - s & p + ir - iq + s \end{pmatrix} \end{aligned} \tag{9.4.18}$$

The property (9.4.14) is invariant under SU(1,1), in fact :

$$\begin{aligned} q'_{\alpha} &= q_{\beta}\, U^{\beta}{}_{\alpha} \\ q'^{\star}_{\alpha} &= q^{\star}_{\beta}\, \left(U^{\beta}{}_{\alpha}\right)^{\star} \\ q'^{\star}_{1} &= q^{\star}_{1}\, \left(U^{1}{}_{1}\right)^{\star} + q^{\star}_{2}\, \left(U^{2}{}_{1}\right)^{\star} = q^{\star}_{1}\, \left(U^{1}{}_{1}\right)^{\star} + q_{1}\, \left(U^{2}{}_{1}\right)^{\star} \end{aligned} \tag{9.4.19}$$

on the other side we have that $q'_2 = q_1\, \left(U^1{}_2\right) + q_2\, \left(U^2{}_2\right) = q_1\, \left(U^1{}_2\right) + q^{\star}_1\, \left(U^2{}_2\right)$ and so if we want that $q'_2 = q'^{\star}_1$ we must have that $\left(U^2{}_1\right)^{\star} = U^1{}_2$ and $\left(U^1{}_1\right)^{\star} = U^2{}_2$, but this is exactly the property of SU(1,1) how we can see in (9.4.18).
The Levi-Civita tensor becomes :

$$\epsilon^{\alpha\beta} = C^{\alpha}{}_{\Lambda}\, C^{\beta}{}_{\Sigma}\, \epsilon^{\Lambda\Sigma} = \begin{pmatrix} 0 & i \\ -i & 0 \end{pmatrix} \tag{9.4.20}$$

and it is very simple to verify that the constrain now is :

$$q_{\alpha}\, p_{\beta}\, \epsilon^{\alpha\beta} = 1 \tag{9.4.21}$$

The covariant formulation of the $\mathcal{N}$ matrix is:

$$\mathcal{N} = i\, \frac{\Lambda^1_- - \Lambda^2_-}{\Lambda^1_- + \Lambda^2_-} = i\, e^{-\phi} + C_{[0]} \;\rightarrow\; \frac{p_{\alpha}\Lambda^{\alpha}_-}{q_{\beta}\Lambda^{\beta}_-} \tag{9.4.22}$$

The condition (9.4.21) says that

$$ps - qr = 1 \tag{9.4.23}$$

Now we can see some examples.

- If $p \neq 0$

$$s = \frac{1}{p} + \frac{qr}{p} \tag{9.4.24}$$

and so

$$p_{\alpha} = \frac{1}{\sqrt{2}}\left(r + i\,\frac{1}{p} + i\,\frac{qr}{p}\, , r - i\,\frac{1}{p} - i\,\frac{qr}{p}\right) = \frac{i}{p\,\sqrt{2}}(1\, , -1) + \frac{r}{p}\, q_{\alpha} \equiv \widetilde{p}_{\alpha} + \frac{r}{p}\, q_{\alpha} \tag{9.4.25}$$

The $\mathcal{N}$ matrix became

$$\mathcal{N} = \frac{r}{p} + \frac{\widetilde{p}_\alpha \Lambda_-^\alpha}{q_\beta \Lambda_-^\beta} \tag{9.4.26}$$

and it is very simple to see that the parameter r disapperaes from the Lagrangian, infact it is the parameter of the invariance (9.4.4). How we said before only two parameters (p and q in this case) are independent and produce SL$(2, \mathbb{R})$ covariant Lagrangianes.

• If $p = 0$

$$q_\alpha = \frac{i\,q}{\sqrt{2}}(1\,,\,-1) \tag{9.4.27}$$

$$p_\alpha = \frac{1}{\sqrt{2}}(r + is\,,\, r - is) \tag{9.4.28}$$

from (9.4.21) or (9.4.23)

$$r = -\frac{1}{q} \tag{9.4.29}$$

and

$$p_\alpha = \frac{1}{\sqrt{2}}\left(-\frac{1}{q} + i\,s\,, -\frac{1}{q} - i\,s\right) = \frac{i\,s}{\sqrt{2}}(1\,,\,-1) - \frac{1}{q\sqrt{2}}(1\,,\,1) \equiv \frac{s}{q}\,q_\alpha + \widetilde{p}_\alpha \tag{9.4.30}$$

The $\mathcal{N}$ matrix now is

$$\mathcal{N} = \frac{s}{q} + \frac{\widetilde{p}_\alpha \Lambda_-^\alpha}{q_\beta \Lambda_-^\beta} = \frac{s}{q} - \frac{1}{q^2\,\mathcal{N}} \tag{9.4.31}$$

and

$$\begin{aligned} q_\alpha\, A^\alpha &= q\, C_{[2]} \\ \widetilde{p}_\alpha\, A^\alpha &= -\frac{1}{q}\, B_{[2]} \end{aligned} \tag{9.4.32}$$

In this case the invariance parameter is s, the only independent parameter is q and the Lagrangian is that we can obtain with the duality:

$$D = \begin{pmatrix} 0 & 1 \\ -1 & 0 \end{pmatrix} \tag{9.4.33}$$

At this point we can ask to ourself what is the relaction between $p \neq 0$ and $p = 0$. If we start from the SL(2,R) vectors $q_\Lambda = (1, 0)$ and $p_\Lambda = (0, 1)$ we can see that the stability group of q_Λ is T of (9.4.4) :

$$\begin{aligned} q'_\Lambda &= (1, 0)\begin{pmatrix} 1 & 0 \\ r & 1 \end{pmatrix} = (1, 0) \\ p'_\Lambda &= (0, 1)\begin{pmatrix} 1 & 0 \\ r & 1 \end{pmatrix} = r\,(1, 0) + (0, 1) = p_\Lambda + r\,q_\Lambda \end{aligned} \tag{9.4.34}$$

Now we transform q_Λ and p_Λ with the duality $U = D$ of (9.4.33) and obtain $q'_\Lambda = (0,1)$ and $p'_\Lambda = (-1,0)$, contemporary :

$$T^\Lambda{}_\Sigma \to \left(U^{-1}TU\right)^\Lambda{}_\Sigma = \begin{pmatrix} 1 & -r \\ 0 & 1 \end{pmatrix} \tag{9.4.35}$$

this transformations are the stability group of $q'_\Lambda = (0,1)$, in fact :

$$\begin{aligned} q'_\Lambda &= (0,1)\begin{pmatrix} 1 & -r \\ 0 & 1 \end{pmatrix} = (0,1) \\ p'_\Lambda &= (-1,0)\begin{pmatrix} 1 & -r \\ 0 & 1 \end{pmatrix} = (-1,0) + r\,(0,1) = p_\Lambda + r\,q_\Lambda \end{aligned} \tag{9.4.36}$$

but as you can see in (9.4.38) $q'_\Lambda = (0,1) \to q_\alpha = \frac{i}{\sqrt{2}}(1,-1)$ and $p'_\Lambda = (-1,0) \to p_\alpha = -\frac{1}{\sqrt{2}}(1,1)$ when we pass to SU(1,1) formalism, and so 9.4.36 become exactly the relations (9.4.27), (9.4.30) if we take $q = 1$.

B.1. Particular choice of q_α and p_α

◇ For :

$$\begin{aligned} q_\Lambda = (1,0) \ &\Rightarrow q_\alpha = q_\Lambda (C^{-1})^\Lambda{}_\alpha = \frac{1}{\sqrt{2}}(1,1) \\ p_\Lambda = (0,1) \ &\Rightarrow p_\alpha = p_\Lambda\,(C^{-1})^\Lambda{}_\alpha = \frac{i}{\sqrt{2}}(1,-1) \end{aligned}$$

we have :

$$\begin{aligned} q_\alpha A^\alpha &= B_{[2]} \\ p_\alpha A^\alpha &= C_{[2]} \\ q_\alpha p_\beta \epsilon^{\alpha\beta} &= 1 \end{aligned} \tag{9.4.37}$$

◇ For :

$$\begin{aligned} q_\Lambda = (0,1) \ &\Rightarrow q_\alpha = \frac{i}{\sqrt{2}}(1\,,-1) \\ p_\Lambda = (-1,0) \ &\Rightarrow p_\alpha = -\frac{1}{\sqrt{2}}(1\,,1) \end{aligned} \tag{9.4.38}$$

we have :

$$\begin{aligned} q_\alpha A^\alpha &= C_{[2]} \\ p_\alpha A^\alpha &= -B_{[2]} \end{aligned} \tag{9.4.39}$$

$$\mathcal{N} \to \mathcal{N}_{Dual} = -\frac{1}{\mathcal{N}} \tag{9.4.40}$$

and

$$Im(\mathcal{N}_{Dual}) = \frac{1}{e^{-\phi} + C_{[0]}^2\, e^{-\phi}}$$

$$Re(\mathcal{N}_{Dual}) = -\frac{e^{\phi}\, C_{[0]}}{e^{-\phi} + C_{[0]}^2\, e^{\phi}} \tag{9.4.41}$$

B.2. Invariant Tensors under SU(1, 1)

The invariant tensors under SU(1, 1) are the metric $\eta^{\alpha\beta}$ and the tensor $\epsilon^{\alpha\beta}$, in fact :

$$U^T \epsilon\, U = \epsilon \tag{9.4.42}$$

$$U^{\dagger} \eta\, U = \eta \tag{9.4.43}$$

where we defined :

$$\eta^{\alpha\beta} = \begin{pmatrix} 1 & 0 \\ 0 & -1 \end{pmatrix} \qquad \epsilon^{\alpha\beta} = \begin{pmatrix} 0 & i \\ -i & 0 \end{pmatrix} \tag{9.4.44}$$

In the SU(1, 1) formalism the scalar product is :

$$\begin{aligned} q^{\alpha}\, p_{\alpha} &= q^{*}_{\alpha}\, \eta^{\alpha\,\beta}\, p_{\beta} = q^{*}_{1} p_1 - q^{*}_{2} p_2 \\ q^{\alpha}\, q_{\alpha} &= q^{*}_{\alpha}\, \eta^{\alpha\,\beta}\, q_{\beta} = |q_1|^2 - |q_2|^2 \\ A^{\alpha}\, q_{\alpha} &= A^{*}_{\alpha}\, \eta^{\alpha\,\beta}\, q_{\beta} \end{aligned} \tag{9.4.45}$$

we have introduced the vectors q_{α}, p_{α}, and the form A^{α} and so the vector A_{α} is :

$$A_{\alpha} = \eta_{\alpha\beta} A^{*\beta} \tag{9.4.46}$$

C. Notations and Conventions

General adopted notations for first order actions are the following ones:

$$\begin{aligned} d &= \text{dimension of the world-volume } \mathcal{W}_d \\ D &= \text{dimension of the bulk space–time } \mathcal{M}_D \\ V^{\underline{a}} &= \text{vielbein 1–form of bulk space–time} \\ \Pi^{\underline{a}}_{\underline{i}} &= D \times d \text{ matrix. 0–form auxiliary field} \\ h^{ij} &= d \times d \text{ symmetric matrix. 0–form auxiliary field} \\ e^{\ell} &= \text{vielbein 1–form of the world-volume} \\ \eta_{\underline{ab}} &= \text{diag}\{+, \underbrace{-, \ldots, -}_{D-1\, times}\} = \text{flat metric on the bulk} \\ \eta^{\underline{ij}} &= \text{diag}\{+, \underbrace{-, \ldots, -}_{d-1\, times}\} = \text{flat metric on the world–volume} \end{aligned} \tag{9.4.1}$$

The supersymmetric formulation of type IIB supergravity we rely on is that of Castellani and Pesando [7] that uses the rheonomy approach [10]. Hence, as it is customary in all the rheonomy constructions, the adopted signature of space–time is the **mostly minus signature**:

$$\eta_{\underline{ab}} = \mathrm{diag}\left\{ +, \underbrace{-, \dots, -}_{9 \text{ times}} \right\} \tag{9.4.2}$$

The index conventions are the following ones:

$$\begin{aligned} \underline{a, b, c, \dots} &= 0, 1, 2, \dots, 9 \quad \text{Lorentz flat indices in } D = 10 \\ i, j, k, \dots &= 0, \dots, d \quad \text{Lorentz flat indices on the world-volume} \quad (9.4.3) \\ \alpha, \beta, \dots &= 1, 2 \quad \mathrm{SU}(1,1) \text{ doublet indices} \\ A, B, C, \dots &= 1, 2 \quad \mathrm{O}(2) \text{ indices for the scalar coset} \\ \Lambda, \Sigma, \Gamma, \dots &= 1, 2 \quad \mathrm{SL}(2, \mathrm{R}) \text{ doublet indices} \end{aligned} \tag{9.4.4}$$

For the gamma matrices our conventions are as follows:

$$\left\{ \Gamma^{\underline{a}}, \Gamma^{\underline{b}} \right\} = 2\, \eta^{\underline{ab}} \tag{9.4.5}$$

The convention for constructing the dual of an ℓ–form ω in D–dimensions is the following:

$$\omega = \omega_{\underline{i_1 \dots i_\ell}} V^{\underline{i_1}} \wedge \dots \wedge V^{\underline{i_\ell}} \quad \Leftrightarrow \quad \star\omega = \frac{1}{(D-\ell)!}\, \epsilon_{\underline{a_1 \dots a_{D-\ell} b_1 \dots b_\ell}} \omega^{\underline{b_1 \dots b_\ell}} V^{\underline{a_1}} \wedge \dots \wedge V^{\underline{a_{D-\ell}}} \tag{9.4.6}$$

Note that we also use ℓ–form components with strength one: $\omega = \omega_{\underline{i_1 \dots i_\ell}} V^{\underline{i_1}} \wedge \dots \wedge V^{\underline{i_\ell}}$ and not with strength $\ell!$ as it would be the case if we were to write $\omega = \frac{1}{\ell!}\omega_{\underline{i_1 \dots i_\ell}} V^{\underline{i_1}} \wedge \dots \wedge V^{\underline{i_\ell}}$ When it is more appropriate to use curved rather than flat indices then the convention for Hodge duality is summarized by the formula:

$$\star\left(dx^{\mu_1} \wedge \dots dx^{\mu_n}\right) = \frac{\sqrt{-\det(g)}}{(10-n)!} G^{\mu_1 \nu_1} \dots G^{\mu_n \nu_n} \epsilon_{\rho_1 \dots \rho_{10-n} \nu_1 \dots \nu_n} dx^{\rho_1} \wedge \dots dx^{\rho_{10-n}} \tag{9.4.7}$$

Acknowledgements

We are grateful to Pietro Fré, Marco Billó, Emiliano Imeroni, Alberto Lerda, Igor Pesando, and Antonio Rago for many important and clarifying discussions about this work.

References

[1] Pietro Fré, Leonardo Modesto A new first order formalism for k-supersymmetric Born Infeld actions: the D3-brane example, *Class. Quant. Grav.* 19 (2002) 5591, [hep-th/0206144].

[2] M. Bertolini, G. Ferretti, P. Fré M. Trigiante, L. Campos, P. Salomonson Supersymmetric 3–Branes on Smooth ALE Manifolds with Flux, *Nucl. Phys.* **B617** (2001) 3-42, [hep-th 0106186].

[3] M. Bertolini, P. Di Vecchia, M. Frau, A. Lerda, R. Marotta, I. Pesando Fractional D-branes and their gauge duals, *JHEP* 0102 (2001) 014 [hep-th 0011077]

[4] P. Di Vecchia, A. Lerda, P. Merlatti N=1 and N=2 Super Yang-Mills theories from wrapped branes, *Nucl. Phys.* **B646** (2002) 43-68, [hep-th 0205204]

[5] L. Modesto and P. Fré to appear

[6] L. Castellani, Chiral D = 10, N=2 Supergravity On The Group Manifold. 1. Free Differential Algebra And Solution Of Bianchi Identities, *Nucl. Phys.* **B294** (1987) 877.

[7] L. Castellani and I. Pesando, The Complete superspace action of chiral D = 10, N=2 supergravity, *Int. J. Mod. Phys.* **A8** (1993) 1125.

[8] P. Fré, Comments on the 6-index photon in d=11 supergravity and the gauging of free differential algebras, *Class. Quant.Grav.* **1**, (1984) L 81

[9] R.D'Auria and P.Fré, D=11 Supergravity and its hidden supergroup, *Nucl. Phys.* **B201** (1982) 101.

[10] L. Castellani, R. D'Auria and P. Fré, *Supergravity and Superstring theory: a geometric perspective*, World Scientific, Singapore (1990).

[11] Polchinski J. *String Theory Volume 1*, *String Theory Volume 2*, Cambridge University Press.

[12] P. Fré, Gaugings and other supergravity tools of p-brane physics, Lectures given at the RTN School **Recent Advances in M-theory**, Paris February 1-8 2001, IHP, [hep-th/0102114].

INDEX

F

G

H

Q

R

S

T

U

V

W